Synthetic Peptides

A User's Guide

UWBC Biotechnical Resource Series

Richard R. Burgess, Series Editor
University of Wisconsin Biotechnology Center
Madison, Wisconsin

G. Grant, ed.	Synthetic Peptides: A User's Guide
M. Gribskov and J. Devereux, eds.	Sequence Analysis Primer

In preparation:

D. Nelson and B. Brownstein, eds	YAC Libraries: A User's Guide

Synthetic Peptides

A User's Guide

Edited by

Gregory A. Grant

Washington University School of Medicine

W. H. Freeman and Company
New York

For
Roger and Clarice
I hope you know.

I would like to thank Carolyn Stock; late of UWBC, and the people at W. H. Freeman, especially Ingrid Krohn and Penny Hull, who helped make the job easier and enjoyable. I would especially like to thank Ronald Niece for his efforts and support from the very beginning. Finally, I would like to thank the contributors for their excellent work; all the credit goes to them.

Gregory A. Grant

Cover illustration by Fineline Illustrations, Inc., based on a concept of the editor.

Library of Congress Cataloging-in-Publication Data

Synthetic Peptides: A User's Guide / edited by Gregory A. Grant
      p.   cm. — (UWBC biotechnical resource series)
    Includes bibliographical references and index.
    ISBN 0-7167-7009-1
    1. Peptides—Synthesis. 2. Protein engineering. I. Grant
    Gregory A., 1949-   II. Series.
    QP552.P4S963   1992
    547.7'56—dc20   91-4436
                                     CIP

Printed in the United States of America

1 2 3 4 5 6 7 8 9 0 VB 9 9 8 7 6 5 4 3 2

Contributors

Ruth Hogue Angeletti, Department of Developmental Biology and Cancer, Albert Einstein College of Medicine, Bronx, New York

George Barany, Department of Chemistry, University of Minnesota, Minneapolis, Minnesota

Nathan Collins, Department of Chemistry, University of Arizona, Tucson, Arizona

Gregg B. Fields, Department of Laboratory Medicine and Pathology, University of Minnesota, Minneapolis, Minnesota

Gregory A. Grant, Department of Molecular Biology and Pharmacology, Washington University School of Medicine, St. Louis, Missouri

Victor J. Hruby, Department of Chemistry, University of Arizona, Tucson, Arizona

Terry O. Matsunaga, Department of Chemistry, University of Arizona, Tucson, Arizona

Michael Moore, Smith-Kline Beecham Pharmaceuticals, New York, New York

K. C. Russell, Department of Chemistry, University of Arizona, Tucson, Arizona

Shubh D. Sharma, Department of Chemistry, University of Arizona, Tucson, Arizona

Zhenping Tian, Department of Chemistry, University of Minnesota, Minneapolis, Minnesota

Advisory Committee Coordinator
Ronald L. Niece
Biotechnology Center
University of Wisconsin–Madison

Contents

Foreword

This volume, *Synthetic Peptides: A User's Guide,* is the second in a series of biotechnical resource manuals being produced by the University of Wisconsin Biotechnology Center and published by W. H. Freeman and Company. Books in this series will cover two broad areas that we refer to as current and emerging technologies. Those books describing current technologies, such as this volume, are directed at the beginning or intermediate user of technology. The books on emerging technologies cover topics at a more advanced level. As we see it, many emerging technologies will soon become current and widely used.

Rapid advances in research have paved the way for rapid changes in the way we do research, leading to more research advances. Once, we relied solely on our laboratories' own power (possibly sharing an ultra-centrifuge with other groups). Now many of us have come to rely on core facilities or contract services to provide us with some of the more powerful research tools. We no longer can be or need to be expert in all areas necessary for modern research, but we do need some familiarity with these services to use them most effectively and efficiently. We hope that books like this one will help provide that familiarity.

We have four main goals we wish to achieve with these books. First, we want to provide information enabling you to understand the principles that underlie the methods. Second, we want to address many of the "common" problems faced by beginners so that you do not repeat these mistakes. In addition, we want to describe relevant applications of the technology to help you in searching for answers to your research problems. Most of all, we hope these books become helpful references to the laboratories using the technologies covered.

This volume begins with a brief overview and history of peptide synthesis, followed by chapters on peptide structure and design, the chemistry of peptide synthesis, evaluation and characterization of synthetic peptides, the diverse uses and applications of synthetic peptides in research, and finally the setup and operation of a peptide synthesis laboratory. An appendix lists some sources offering custom peptide synthesis and related analytical services.

I would like to thank the volume editor, Greg Grant, for developing the overall organization of the volume and for his careful work in improving each chapter and helping the individual sections fit together. The chapter authors provided knowledgeable and thoughtful contributions. Early ideas on what the volume should contain and who might be suitable contributors were provided by an advisory committee drawn from members of the Association of Biomolecular Resource Facilities consisting of Ruth Hogue Angeletti (Albert Einstein College of Medicine), Mel Billingsley (University of Wisconsin–Madison), Greg Grant (Washington University School of Medicine), Mary Padgett (National Institutes of Health), and Roger Perlmutter (University of Washington), and led by Ronald Niece (University of Wisconsin–Madison). Finally I want to thank Carolyn Stock, series developmental editor at the University of Wisconsin Biotechnology Center, for her excellent efforts to coordinate the whole endeavor.

Richard Burgess, Series Editor
Madison, Wisconsin
March 10, 1992

1

Overview

Gregory A. Grant

SYNTHETIC PEPTIDES: SPANNING THE TWENTIETH CENTURY

... every synthesis of a complicated polypeptide represents the sum of prolonged intellectual and experimental exertions which, however, often find their reward in the interesting physical, chemical, and biological properties of the final product.

(Theodor Wieland, in the foreword to *Peptide Synthesis*
by M. Bodanszky and M.A. Ondetti, 1966.)

From the first decade of the twentieth century, when peptide chemistry had its beginnings, to the last decade, which we are just beginning, the evolution of the development and use of synthetic peptides has been dynamic and exciting.

Emil Fischer (1902, 1903, 1906) introduced the concept of peptides and polypeptides and presented protocols for their synthesis in the early 1900s. Although others also made contributions in those days, most notably Theodor Curtius, the work of Fischer and his colleagues stands out, and he is generally regarded as the father of peptide chemistry. An excellent account of the history of peptide synthesis, which treats the subject much more comprehensively than can be attempted here, can be found in a recent book by Wieland and Bodanszky (1991). Fischer's place in the history of synthetic peptides is eloquently summed up in one sentence from that book, which simply states, "To Emil Fischer we owe the systematic attack on a field of natural substances that had previously been avoided by chemists."

In those early days, the chemistries developed by Fischer and others led to the production of molecules containing as many as 18 amino acids, such as Leucyl (triglycyl) leucyl (triglycyl) leucyl (octaglycyl) glycine (Fischer, 1907). Nonetheless, the syntheses performed were difficult and limited to simple amino acids, and progress was slow for many years. Then, the discovery of an easily removable protecting group, the carbobenzoxy group, by Bergmann and Zervas in 1932 provided new impetus by opening the way to the use of polyfunctional amino acids. As a result, the synthesis of naturally occurring small peptides such as carnosine (Sifferd and du Vigneaud, 1935) and glutathione (Harington and Mead, 1935) were soon achieved. Almost 20 years later, the synthesis of an active peptide hormone, the octapeptide oxytocin, by du Vigneaud et al. (1953) was acclaimed as a major accomplishment and spurred the advancement of peptide synthesis once again. The synthesis of the 39-residue porcine adrenocorticotropic hormone in 1963 (Schwyzer and Sieber, 1963), using solution phase segment condensation methods, was viewed as no less than sensational at the time; and in 1967, Bodanszky and colleagues succeeded in synthesizing the 27-residue secretin peptide by solution phase stepwise addition methods (Bodanszky and Williams, 1967; Bodanszky et al., 1967). These accomplishments in peptide synthesis were considered to be monumental at the time in that they were exceptionally difficult undertakings that pushed the prevailing technology to its limits.

In 1963, Bruce Merrifield published a landmark paper describing the development of solid-phase peptide synthesis. This technique, for which he was awarded the Nobel Prize in Chemistry in 1984, was responsible, more than anything else, for opening the way to the widespread use of synthetic peptides as reagents in chemical and biomedical investigations. Theodor Wieland (1981) once described the advances in peptide chemistry in the following way: " . . . the synthesis of glutathione opened the door to peptide synthesis a crack, and the synthesis of oxytocin pushed the door wide open." To extend that analogy, the introduction of the solid-phase method "blew the door off its hinges." Since that time, many other developments, such as improved synthesis chemistry, automated instrumentation for the unattended production of peptides, and improved purification and analytical methods have contributed to our ability to exploit the potential of synthetic peptides. Not only has

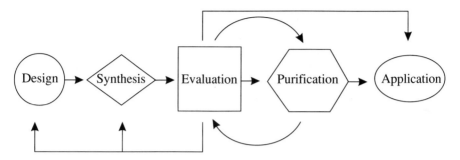

FIGURE 1   Diagram of the steps involved in obtaining a synthetic peptide.

there been an explosion in the number of synthetic peptides being produced, but reports of the synthesis of larger and larger peptides (Gutte and Merrifield, 1969, 1971; Clark-Lewis et al., 1986; Nutt et al., 1988; Schneider and Kent, 1988), some exceeding 100 residues in length, are becoming more commonplace.

Now, in the last decade of the twentieth century, more than ever before, synthetic peptides are available not only to those actively involved in developing the field, but to virtually every investigator in any field who perceives a need for them and wishes to pursue their use.

This book is intended primarily for those investigators who wish to utilize synthetic peptides in their research but who themselves are not already intimately involved in the field. This group would encompass researchers who either simply want access to synthetic peptides as tools or those who wish to become actively involved in the production of the peptides. As such, it strives to provide practical background information, answer basic questions, and address common problems relating to all aspects of synthetic peptides. At the same time, though, it is intended to be current, comprehensive, and sophisticated in its treatment of the subject and should be of general interest to researchers and educators at all levels.

The process of obtaining a synthetic peptide for use in biochemical or biomedical research involves several discrete steps that can be represented diagrammatically, as shown in Figure 1. The process starts with the design of the peptide; it follows through with the chemical synthesis, evaluation, and purification of the product; and finally, the application, or the actual use, of the peptide is demonstrated in an experimental situation. As indicated in the diagram, the process can be viewed as pivoting on the evaluation step, which actually is part of a more comprehensive evaluation-purification cycle. It is at this point that the product is characterized and its disposition determined. That disposition can include (1) use of the peptide as is, (2) additional purification and subsequent evaluation, eventually leading to experimental use, or (3) changes in either the design or synthesis protocols, or both, followed by reiteration of the latter steps of the process.

Each subsequent chapter of this book deals with one major aspect of this process, that is, design, synthesis, evaluation, and application. In addition, a separate chapter is devoted to the practical aspects of the melding of several of these individual steps, the setup and operation of a peptide synthesis core laboratory.

Chapter 2 begins with a discussion of the fundamental aspects of peptide structure and uses that structure as a foundation for both basic and more advanced considerations in the design of synthetic peptides. The design process is constantly evolving and it is something of an art itself. In many cases, there are no hard and fast rules, and function usually dictates design. Perhaps the most common concern expressed by general users of peptide synthesis facilities is how to choose a sequence for the production of antibodies. Some general rules for this procedure are presented, and the basic approach is discussed in this chapter. The user must keep in mind , however, that although some general guidelines are available, they are not foolproof and a sequence cannot be guaranteed a priori to be antigenic. A relatively new and exciting area of peptide design that is discussed in Chapter 2 is that of peptidomimetics. This type of design involves peptides whose structures are altered in some way to mimic the natural product in activity, usually by using structures that do not occur naturally in the system to be used, but which have beneficial attributes, such as an increased biological lifetime. This area is probably the most sophisticated and demanding and, at the same time, the most creative area of peptide design today. In some respects it may go beyond the need of many readers, but it is a fast-growing field that is of great importance, especially to pharmaceutical research. Many readers may find it relevant and useful as they become increasingly involved with synthetic peptides.

Chapter 3 deals with the synthesis chemistry itself. It presents a logical and straightforward explanation of the solid-phase method of peptide synthesis and discusses the overall state of the chemistry today. This chapter places particular emphasis on the Boc and Fmoc protection strategies, which are the two most useful approaches available, and provides specific recommendations for undertaking routine synthesis. It is immediately evident that the chemistry is very sophisticated and com-plicated, yet it is relatively straightforward in the overall approach and amenable to routine procedures. One must always be mindful that al-though routine synthesis using a uniform set of reagents is commonly performed successfully, a vast array of problems can occur. At the same time, a large number of alternatives can be utilized to prevent or circum-vent these problems. This chapter also contains a useful section on the automation of solid-phase synthesis. It gives an informative overview of the range of instruments available and illustrates the choice of capabilities, from the synthesis of single peptides to that of as many as 96 at one time and from micrograms to grams.

Chapter 4 deals with the evaluation of the finished peptide. It presents a discussion of the latest methods available for the routine characterization of the synthetic product and illustrates its use through examples of the evaluation of actual peptides. The message that should be very clear from this chapter is that although the chemistry can, in many cases, be preprogrammed on modern peptide synthesizers, and these machines do most of the work that used to be done laboriously on the benchtop, the process is by no means foolproof or trivial. Many problems from a wide variety of sources can be manifest even in a "routine" synthesis, and the final product can never be taken for granted. Every peptide produced must be rigorously evaluated with the ultimate goal being proof that the peptide obtained is the one intended. In addition, problematic residues and peptide solubility are very important considerations in the evaluation and use of the peptides, and they are often overlooked. Whenever possible, an attempt should be made to avoid problems arising from these two aspects in the design process. Chapter 4 contains general guidelines and useful information concerning these two areas as well as helpful suggestions for identifying and dealing with them if a problem does occur.

Chapter 5 presents an excellent survey of the use and application of synthetic peptides in research. This chapter not only illustrates the diversity in the use and application of synthetic peptides in modern research, but, perhaps, it also serves as a basis from which new applications and ideas may develop. These applications include the antigenic and immunogenic use of synthetic peptides, the use of peptides as enzyme inhibitors, structure/function studies involving synthetic peptides, and the use of peptides for affinity labeling of receptors, or "acceptors," and their use in structure/function studies of receptors. Undoubtedly, some uses for synthetic peptides have not been included, but the areas that are discussed are among the more exciting and successful applications being investigated today. Furthermore, many of the concepts and approaches discussed can be easily adapted to other systems and areas of exploration. Also, because use very often dictates design, many aspects of peptide design are discussed in this chapter and, as such, it complements Chapter 2 very nicely.

Finally, Chapter 6 presents both a philosophical and a practical discussion of the requirements and considerations that should go into the setup and operation of a peptide synthesis laboratory. It presents a question-oriented approach that may be of particular interest to those who are attempting peptide synthesis for the first time as well as to those who are faced with the ongoing task of operating an already established synthesis laboratory. Safety, which is a very important aspect of peptide synthesis, is dealt with here and is something that must take priority in the operation of a synthesis laboratory.

Peptide synthesis has come a long way since the beginning of this century. Research has occurred throughout the twentieth century, which has enhanced the passing of that period. As the century comes to a close and we head toward the beginning of the next millenium, it is tempting to speculate on what may lie ahead. However, history has told us that such attempts invariably fall far short of the eventual reality. After all, even Merrifield admitted in a 1986 article that he did not forsee the impact of his technique:

> From the accumulated data presented, I conclude that the solid phase synthesis of peptides up to 50 or somewhat more residues can be readily achieved in good yield and purity; this is a far better situation than I could have expected when this technique was first proposed.

Undoubtedly, great things will be accomplished in the field of peptide synthesis and in the use of the peptides in the future. It is, perhaps, sufficient to have had the privilege of taking part in a small bit of that process.

REFERENCES

Bergman, M. and Zervas, L. 1932. Über ein allgemeines Verfahren der Peptidsynthese. Ber. Dtsch. Chem. Ges. 65: 1192–1201

Bodanszky, M., and Ondetti, M.A. Peptide Synthesis. John Wiley and Sons, New York, 1966.

Bodanszky, M., Ondetti, M.A., Levine, S.D., and Williams, N.J. 1967. Synthesis of secretin II. The stepwise approach. J. Am. Chem. Soc. 89: 6753–6757.

Bodanszky, M., and Williams, N.J. 1967. Synthesis of secretin I. The protected tetradecapeptide corresponding to sequence 14-27. J. Am. Chem. Soc. 89: 685–689.

Clark-Lewis, I., Aebersold, R., Ziltener, H., Schrader, J.W., Hood, L.E., and Kent, S.B.H. 1986. Automated chemical synthesis of a protein growth factor for hemopoietic cells, Interleukin-3. Science 231: 134–139.

Fischer, E. 1902. Ueber einige Derivate des Glykocolls Alanins und Leucins. Ber. Dtsch. Chem. Ges. 35: 1095–1106.

Fischer, E. 1903. Synthese von Derivaten der Polypeptide. Ber. Dtsch. Chem. Ges. 36: 2094–2106.

Fischer, E. 1906. Untersuchungen über Aminosäuren, Polypeptide, und Proteïne. Ber. Dtsch. Chem. Ges. 39: 530–610.

Fischer, E. 1907. Synthese von Polypeptiden XVII. Ber. Dtsch. Chem. Ges. 40: 1754–1767.

Gutte, B., and Merrifield, R. B. 1969. The total synthesis of an enzyme with ribonucleasae A activity. J. Am. Chem. Soc. 91: 501.

Gutte, B., and Merrifield, R. B. 1971. The synthesis of ribonuclease A. J. Biol. Chem. 246: 1922.

Harington, C. R., and Mead, T.H. 1935. Synthesis of glutathione. Biochem. J. 29: 1602–1611.

Merrifield, R.B. 1963. Solid phase peptide synthesis I. The synthesis of a tetrapeptide. J. Am. Chem. Soc. 85: 2149–2154.

Merrifield, R.B. 1986. Solid phase synthesis. Science 232: 341–347.

Nutt, R.F., Brady, S.F., Darke, P.L., Ciccarone, T.M., Colton, C.D., Nutt, E.M., Rodkey, J.A., Bennett, C.D., Waxman, L.H., Sigal, I.S., Anderson, P.S., and Veber, D.F. 1988. Chemical synthesis and enzymatic activity of a 99 residue peptide with a sequence proposed for the human immunodeficiency virus protease. Proc. Natl. Acad. Sci. U.S.A. 85: 7129–7133.

Schneider, J., and Kent, S.B.H. 1988. Enzymatic activity of a synthetic 99 residue protein corresponding to the putative HIV-1 protease. Cell 54: 363–368.

Schwyzer, R., and Sieber, P. 1963. Total synthesis of adrenocorticotrophic hormone. Nature 199: 172–174.

Sifferd, R.H., and du Vigneaud, V. 1935. A new synthesis of carnosine, with some observations on the splitting of the benzyl group from carbobenzoxy derivatives and from benzylthio ethers. J. Biol. Chem. 108: 753–761.

Vigneaud, V. du, Ressler, C., Swan, J.M., Roberts, C.W., Katsoyannis, P.G., and Gordon, S. 1953. The synthesis of an octapeptide amide with the hormonal activity of oxytocin. J. Am. Chem. Soc. 75: 4879–4880.

Wieland, T. From Glycylglycine to Ribonuclease: 100 Years of Peptide Chemistry. In Perspectives in Peptide Chemistry, Eberle, A., Geiger, R., and Wieland, T., eds., Karger, Basel, 1981, pp. 1–13.

Wieland, T., and Bodanszky, M. The World of Peptides: A Brief History of Peptide Chemistry. Springer-Verlag, Berlin, 1991.

2

Peptide Design Considerations

Michael L. Moore

Peptides have become an increasingly important class of molecules in biochemistry, medicinal chemistry, and physiology. Many naturally occurring, physiologically relevant peptides function as hormones, neurotransmitters, cytokines and growth factors. Peptide analogs that possess agonist or antagonist activity are useful as tools to study the biochemistry, physiology, and pharmacology of these peptides, to characterize their receptor(s), and to study their biosynthesis, metabolism, and degradation. Radiolabeled analogs and analogs bearing affinity labels have been used for receptor characterization and isolation. Peptide substrates of proteases, kinases, phosphatases, and aminoacyl or glycosyl transferases are used to study enzyme kinetics, mechanism of action, and biochemical and physiological roles and to aid in the isolation of enzymes and in the design of inhibitors. Peptides are also used as synthetic antigens for the preparation of polyclonal or monoclonal antibodies targeted to specific sequences. Epitope mapping with synthetic peptides can be used to identify specific antigenic peptides for the preparation of synthetic vaccines, to determine protein sequence regions that are important for biological action, and to design small peptide mimetics of protein structure or function.

A number of peptide hormones or analogs thereof, including arginine vasopressin, oxytocin, luteinizing hormone releasing hormone (LHRH), adrenocorticotropic hormone (ACTH), and calcitonin, have already found use as therapeutic agents, and many more are being investigated actively. Peptide-based inhibitors of proteolytic enzymes, such as angiotensin converting enzyme (ACE), have widespread clinical use, and inhibitors of renin, human immunodeficiency virus (HIV) protease, and elastase are also being investigated for therapeutic use. Finally, peptides designed to block the interaction of protein molecules by mimicking the combining site of one of the proteins, such as the fibrinogen receptor antagonists, show great therapeutic potential as well.

With the development of solid-phase peptide synthesis by Bruce Merrifield (1963) and the optimization of supports, protecting groups, and coupling and deprotection chemistries by a large number of researchers, it has become possible to obtain useful amounts of peptides on a more or less routine basis. With the increasing ease of synthesizing peptides, it has become all the more important to understand the underlying principles of peptide structure and physical chemistry that govern solubility, aggregation, proteolytic resistance or susceptibility, secondary structure stabilization or mimicry, and interaction with nonpolar environments like lipid membranes.

The design of any specific peptide depends primarily on the use for which it is intended as well as on synthetic considerations. A number of aspects of the peptide sequence and structure can be manipulated to affect solubility, proteolytic resistance or stability, and the predilection to adopt specific secondary structures to produce peptides with specific properties. It is this topic that will be addressed in this chapter.

The distinction between what constitutes a peptide and what constitutes a protein becomes increasingly fuzzy as peptides increase in length. An operational definition in the context of this chapter might be that a peptide is any sequence that the researcher can conveniently synthesize chemically. Although such proteins as ribonuclease A (124 amino acids) (Hirschmann et al., 1969; Yajima and Fujii, 1981), acyl carrier protein (74 amino acids) (Hancock et al., 1972), and the HIV protease (99 amino acids) (Nutt et al., 1988; Wlodawer et al., 1989) have been chemically synthesized, most peptide synthesis generally involves peptides of 30 amino acids or less. The underlying principles applied to peptide design will be dependent in some degree on peptide length, especially those relating to peptide secondary structure. For example, shorter peptides do not tend to exhibit preferred solution conformations and possess a large amount of segmental flexibility. If secondary structure is important in a small peptide's design, it must be approached by chemical modifications designed to decrease conformational flexibility. As peptides increase in length, they have a greater tendency to exhibit elements of secondary structure, with a consequent decrease in segmental flexibility. In such cases, secondary structure can often be induced by optimizing the peptide sequence.

PROTEIN AND AMINO ACID CHEMISTRY

The physical and chemical properties of proteins and peptides are determined by the nature of the constituent amino acid side chains and by the polyamide peptide backbone itself. Twenty protein amino acids are coded for by DNA, which are translationally incorporated into proteins. Several amino acids can be modified in the protein post-translationally to yield new amino acids. Also, many peptides are synthesized enzymatically rather than ribosomally, especially in lower eukaryotes, and those peptides often contain highly unusual amino acids.

Protein Amino Acids

The structures of the 20 primary protein amino acids (those coded for by DNA) are given in Table 1 along with their three-letter abbreviations, one-letter codes, and a general grouping by physical properties. Amino acids generally can be divided into hydrophobic and hydrophilic residues. The hydrophobic residues include those with aliphatic side chains, such as alanine, valine, isoleucine, leucine, and methionine, and those with aromatic side chains, such as phenylalanine, tyrosine, and tryptophan. The hydrophilic residues include amino acids with (1) neutral, polar side chains, such as serine, threonine, asparagine, and glutamine; (2) those with acidic side chains, such as aspartic acid and glutamic acid; and (3) those with basic side chains such as histidine, lysine, and arginine. It can be appreciated that these categories are not entirely exclusive. Alanine, with its small aliphatic side chain, and glycine can be found in hydrophilic regions of peptides and proteins. Conversely, the long alkyl chains of lysine and arginine can give those residues an overall hydrophobic character with just the terminal charged group being hydrophilic.

Two amino acids, cysteine and proline, have special properties that set them apart. Cysteine contains a thiol moiety that can be oxidatively coupled to another cysteine thiol to form a disulfide linkage. Disulfides are the principal entities by which peptide chains are covalently linked together to stabilize secondary or tertiary structure or to hold two different peptide chains together. Although the disulfide form is the most stable form under normal aerobic conditions, free thiols are also present in some proteins, where they often serve as ligands for metal chelation,

Table 1   Amino Acid Structures and Properties

	H_2N ⋮ CO_2H (R, H)	
H	CH_3	CH_3 CH_3
Glycine Gly G	Alanine Ala A	Valine Val V
Neutral, hydrophobic, aliphatic		

CH_3 H CH_3	CH_3 CH_3	S CH_3
Isoleucine Ile I	Leucine Leu L	Methionine Met M
Neutral, hydrophobic, aliphatic		

	OH	N
Phenylalanine Phe F	Tyrosine Tyr Y	Tryptophan Trp W
Neutral, hydrophobic, aromatic		

Table 1   (continued)

$\diagup^{OH}$	$CH_3 \diagdown \overset{H}{\underset{}{\diagup}} OH$	$\diagup^{CONH_2}$
Serine Ser S	Threonine Thr T	Asparagine Asn N
Neutral, hydrophilic		

$\diagup^{CONH_2}$	$\diagup^{CO_2H}$	$\diagup^{CO_2H}$
Glutamine Gln Q	Aspartic Acid Asp D	Glutamic Acid Glu E
Neutral, hydrophilic	Acidic, hydrophilic	

	$\diagup^{NH_2}$	$H_2N \diagdown \diagup NH$ $\diagdown NH$
Histidine His H	Lysine Lys K	Arginine Arg R
Basic, hydrophilic		

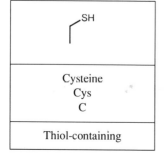

$\diagup^{SH}$
Cysteine Cys C
Thiol-containing

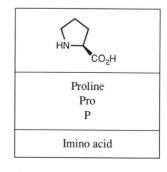

$HN \diagdown \diagup^{CO_2H}$
Proline Pro P
Imino acid

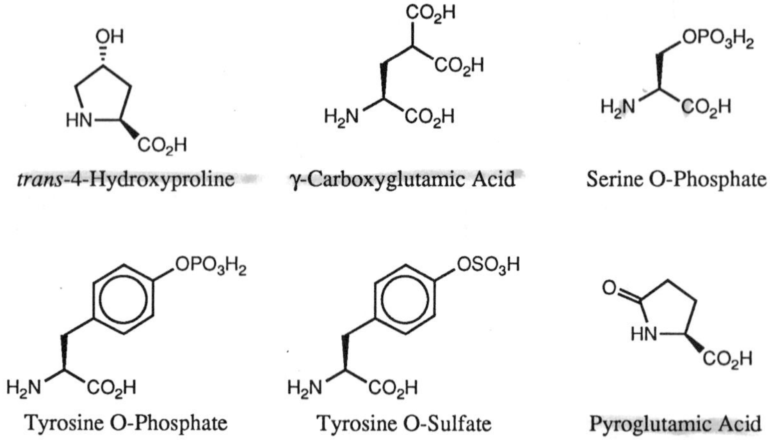

trans-4-Hydroxyproline   γ-Carboxyglutamic Acid   Serine O-Phosphate

Tyrosine O-Phosphate   Tyrosine O-Sulfate   Pyroglutamic Acid

FIGURE 1   Some amino acids formed by post-translational modification.

as nucleophiles in proteolytic enzymes, such as papain, or as carboxyl activators in acyl transferases. The secondary amino acid proline has specific conformational effects on the peptide or protein backbone both because of its cyclic structure and because of the alkylation of the amino group. It often plays an important role in stabilizing or influencing the secondary structure of proteins, as will be discussed later.

Some amino acids can be modified enzymatically after incorporation into certain proteins to give rise to new amino acids, some of which are shown in Figure 1. In collagen, for example, proline can be hydroxylated to yield *trans*-4-hydroxyproline (Hyp). Glutamic acid is carboxylated by a vitamin K-dependent carboxylase to yield γ-carboxyglutamic acid (Gla) in a number of proteins involved in blood coagulation, where the malonyl moiety of the Gla residue is thought to be important in providing a bidentate ligand for calcium ions. The hydroxyl functions of tyrosine, serine, and threonine can be reversibly phosphorylated by kinases and phosphatases; that process is thought to be an important regulator of biological activity in the target proteins. The hydroxyl of tyrosine may also be sulfated in peptide hormones such as gastrin and cholecystokinin.

One important non-enzymatic transformation of a protein amino acid occurs with glutamine, which is chemically unstable at the amino terminus of a peptide or protein in aqueous solution. Glutamine will spontaneously cyclize to form pyroglutamic acid (pyrrolidone carboxylic acid, also shown in Figure 1). This transformation typically occurs when an internal glutamine residue is exposed by proteolysis of an X-Gln bond (where X represents any amino acid) or during protein sequencing.

All amino acids (except for glycine) are chiral molecules with an asymmetric center at the α-carbon. The protein amino acids all have the L absolute configuration at this center, as shown in the structure at the top of Table 1. The L does not refer to the direction of optical rotation but, rather, to having the same stereochemical arrangement as L(−)-glyceraldehyde. In the more unambiguous Cahn-Ingold-Prelog convention, L-amino acids have the S absolute configuration (except cysteine, for which this configuration is defined as R).

Two amino acids, threonine and isoleucine, have a second asymmetric center at the β-carbon, as shown in Table 1. Threonine has the R absolute configuration at the β-carbon; that is, it is 2-(S)-amino-3-(R)-hydroxybutanoic acid. Isoleucine has the S absolute configuration at the β-carbon; that is, it is 2-(S)-amino-3-(S)-methylpentanoic acid. Because there are two asymmetric centers in the molecule, four stereoisomers exist for each of these amino acids. The D-amino acid has the opposite configuration at both asymmetric centers; D-isoleucine is the 2-(R)-3-(R) isomer, for example. The trivial prefix *allo* is used to denote inversion at only one asymmetric center. Thus D-*allo*-isoleucine is the 2-(R)-3-(S) isomer. D-*allo*-isoleucine occasionally forms as an artifact during hydrolysis of isoleucine-containing peptides or proteins. Because it is a diastereomer of isoleucine rather than an enantiomer, it will appear as a separate peak in amino acid analyses.

Non-protein Amino Acids

Literally hundreds of naturally occurring amino acids exist that are not found in proteins. Some of them, like γ-aminobutyric acid, have important functions as neurotransmitters. Others, like ornithine, appear as intermediates in metabolic pathways or, like dihydroxyphenylalanine, are precursors to amino acid-derived products, catecholamines in this case.

By far the largest number of non-protein amino acids are found in prokaryotes and lower eukaryotes, especially in algae, sponges, yeasts, and fungi, although peptides with unusual amino acids have been isolated from chordates, such as tunicates, as well. These amino acids are incorporated into peptides by enzymatic synthesis rather than ribosomally. More than 700 of these non-protein amino acids are known, and the structural variations are immense (Hunt, 1985). Those amino acids and the peptides containing them are usually the products of secondary metabolism, and their function in the producing organism is often obscure. Those unusual amino acids can confer unusual biological activities on peptides that contain them. Such peptides have been the basis either of biochemical tool molecules or of antibiotics, immunomodulators or antineoplastic agents that have therapeutic utility.

Non-protein amino acids can be roughly categorized into a few basic structural types, representative members of which are depicted in Figure

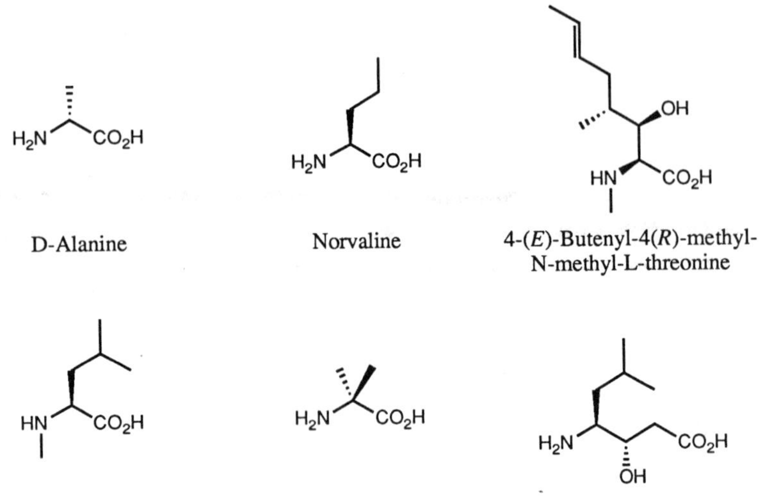

D-Alanine

Norvaline

4-(*E*)-Butenyl-4(*R*)-methyl-N-methyl-L-threonine

N-Methyl-L-leucine

Aminoisobutyric Acid

Statine

FIGURE 2 Some non-protein amino acids.

2. Some non-protein amino acids are simply the enantiomeric D-amino acid analog of a protein L-amino acid, e.g., D-alanine and D-glutamic acid, which are important constituents of the proteoglycan bacterial cell wall. Others have a normal, α-amino acid structure but with a novel side chain. The side chain can be a simple alkyl group, such as in norvaline, or it can be quite unusual, such as in 4-(*E*)-butenyl-4(*R*)-methyl-N-methyl-L-threonine (MeBmt), a critical constituent of the immunosuppressive peptide cyclosporin A (Rüegger et al., 1976). Some non-protein amino acids deviate from the normal α-amino acid structure and are methylated on their amine function, such as the N-methylleucine (MeLeu) of the tunicate-derivated antineoplastic peptide didemnin B (Rinehart et al., 1981), or on their α-carbon, such as the aminoisobutyric acid (α-methylalanine, Aib) of the ionophoretic antibiotic peptide alamethicin (Payne et al., 1970; Pandey et al., 1977). In addition, there are large numbers of amino acids in which the amino group is not on the α-carbon but at some other position in the molecule. Sometimes several of these features are combined in one amino acid, such as statine [3-(*S*)-hydroxy-4-(*S*)-amino-6-methylheptanoic acid, Sta], the critical residue in the fungal protease inhibitor pepstatin (Morishima et al., 1970).

Side-Chain Interactions

Side chains interact with each other, with the amide peptide backbone, with bulk solvent, and through non-covalent interactions, such as hydrogen bonds, salt bridges, and hydrophobic interactions. Cysteine also participates in a covalent interaction—disulfide bond formation.

Hydrogen Bonds and Salt Bridges

In proteins, polar side chains tend to be extensively solvated. Acidic (Asp and Glu) and basic (Lys, Arg, and His) residues generally are found on the protein surface with the charged ends of the side chains projecting into the bulk solvent, although the alkyl portion of the Lys and Arg side chains is usually buried. Internal charged residues are almost invariably involved in salt bridges, where acidic and basic side chains are either directly bonded ionically to each other or connected by a single inter-mediary water molecule (Baker and Hubbard, 1984). Non-ionic polar residues (Ser, Thr, Asn, Gln, and Tyr at the phenolic hydroxyl) are also extensively hydrogen-bonded, either to bulk solvent or to backbone, other side-chain groups, or to specifically bound water molecules (Thanki et al., 1990). In helices, the side chains of Ser, Thr, and Asn often make specific hydrogen bonds to the carbonyl oxygen of the third or fourth residue earlier in the sequence, which may to help stabilize helical segments (Gray and Matthews, 1984). In shorter peptides, side-chain hydration occurs mostly through the bulk solvent, although the ability to form low-energy intramolecular hydrogen-bonded structures or salt bridges may be an important factor in the association of peptides with macromolecular targets or receptors.

Consideration of solvation and desolvation effects is a potentially critical, but often overlooked, aspect of peptide design and structure–activity relationships. One case will serve to illustrate the importance of solvation in interpreting structure–activity studies. Bartlett and cowork-ers prepared a series of thermolysin inhibitors that were based on the peptide substrate sequence Cbz-Gly-Leu-Leu-OH but contained a phos-phoryl moiety in place of the Gly carbonyl function (Morgan et al., 1991). The phosphoryl moiety was designed to mimic the tetrahedral transition state for amide bond hydrolysis and was suggested by a naturally occurring glycopeptide inhibitor of thermolysin, phosphor-amidon, which contains a similar phosphoryl group. The inhibitors that were prepared are shown in Table 2. Compound 1, which contains a phosphoramidate linkage (PO_2NH), was found to be a potent inhibitor. X-ray crystal structure analysis of the enzyme-inhibitor complex showed a hydrogen bond between the phosphoramidate NH and a backbone car-bonyl in the enzyme. Compound 2 contained a phosphonate linkage (PO_2O). It was unable to form the same hydrogen bond as compound 1 because it lacks the corresponding hydrogen. It was found to be three orders of magnitude less potent than compound 1, but it bound to the enzyme in an identical fashion (except for the absence of the hydrogen bond) by X-ray crystal analysis. Compound 3 contained a phosphinate

Table 2   Phosphorus-Containing Thermolysin Inhibitors

Compound	X	Ki(nM)
1	NH	9.1
2	O	9000
3	CH_2	10.6

linkage (PO_2CH_2). Like compound 2, it could not form that same hydrogen bond. Unlike compound 2, however, it was essentially equipotent with the phosphoramidate 1. The explanation includes not only its ability to form a hydrogen bond with the enzyme but also its solvation and desolvation effects. Both the phosphoramidate 1 and the phosphonate 2 will be solvated in aqueous solution and require desolvation to bind to the enzyme active site. The phosphoramidate 1 can recover some of the energy required for desolvation by forming a hydrogen bond with the enzyme, but the phosphonate 2 cannot. The phosphinate 3 is somewhat less polar due to the presence of a methylene group instead of an amine or oxygen moiety. It is correspondingly less solvated in aqueous solution, and, therefore, it requires less desolvation when it binds to the enzyme. The net change in free energy in going from solution to enzyme-bound states is roughly comparable for compound 1 (with the PO_2X group going from solvated to solvated states) and compound 3 (going from unsolvated to unsolvated states) and is more favorable than for compound 2 (going from solvated to unsolvated states.)

Hydrophobic Interactions

Just as hydrophilic residues tend to be solvated, hydrophobic side chains have an equally strong tendency to avoid exposure to the aqueous environment. This effect is largely entropic, reflecting the unfavorable free energy of forming a water–hydrocarbon interface, where the side chain would penetrate the aqueous solvent (Tanford, 1973; Burley and Petsko, 1988). It has long been recognized that the interiors of soluble proteins are highly hydrophobic (Kauzmann, 1959) and that proteins fold in such a way as to minimize the exposure of hydrophobic side chains on the protein surface. It has been proposed that protein folding proceeds first through the formation of hydrophobic clusters, which direct further folding of the peptide chain into the various low-energy secondary structures such as helices and β-structure (Rose and Roy, 1980). The importance of hydrophobic interactions to protein stability has been studied by comparing the susceptibility to denaturation of a series of proteins in which site-specific mutagenesis was used to modify single residues in the protein sequence (Kellis et al., 1988). It was found that absence of even a single methyl group (Ile to Val substitution) destabilized the protein by 1.1 kcal/mol, underscoring the important cumulative effect of hydrophobic interactions on overall protein structure and stability.

Although the sequestering of nonpolar residues away from the aqueous environment is largely an entropy-driven process, specific interactions of hydrophobic side chains occur as well. These are typically induced dipole-induced dipole interactions, known as van der Waals or London interactions (Burley and Petsko, 1988). Although they are much weaker than the salt bridges and hydrogen bonds involving polar residues, they can be important in local secondary structure and protein interactions. For example, a number of DNA-binding proteins that function biologically as dimers share an unusual sequence, featuring a leucine every seventh residue in a 30-residue sequence and having a relatively high probability of helical structure. This sequence would create a helix in which the leucine residues occupied every other turn of the same side of the helix. The structure was termed a "leucine zipper" based on the hypothesis that the leucines from such a helix in two monomers could interdigitate like a zipper, holding the monomers together (Landschutz et al., 1988). Although it was subsequently found that the Leu residues do not interdigitate but rather align themselves parallel to each other in pairs along the helical interface (O'Shea et al., 1989), the term leucine zipper has remained, and it is an important structural motif in hydrophobic protein–protein interaction.

Aromatic residues have an inherent dipole with the electron-rich π-cloud lying parallel to and above and below the plane of the ring and with positively polarized hydrogen atoms in the plane of the ring (Burley and Petsko, 1988). Although aromatic rings in proteins do interact with each other, the arrangement is not typified by the parallel stacking of base pairs in DNA helices. Rather, the edge of one ring interacts with the face of the other in a roughly perpendicular arrangement (Burley and Petsko, 1985), allowing a favorable interaction between the positively polarized hydrogens and the negatively polarized π-cloud.

Peptides, in general, are not long enough to allow the hydrophobic residues to arrange themselves in such a way that they are totally shielded from solvent. This occurrence undoubtedly contributes to the poor aqueous solubility of many peptides compared to proteins. Many peptides require the presence of strong organic cosolvents like dimethylsulfoxide (DMSO), dimethylformamide (DMF), or ethanol to achieve

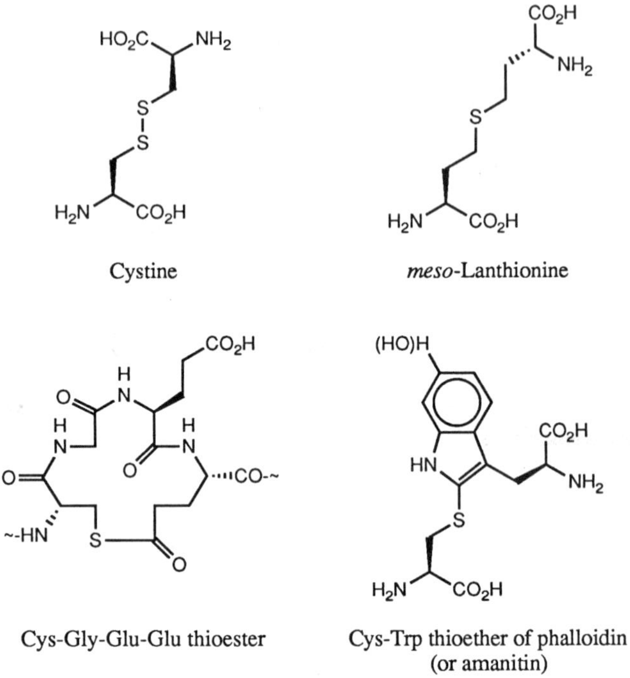

Cystine

meso-Lanthionine

Cys-Gly-Glu-Glu thioester

Cys-Trp thioether of phalloidin
(or amanitin)

FIGURE 3   Disulfide, thioester, and thioether linkages.

sufficient solubility for biological testing, which may be a limiting factor in some biological test systems. Peptides that have substantial hydrophobic character also tend to aggregate with increasing concentration. Solubility generally increases with peptide length because of the peptide's consequent ability to adopt stable secondary structures and to segregate non-polar residues.

Disulfide Bonds, Thioesters, and Thioethers

Unlike side-chain interactions of other residues that are non-covalent in nature, cysteine residues form a number of covalent linkages with other amino acid side chains. The most common of these is the disulfide, which involves oxidative coupling of two cysteine thiol groups to form cystine (Figure 3). Like the amide bond, the sulfur–sulfur bond in a disulfide is not freely rotatable. Rather, it exists in one of two rotamers, with torsional angles in the vicinity of either + 90° or –90°. The entire disulfide moiety, CH_2—S—S—CH_2, can rotate as a unit by simultaneous rotation about the side-chain angles χ_1 (NH—CαH—CH_2—S) and χ_2 (CαH—

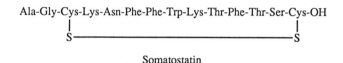

Cys-Tyr-Ile-Gln-Asn-Cys-Pro-Leu-Gly-NH₂
  | |
  S————————S

Oxytocin

Cys-Tyr-Phe-Gln-Asn-Cys-Pro-Arg-Gly-NH₂
  | |
  S————————S

Arginine Vasopressin

Ala-Gly-Cys-Lys-Asn-Phe-Phe-Trp-Lys-Thr-Phe-Thr-Ser-Cys-OH
      | |
      S————————————————S

Somatostatin

FIGURE 4   Disulfide-containing peptides.

CH_2—S—S). In proteins, disulfide bonds are important in the stabilization of tertiary structure. In disulfide-containing peptides like oxytocin, vasopressin, and somatostatin (see Figure 4), the disulfide has proportionally greater importance in maintaining a biologically active conformation because the peptides are too small to maintain a stable conformation otherwise. Replacement of the cysteine residues with isosteric alanine residues results in a dramatic loss in biological activity (Walter et al., 1967; Polácek et al., 1970; Sarantakis et al., 1973). A great deal of conformational flexibility still remains in cyclic disulfide-containing peptides like vasopressin, and it does not exhibit a single preferred conformation (Hagler et al., 1985). The disulfide does constrain the peptide to folded conformations that otherwise would have a low probability for linear peptides.

Cysteine residues can also form covalent thioester and thioether linkages. The thioester is found as the activated form of acyl groups in acyl transferases and acyl enzyme intermediate in thiol proteases such as papain. Thioesters also are found in the reactive binding sites of complement C3b and α_2-macroglobulin, where the tetrapeptide sequence Cys-Gly-Glu-Glu contains a thioester involving the Cys[1] thiol and the Glu[4] side-chain carboxylate (Figure 3) (Sothrup-Jensen et al., 1980; Tack et al., 1980; Howard, 1981). Cysteine forms aliphatic thioether linkages, such as in lanthionine, a constituent of the peptide antibiotics, subtilin (Alderton and Fevold, 1951), and nisin (Berridge et al., 1952). Aromatic thioether linkages also occur, such as in the Cys-Trp conjugate found in mushroom toxins, for example amanitin and phalloidin (Wieland, 1968).

(a)   Resonance forms of amide bond

trans amide bond
$\omega = 180°$

cis amide bond
$\omega = 0°$

(b)   *trans* and *cis* amide bonds

FIGURE 5   Amide bonds.

A novel thioether Cys-Tyr conjugate recently has been identified in the enzyme galactose oxidase (Ito et al., 1991).

The Amide Bond

The polyamide peptide backbone is also an important contributor to overall protein and peptide structure. A substantial double-bond character is found in the carbon–nitrogen peptide bond due to the resonance structure shown in Figure 5(a). This structure gives the amide bond several characteristics that are important in peptide and protein structure. The amide bond is flat, with the carbonyl carbon, oxygen, nitrogen, and amide hydrogen all lying in the same plane. No free rotation occurs about the carbon–nitrogen bond because of its partial double-bond character (the barrier to rotation is about 25 kcal/mol). The torsional angle of that bond, ω, is defined by the peptide backbone atoms C_α—$C(O)$—N—C_α. Because of the partial double-bond character, there are two rotational isomers for the peptide bond: *trans* ($\omega = 180°$) and *cis* ($\omega = 0°$), as shown in Figure 5(b). The lower-energy isomer is the *trans* peptide bond, which is the isomer generally found for all peptide bonds not involving proline. In the case of amide bonds involving proline, the energy of the *trans* X-Pro bond is somewhat elevated, and both the difference in energy between *cis* and *trans* isomers and the barrier to rotation is lowered. Proline-containing peptides thus will often exhibit *cis-trans* isomerism about the X-Pro bond. This can be detected by NMR studies because the chemical shifts of some hydrogens, especially those on the proline δ-carbon, are often different. The equilibrium ratio of *cis* to *trans* isomers and the rate of isomerization are highly dependent on the exact peptide sequence. Empirically, it has been found that the *cis* content of X-Pro peptides generally increases when X is a bulky, hydrophobic amino acid (Harrison and Stein, 1990).

It has been appreciated only recently that *cis* X-Pro peptide bonds may be important in proteins as well. The X-ray crystal structure of several proteins has revealed specific *cis* X-Pro peptide bonds (Frömmel and Preissner, 1990). Proline isomerization has also been implicated as a slow step in protein refolding (Kim and Baldwin, 1982). Most recently, it has been shown that cyclophilin, the binding protein for the immunosuppressive peptide cyclosporin A, catalyzes peptidyl proline *cis-trans* isomerization (Fischer et al., 1989; Takahashi et al., 1989). Intriguingly, unrelated immunosuppressive macrolides, such as FK506, which apparently have a different mechanism of action from cyclosporin A, also have cellular binding proteins (immunophilins) that possess proline isomerase activity (Harding et al., 1989; Siekierka et al., 1989). It is unclear how proline isomerase activity is related to immunosuppressive activity, since the two activities can be dissociated using synthetic analogs of cyclosporin or FK506 (Bierer et al., 1990; Sigal et al., 1991). Nor is there a known physiological role for cyclophilin, although it is suspected of being involved in protein folding (Steinmann et al., 1991). It may turn out that proline isomerization plays an important regulatory role in cellular biochemistry as well.

The resonance forms of the amide bond give it one other characteristic that is extremely important in peptide and protein structure. The amide bond is quite polar and has a significant dipole moment, which makes the amide carbonyl oxygen a particularly good hydrogen-bond acceptor and the amide NH a particularly good hydrogen-bond donor. Hydrogen bonds involving the peptide backbone are an important stabilizing factor in protein secondary structures. Peptide bonds have a strong tendency to be solvated, by either bulk solvent or specifically bound water molecules, or by internal hydrogen bonds. This is especially true in the hydrophobic interior of soluble protein molecules. Peptide bonds of regular secondary structures, such as helices and β-sheets, are internally solvated by the hydrogen bonds that stabilize those structures. The peptide bonds of the so-called random coil, or irregular structures, also participate in an extensive network of hydrogen bonds involving internal polar side chains, bound water molecules, and backbone interactions.

The polarity of the amide bond also can impart a net dipole to regular structures containing peptide bonds, such as helices. The overall dipole points from the amino terminus (positive partial charge) to the carboxyl terminus (negative partial charge) in a helix, which can be an important factor in protein tertiary structure as well as a contributor to catalytic activity in enzymes. The dipole can be used to stabilize interactions with substrates or to modify the pK_a, or nucleophilicity, of catalytically active residues (Knowles, 1991).

PROTEIN STRUCTURE

Protein structure is organized in several hierarchical levels of increasing structural complexity. The most basic level is the primary structure, or amino acid sequence, of the protein. Secondary structure deals with the folding up of short segments of the peptide, or protein chain, into regular structures such as α-helices, β-sheets, and turns. Tertiary structure describes how the secondary structural elements of a single protein chain interact with each other to fold into the native protein structure. Quaternary structure involves the interaction of individual protein subunits to form a multimeric complex.

In the design of peptides, we are concerned mainly with the primary and secondary structures. The primary structure, or sequence, of a protein, is readily obtainable by chemical sequence analysis of the isolated, purified protein or by translation of the corresponding DNA coding sequence. It is easier to sequence DNA than to sequence protein, and "fishing out" genomic DNA fragments from a DNA library has become quite routine in many laboratories. Thus, most new protein sequences are now obtained in this way. Computer programs are available, which will be discussed later (see Prediction of Protein/Peptide Structure), to help analyze and compare protein primary structures. Protein secondary structure has also become increasingly well understood, to the point that it can be predicted or optimized on the basis of the primary sequence reasonably well by using computer algorithms. In addition, several techniques have been applied to either induce or stabilize secondary structure in peptides, which will be discussed in more detail later (see Conformational Design and Constraint).

Protein Secondary Structure

Chains of amino acids can fold into several types of regular structures that are stabilized by intrachain or interchain hydrogen bonds in the amide backbone. Helices and turns are formed from continuous regions of protein sequence and are stabilized by intrachain hydrogen bonds. Sheets are formed from two or more chains, which are separated by intervening sequences (regions of secondary structure as well), and are stabilized by interchain hydrogen bonds. The parts of a protein sequence that do not appear in helices, sheets, or turns are said to be in random coil, which is meant to imply only that no regular, repeating structure exists. It should not be taken to mean that the conformation of the protein is random in those regions, because in any individual protein, the random coil sequences are just as highly ordered and reproducibly formed as other regions of secondary structure.

The conformation of the peptide backbone can be described by three torsional angles: ϕ, which is the angle defined by C(O)—N—C$_\alpha$—C(O); ψ, which is defined by N—C$_\alpha$—C(O)—N; and the amide torsional angle ω, which was discussed previously (see The Amide Bond). These angles are shown graphically in Figure 6(a). The convention in protein chemistry is that these angles are defined with respect to the peptide backbone. The angle ϕ is 180° when the two carbonyl carbons are *trans* to each other, and the angle ψ is 180° when the two amide nitrogens are *trans* to each other as in Figure 6(b). The angle ψ, therefore, is +180° compared to the usual chemical definition. A fully extended peptide chain, however, would have backbone torsional angles all of 180°, just like a fully extended carbon chain.

Helices

One of the most common types of secondary structure in proteins is the helix in which amino acid residues are wrapped around a central axis in a regular pattern. Given planar, *trans* peptide bonds, helices will be uniquely defined by their ϕ and ψ angles. Several types of helices are found in protein structures, the α-helix being the most common. These helices and their characteristic torsional angles are listed in Table 3. Helices are also characterized by the number of residues per turn and the hydrogen-bonding pattern, specifically the number of atoms in the cyclic structure formed by the hydrogen bond. The α-helix is a 3.6_{13} helix, meaning there are 3.6 residues per turn of the helix, and 13 atoms are included in each repetitive hydrogen-bonding structure. Likewise, a 3_{10} helix has 3 residues per turn and 10 atoms in the cyclic hydrogen-bonded structure. These are the two most relevant helices for protein and peptide structure. There is also a handedness to helical structure, which is defined by the screw sense of the helix. Looking down the helical axis from the amino terminal end, the peptide backbone can be traced from amino terminus to carboxyl terminus in a clockwise sense in a right-handed helix, which is the form that generally occurs.

Helical structures are stabilized by intrachain hydrogen bonds. In a helix, the carbonyl bonds and amide NH bonds lie parallel to the helix's axis with the carbonyls pointing downward, in the direction of the carboxyl terminus of the helix, and the NHs pointing upward, in the direction of the amino terminus. In an α-helix, the carbonyl of any residue i is hydrogen-bonded to the amide NH of the $i+4$ residue; that is, the residue 4 amino acids farther down along the peptide chain. In a 3_{10} helix, the ith residue carbonyl is hydrogen-bonded to the NH of the $i+3$ residue. These relationships are shown graphically in Figures 7 and 8 for six-residue

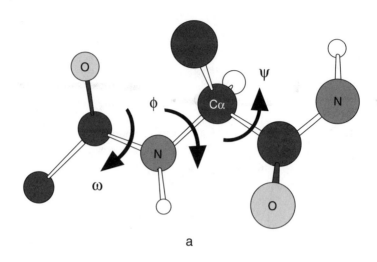

a

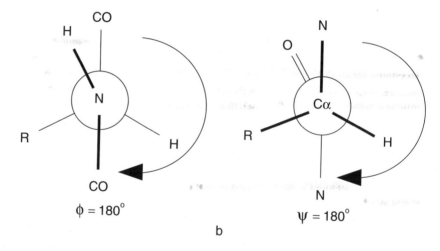

$\phi = 180°$ $\psi = 180°$

b

FIGURE 6   Torsional angles of the peptide backbone. (a) Backbone torsional angles ϕ, ψ, and ω. (b) Newman projections of torsional angles ϕ and ψ.

segments of α and 3_{10} helices as viewed from the side, perpendicular to the helical axis. For clarity, only the peptide backbone atoms are shown.

The helix forms a polyamide cylinder. The uniform arrangement of carbonyl and NH groups imparts a strong dipole moment to the helix,

TABLE 3 Torsional Angles for Helices ϕ^{hi} ρ^{si}

	φ	ψ
α Helix (right-handed)	−57°	−47°
α Helix (left-handed)	57°	47°
3_{10} helix	−60°	−30°
Collagen helix	−51°	153°
	−76°	127°
	−45°	148°
Polyproline	−78°	149°

Angles taken from the IUPAC-IUB Commission on Biochemical Nomenclature, 1970.

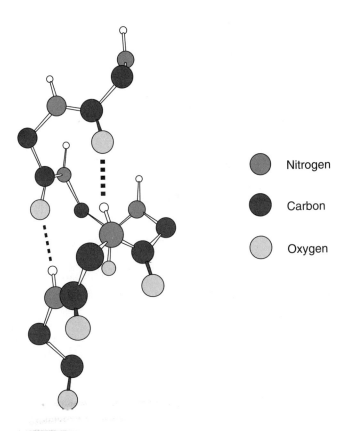

Nitrogen

Carbon

Oxygen

FIGURE 7   Backbone structure of an α-helix. Note that the NH and carbonyl groups are aligned parallel to the axis of the helix. Each hydrogen-bonded segment contains 13 atoms.

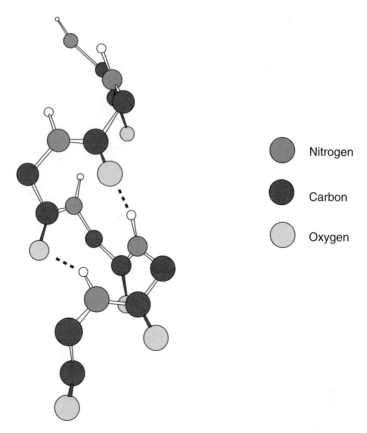

Nitrogen

Carbon

Oxygen

FIGURE 8   Backbone structure of a 3_{10} helix. It is slightly elongated compared to the α-helix, with 3 residues per turn of the helix instead of 3.6. Each hydrogen-bonded segment contains 10 atoms.

with the positive end at the amino terminus and the negative end at the carboxyl terminus. The side chains of the residues are arranged radially outward from the helix, looking down the helix axis (Figure 9). Because all the backbone amide groups are involved in intrachain hydrogen bonds, the interactions of helices with other peptide chains or small molecules occurs predominantly through side-chain interactions (hydrophobic interactions, salt bridges, and hydrogen-bonding interactions; see Side-Chain Interactions) and interactions with the helix dipole itself. Because of this arrangement of intrachain hydrogen bonds, the amide bonds of the helix can be thought of as being internally solvated, resulting in the helix being more readily accommodated in a non-polar environment, such as a lipid bilayer. The membrane-spanning regions of transmembrane proteins are generally thought to be helical, with the axis

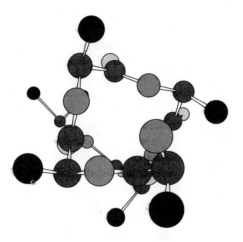

FIGURE 9   View of an α-helix looking down the helical axis. The residue side chains, shown in black, all project radially outward from the helix.

of the helix perpendicular to the plane of the membrane (Eisenberg, 1984; Wickner and Lodish, 1985). The membrane pore-forming peptide antibiotics, such as alamethicin, also have a highly helical structure, and a high degree of helix potential exists in the hydrophobic leader sequences of proteins that are transported through membranes.

β-Structure

Although helices are a form of secondary structure resulting from one strand of peptide chain, β-structure requires two strands. The conformation of each peptide chain is generally extended, and interchain hydrogen bonds occur between the carbonyl and amide NHs of every other residue in the backbone. Two possible arrangements of adjacent strands occur: parallel, in which each peptide strand runs in the same direction, and antiparallel, in which the peptide strands run in opposite directions. Both can be found in protein structures. These two arrangements are shown diagrammatically in Figure 10.

The β-structure can be thought of as forming a surface or sheet, although there is a slight twist to it because the peptide backbone is not fully extended (typical angles are given in Table 4). The side chains extend above and below the rough plane of the sheet, with every other side chain on one surface. Because the interchain hydrogen bonds between two adjacent chains involve only every other residue, there is the possibility of larger structures forming that involve many strands. Such extended β-structures are common features in proteins. The twist of the

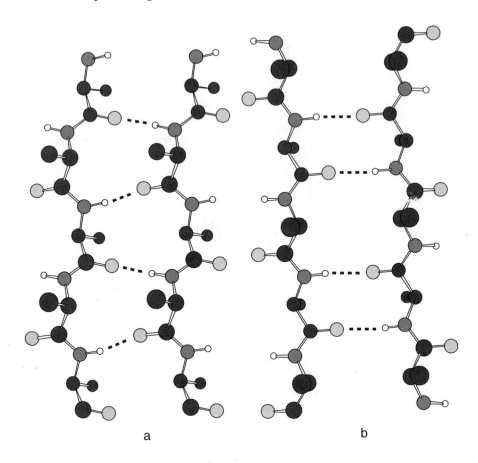

a   b

FIGURE 10   Structure of β-sheet. Amide bonds lie in the plane of the sheet, with side chains alternately projecting above and below the plane of the sheet. (a) Parallel sheet, with both strands running in the same direction. (b) Antiparallel sheet, with strands running in opposite directions.

TABLE 4   Torsional Angles for β Structure

	ϕ	ψ
Parallel β sheet	−119°	113°
Antiparallel β sheet	−139°	135°

β-structure also allows sheets to fold into cylindrical structures, called β-barrels, which is another common structural motif.

Turns

Strands of α-helix or β-structure do not extend indefinitely in proteins but rather fold back on themselves. Several regular structures, called turns, are involved in changing the direction of the peptide chain. Turns are classified by the number of residues that are involved in the regular structure; β-turns comprise four amino acid residues, while γ-turns contain three residues. Each structure is stabilized by a hydrogen bond extending across the turn, in effect holding the two ends together. The first residue of a turn is usually designated as i. In a β-turn, the hydrogen bond is between the carbonyl of the i residue and the NH of the $i + 3$ residue, giving the equivalent of a ten-membered ring. There may be an additional hydrogen bond between the NH of the i residue and the carbonyl of the $i + 3$ residue, as if they were an incipient antiparallel β-structure. In a γ-turn, the hydrogen bond is between the carbonyl of the i residue and the NH of the $i + 2$ residue, giving the equivalent of a seven-membered ring. (Turns are sometimes named for the number of atoms in the ring structure formed by the hydrogen bond. Thus a β-turn would be a C_{10} turn and a γ-turn a C_7 turn.)

Several types of β-turns are possible. Given planar amide bonds, β and γ turns are uniquely defined by the torsional angles of the internal residues $i + 1$ and, in the case of β-turns, $i + 2$. Table 5 lists the torsional angles associated with the various types of β-turns and γ-turns. Note that

TABLE 5 Torsional Angles for Turns

	$i + 1$		$i + 2$	
	ϕ	ψ	ϕ	ψ
Type 1β	$-60°$	$-30°$	$-90°$	$0°$
Type I′β	$60°$	$30°$	$90°$	$0°$
Type IIβ	$-60°$	$120°$	$80°$	$0°$
Types II′β	$60°$	$-120°$	$-80°$	$0°$
Type IIIβ	$-60°$	$-30°$	$-60°$	$-30°$
Type III′β	$60°$	$30°$	$60°$	$30°$
γ turn	$70 - 85°$	$-60 - -70°$		
Inverse γ turn	$-70 - -85°$	$60 - 70°$		

Angles taken from Smith and Pease (1980).

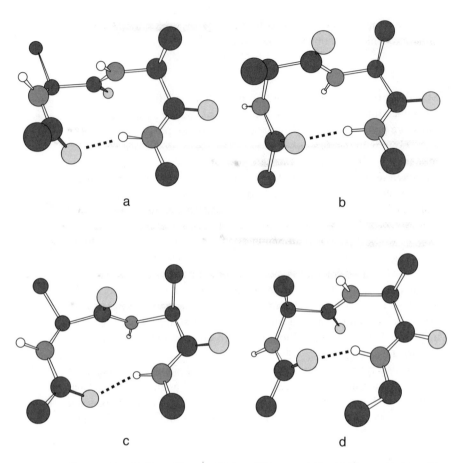

FIGURE 11   (a) Type I β-turn. (b) Type I′ β-turn. (c) Type II β-turn. (d) Type II′
β-turn. The turn is stabilized by a hydrogen bond from the carbonyl of the ith
residue to the NH of the $i+3$ residue. The amide bond between the $i+1$ and $i+2$
residues is perpendicular to the plane of the turn. The side chain of the $i+1$
residue in a type I′ and II′ turn is in a psuedo-axial orientation. Substitution of a
D amino acid would give the more stable pseudo-equatorial orientation. The
NH and side chain of the $i+2$ residue of the type II′ turn are both on the same
side of the ring, an arrangement in which a proline residue could be accommodated as well. The NH and side chain of the $i+2$ residue of a type I′ turn are
on opposite sides of the ring, which would exclude the possibility of a proline
residue at that position.

there is an inverse form for each turn in which the sign of the internal
torsional angles is reversed. Note also that the type III β-turn is
equivalent to a 3_{10} helix structure. Figure 11 depicts the backbone of
types I, I′, II, and II′ β-turns and Figure 12 depicts that of γ- and inverse
γ-turns.

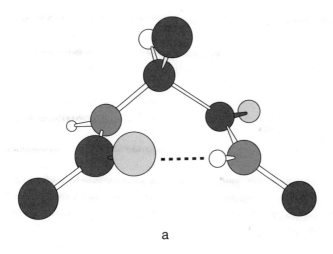

a

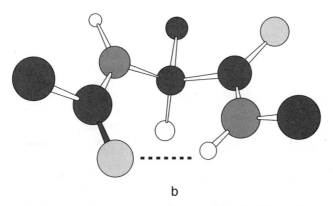

b

FIGURE 12   Structure of a γ-turn (a) and inverse γ-turn (b). The turns are stabilized by a hydrogen bond from the carbonyl of the *i*th residue to the NH of the
i+2 residue. The side chain of the *i*+1 residue adopts a pseudo-axial orientation
in the γ-turn and a pseudo-equatorial orientation in the inverse γ-turn.

Turns have several important structural features. The peptide backbone in a turn forms a rough plane that contains the intramolecular hydrogen bond. In a β-turn, the amide bond between the $i + 1$ and $i + 2$
residues lies perpendicular to this plane. Since it is not part of the hydrogen-bonding structure of the turn, its hydrogen bonding needs must be
satisfied elsewhere. In proteins, β-turns are most often found on the
protein surface, where the $i + 1 - i + 2$ peptide bond can be solvated by
bulk solvent. When such a turn is located in the (generally hydrophobic)

interior of a protein, it is often solvated by a bound water molecule (Smith and Pease, 1980). Besides the amide bond, the side chains of the $i + 1$ and $i + 2$ residues, which form the "corners" of the turn, are also particularly exposed. In small peptides, such as vasopressin or somatostatin, where a turn has been postulated based on the cyclic nature of the peptide chain, the corner residues are implicated as important elements in the expression of the peptide's biological activity (Rivier et al., 1976; Walter et al., 1977).

Side chains will adopt either a pseudo-axial or pseudo-equatorial arrangement around the cyclic hydrogen-bonded turn structure. One diagnostic criterion for the importance of a turn involves changing the chirality of a corner residue to allow the side chain to be in the more stable equatorial configuration without otherwise altering the peptide backbone structure. In a type II' β-turn, for example, the $i + 1$ side chain is in a pseudo-axial configuration (see Figure 11). Replacement of this residue with a D amino acid should allow for equatorial placement of the side chain while the peptide backbone remains unchanged, lowering the overall free energy of that structure and making it more stable. Replacing the $i + 1$ residue with a glycine, with no side chain, should have a similar effect.

Amphiphilic Structures

Given the regular placement of side chains in helical or β-structures, it is possible to obtain forms in which the polar side chains are segregated on one face of the structure and hydrophobic side chains are on the other. This type of arrangement is termed an *amphiphilic,* or *amphipathic,* structure.

In an ideal amphiphilic helix, the hydrophobic residues are all found on one side of the helix cylinder, and the hydrophilic residues are found on the other side of the cylinder. In the linear sequence, these hydrophilic and hydrophobic residues will be interdispersed, so it can achieve the desired arrangement when folded into a helix. One idealized amphipathic helix, designed by Kaiser and coworkers as a model for apolipoprotein A, has the helical sequence Lys-Leu-Glu-Glu-Leu-Lys-Glu-Lys-Leu-Lys-Glu-Leu-Leu-Glu-Lys-Leu-Lys-Glu-Lys-Leu (Fukushima et al., 1980). The arrangement of side chains can best be shown schematically in a helix wheel projection, which is depicted looking down the helix axis with the side chains arranged radially outward, as shown in Figure 13(a).

In β-structure, the side chains alternate above and below the plane of the sheet. In an amphiphilic strand of β-structure, every other residue in the linear sequence would be hydrophobic and the remaining residues would be hydrophilic. This arrangement would give a structure in which one face of the sheet would be hydrophobic and the other face would be hydrophilic, as in Figure 13(b).

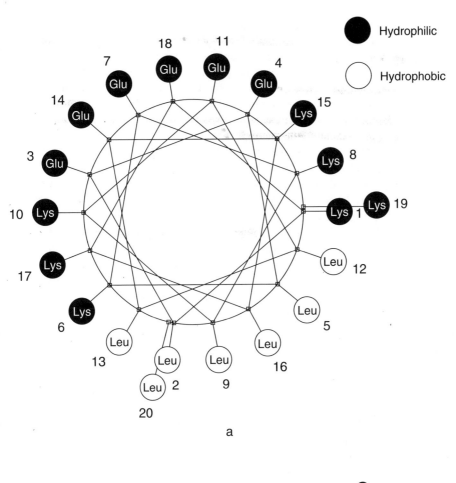

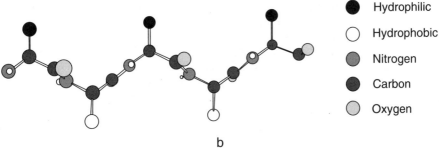

FIGURE 13   (a) Helical wheel projection of a designed amphipathic helix. The residues with hydrophilic side chains are segregated on one face of the helix, while the hydrophobic residues are segregated on the other. (b). Side view of one strand of an amphipathic β-sheet. The hydrophilic side chains project on one side of the sheet, while the hydrophobic side chains project on the other side. The puckering of the backbone, which gives rise to the often-used name β-pleated sheet, is clearly evident in this side view.

In proteins, amphiphilic structures can be important structural elements, allowing the hydrophobic residues to be sequestered in the interior of the protein while a hydrophilic face is presented to the aqueous environment at the protein surface. This arrangement can be reversed in transmembrane proteins, with amphiphilic helices clustered so that the hydrophobic face is exposed to the lipid environment and the hydrophilic faces are sequestered in the interior of the protein, forming a hydrophilic pore, or channel, in the membrane.

The recognition that amphipathic structures are important to the biological activity of smaller peptides is due in large part to the pioneering work of E. T. Kaiser. He recognized potential amphiphilic helices in such membrane-active peptides as apolipoprotein A and mellitin, as well as peptide hormones such as calcitonin, corticotropin releasing factor, and β-endorphin, which interact with membrane-bound receptors. Structure–activity studies of synthetic analogs of these peptides have demonstrated the importance of their amphiphilic structures by designing sequences optimizing amphipathicity or replacing it with nonhomologous sequences adopting the same amphiphilic structure (Kaiser and Kézdy, 1984). Examples of the design of amphiphilic peptides will be discussed later (see Hydrophobicity Measurements).

Prediction of Protein / Peptide Structure

Protein Sequence Homology

The primary sequence of a peptide, or protein, is its most fundamental structural element. It has long been presumed that all the information content of a protein—folding, translocation, and function—are encoded within the amino acid sequence of the primary translation product. Many proteins can be denatured reversibly and renatured with complete recovery of biological activity, lending credence to this concept. The situation in protein biosynthesis is more complex, because nascent peptides can begin to interact with cellular proteins, such as signal recognition particles (which mediate protein translocation across biological membranes), chaperonins, and heat shock proteins (which seem to be involved in protein folding and assembly) even before the full-length translation product has come off the ribosome (Landry and Gierasch, 1991). The elements involved in the recognition of the nascent polypeptide chain by these signal recognition particles and chaperonins still must derive from the primary sequence of the peptide chain.

Although the rules relating protein sequence to folding, translocation, and function are not well understood, the ability to compare protein sequences and determine homology is quite useful. Homology can help to suggest protein class or function in newly identified protein sequences. Identification of a homologous protein for which structural information is known, particularly X-ray crystal structures, can aid in the construction of three-dimensional models for the protein structure. A case in point was the early identification of the HIV-1 protease as an aspartic protease, based on sequence homology to other aspartic proteases, and the construction of a model structure, based on the known X-ray crystal structures of fungal aspartic proteases, such as penicillopepsin and rhizopus pepsin (Pearl and Taylor, 1987).

Our ability to search protein databases for protein similarities is based on the pioneering effort of Margaret Dayhoff at the National Biomedical Research Foundation (NBRF), who began publishing compilations of protein sequences during the 1960s (Dayhoff et al., 1965). The NBRF protein database is now available from the Protein Identification Resource (PIR)[1] along with a sequence searching and comparison software suite (Protein Sequence Query, PSQ), which is available as standalone software for the Digital VAX environment. Another software package for analyzing the NBRF database is available in VAX format from Genetics Computer Group, Inc. (GCG).[2] Sequence analysis software on PC platforms is available from sources including International Biotechnologies, Inc.,[3] and IntelliGenetics, Inc.[4] With the tremendous advances in nucleotide sequencing, many new protein sequences are available from a translation of the nucleotide sequence rather than from conventional protein sequencing techniques. There are several nucleotide sequence databases, including GenBank[4] (funded by the National Institutes of Health and compiled by IntelliGenetics and the Los Alamos National Laboratory) and the European Molecular Biology Laboratory (EMBL)[5] database. GenBank and EMBL (and the DNA Data Bank of Japan) have been exchanging nucleotide sequence data since 1982, but the content of these databases does not entirely overlap.

Two basic operations exist for sequence homology searching. One is a matching function that allows searching for occurrences of a specific sequence. For example, one could search the protein database for all occurrences of the sequence Arg-Gly-Asp-Ser. The number of mismatches allowed in the sequence can be specified, allowing retrieval of such sequences as Arg-Ala-Asp-Ser. It is also usually possible to specify a list of amino acids that would be considered a match at a given position

---

1 PIR Technical Services Coordinator, National Biomedical Research Foundation, 3900 Reservoir Road N.W., Washington, D.C. 20007; 202-687-2121.

2 Genetics Computer Group, Inc., 575 Science Drive, Madison, Wisconsin 53711; 608-231-5200.

3 International Biotechnologies, Inc., P.O. Box 9558, 25 Science Park, New Haven, Connecticut 06535; 800-243-2555.

4 IntelliGenetics, Inc., 700 East El Camino Real, Mountain View, California 94040; 415-962-7300.

5 EMBL Database, Postfach 10.2209, 6900 Heidelberg, Federal Republic of Germany; 49-6221-3870.

in the sequence, e.g. Arg-Gly-(Asp or Glu)-Ser. Finally, one can also search for patterns like Ser-X-X-Tyr-Pro when one is interested in sequences in which a Ser is found three residues upstream from a Tyr-Pro sequence and where the nature of the intervening amino acids (X) is not important.

Although matching is useful for identifying regions of contiguous sequence, many proteins may have homologous regions separated by sequences of varying lengths or may have small gaps or insertions within a region of homology. For this reason, an alignment algorithm is used to compare two entire protein sequences and to determine the best fit of the two sequences. Several variations occur on sequence alignment algorithms: one is based on methodology originally derived by Needleman and Wunsch (1970), and another (FASTA), developed by Pearson and Lipman (Lipman and Pearson, 1985; Pearson and Lipman, 1988), is optimized for speed, something of concern when one wants to check a new sequence against the entire protein database. Both methods rely on the use of a scoring table or matrix that assigns a numerical value to each pair of amino acids. The simplest scoring table would have 1 for identical amino acids and 0 for any other combination. If you were to create a matrix with the sequence of one protein running along the columns and that of the other protein running along the rows, each cell (i, j) would be either 0 or 1 depending on whether residue i from the first protein matched residue j from the second protein. If the two proteins were identical, there would be 1's all along the diagonal of the matrix. If there were more limited regions of homology, they would appear as stretches consisting mainly of 1's along or parallel to the diagonal. Any offset from the diagonal reflects gaps or insertions of amino acids into one of the protein sequences. This often indicates the presence of variable-length loops of sequence that connect conserved regions of secondary structure in homologous proteins.

It is readily apparent that the scoring table does not have to be limited to values of 0 and 1. The comparisons can be graded so that a conservative substitution (e.g., valine for isoleucine) would have a high score, and substitution of a very dissimilar amino acid (such as aspartic acid for isoleucine) would have a very low score. The most commonly used scoring table (Dayhoff et al., 1978) takes into account conservative substitutions, the frequency of appearance of amino acids, the frequency of single base mutations to change the genetic three-base code from one specific amino acid to another, and the conservation of rare amino acids (like tryptophan) or structurally significant amino acids (like cysteine). Scoring values range from a high of 17 (conservation of the relatively uncommon tryptophan in a sequence) to 2 (conservation of the commonly occurring and not generally structurally significant alanine) to lows of −6 to −8 (substitution of the structurally significant cysteine). This sort of table is ideal for studying the evolutionary relatedness of various proteins. If one were interested in convergent evolution, where a peptide or protein might have independently evolved to a similar structure and function, one could construct a scoring matrix which emphasized chemically conservative substitutions.

The regions of sequence homology are not the only regions of interest or importance in protein function. For example, if there are species differences in the activity of homologous proteins, the structural basis for these differences may lie in the regions of non-homologous amino acid sequence. In addition, loops, such as those connecting conserved regions of secondary structure, have a high probability of lying on the protein surface and may serve as important sites for post-translational modification or as antigenic epitopes.

Secondary Structure Prediction

Protein sequence alignment alone cannot suggest any of the structure of a protein unless a homologous protein has been studied by X-ray crystallography, nuclear magnetic resonance, circular dichroism or any of the other physical methods of protein structure determination. Structural information of this type is not available for most protein sequences. However, algorithms have been developed that allow prediction of secondary structures based on the primary sequence of the protein. Secondary structure prediction algorithms are generally available as part of the protein sequence analysis software packages. They are based on statistical analyses of the frequency of residues or sequences being found in particular secondary structures in a dataset of known protein structures. Prediction of overall secondary structure is generally good, but specific details, such as exact helix boundaries, may be less reliable.

The most widely used secondary structure prediction scheme is that of Chou and Fasman (1978). It is based on a statistical analysis of the frequencies of individual amino acid residues being found in specific secondary structures (α-helix, β-structure or β-turn) compared to the overall frequency in which they appear in protein sequences. This allows each amino acid to be assigned a numerical probability of being found in a given secondary structure and to be classified as a strong helix-former, helix-breaker, weak β-former, β-breaker, and so on. Since an α-helix or β-structure is a regular repeating one, each position in it is essentially equivalent. To a first approximation, then, the probability of a given amino acid appearing in a helix or a β-structure is not dependent on the position it would occupy in that structure. Each amino acid, therefore, has a single probability value for α-helix or β-structure. Since the conformational requirements for each position in a β-turn are different, however, each amino acid will have a different probability of appearing in each of the four positions of a β-turn.

Secondary structure is predicted by applying a series of empirically determined rules to the primary sequence. Helices are predicted by searching for nucleation sites—clusters of residues with a high helix-forming potential. The helix propagates out in both directions until a cluster of helix-breaking residues is encountered. There are other rules for locating helix boundaries, such as the general tendency for negatively charged amino acids (Asp and Glu) at the amino terminal end (where they help stabilize the helix dipole) and positively charged amino acids (His, Lys and Arg) at the carboxyl terminal end. β-sheet structure is predicted in a similar fashion. Turns are predicted for a sequence of amino acids i to $i+3$ by considering the sum of their position-dependent probability values. In practice, algorithms supplied with protein database software generally evaluate helix potential, β-sheet potential, and β-turn potential (also, sometimes helix and sheet boundary conditions) separately and plot each against residue number. Helix and sheet potential are averaged over a user-defined window of residues (typically 4) and normalized. Average probabilities greater than one indicate a propensity for that sequence to adopt that type of secondary structure. The graphical output allows regions with a high probability of adopting helix, β-structure, or β-turns to be readily identified, and plotting of residues often found at helix or sheet boundaries helps to define the extent of secondary structures.

It often happens that there are ambiguities in the assignment of secondary structure to regions that may have high probabilities of adopting either helix or β-structure. There are additional rules by which these ambiguities can be resolved, including considering which type of secondary structure has the highest numerical probability for that sequence, the length of the sequence (α-helices, being more compact structures, typically are longer than regions of β-structure), and so forth. Garnier et al. (1978) developed a method, again based on a statistical analysis using a known set of protein structures, in which position-dependent probability values are used to generate a numerical value which specifies helix, β-structure, β-turn, or random coil for each residue of the protein. An algorithm based on this method is also often included in protein sequence analysis software packages.

It is not currently possible to predict the secondary structure of a protein from the sequence alone with absolute certainty. These predictive methods generally have an overall accuracy of about 70% when applied to proteins of known structure that are not in the "knowledge set" used to generate the predictive rules; i.e., about 70% of the secondary structure present is predicted in an overall sense. Helical and turn structures are reasonably well predicted, probably in part because they require only one strand of peptide, while β-structures, which require the presence of at least one complementary strand, are not as well predicted. It must be remembered that these methods rely on statistical analyses and empirically determined rules. Predictions of overall helical content or β-sheet content may be reasonably accurate, but details such as boundaries are generally much less precise. There are many sequences that are ambiguous as to secondary structure preferences and, indeed, many peptide sequences are able to adopt different conformations or structures depending on their environment, solvent, and so on. Also, while random coil or irregular structure can be inferred from the absence of any propensity to form a regular secondary structure, there is no conformational or structural information available for such sequences to aid in model building. Finally, prediction of secondary structure in and of itself does not necessarily yield much information on tertiary structure or protein topology. Secondary structure prediction is best utilized as a guide in model building or in selection of peptide targets in conjunction with other techniques, such as sequence homology, chemical modification, proteolytic digestion, site-specific mutagenesis, and so on.

Hydrophobicity Measurements

Given that soluble proteins fold in such a way as to minimize the exposure of hydrophobic side chains to the aqueous medium, it is possible to infer portions of the protein sequence that are likely to be buried or to be exposed on the protein surface by examining the relative hydrophilicity or hydrophobicity of the protein sequence. The method most widely available in protein sequence analysis software is that of Kyte and Doolittle (1982). They devised a set of values representative of the free energy change of each amino acid side chain going from a polar to a non-polar environment. These values are averaged over a window (generally six to nine residues), and these average hydropathy values are plotted against the residue number of the first residue in the window. Positive values indicate hydrophobic segments of the protein, and negative values indicate hydrophilic regions. Segments of the protein chain with a strongly hydrophobic character correlate well with the interior regions of the protein, and those with a strongly hydrophilic character correlate well with regions on the surface.

Hydrophilicity analysis of protein sequences has been used to predict antigenic determinants or epitopes in proteins. As one might intuitively expect, most antigenic determinants in proteins (though not all) are found in regions exposed on the surface of the protein molecule. Hopp and Woods (1981) modified a set of hydrophilicity values for amino acids ultimately derived from octanol–water partition coefficients, used them to calculate the hydrophilicity of protein sequences over a six-residue window, and plotted the results against the residue number. They found a good correlation between peaks of hydrophilicity and the known antigenic sequences of several proteins, although not all highly hydrophilic segments corresponded to known antigenic regions, and not all antigenic determinants corresponded to the highest hydrophilic peaks.

The region of highest hydrophilicity did always correspond to an antigenic determinant in the set of 12 proteins examined. This method is useful for examining proteins of unknown structure, both to suggest peptides for epitope mapping studies or antibody production and also to look for other probable surface features, such as receptor binding sites, enzyme cleavage sites, or sites for post-translational modification.

From the earlier discussion on amphipathic secondary structures (see Amphipathic Structures), it is clear that sequences can exhibit an overall hydrophobic or hydrophilic profile that is rather flat and yet fold into structures that present very hydrophobic or hydrophilic faces. In order to evaluate this type of behavior, Eisenberg et al. (1982) developed the concept of a hydrophobic moment, which can be calculated for a protein sequence having a given secondary structure. For a helix, the hydrophobicity of a side chain is represented by a vector projecting outward from the helix, and its magnitude reflects the magnitude of the hydrophobicity. Its direction is determined by its position in the helix; successive side chains will be offset from each other by 100° in an α-helix. Hydrophilic side chains give rise to a hydrophobic vector of the same magnitude but pointing 180° back from the side chain projection. A hydrophobic moment can be calculated by averaging the magnitude and direction of these vectors over a window of residues. A large moment reflects the segregation of hydrophobic residues along one face of the helix and hydrophilic on the other. A similar calculation can be applied to β-sheet structure. In this case, the angular offset for succeeding residues is 160°.

Eisenberg et al. (1982) calculated hydrophobic moments for a series of helices in 26 different proteins of known structure and compared those values to the average hydrophobicity. The moment and average hydrophobicity were calculated for the entire helix sequence, so each different helix sequence gave a single value for moment and hydrophobicity. Buried or membrane-spanning helices tended to have large hydrophobicity and low hydrophobic moment, as one might intuitively expect, since they exist in a uniformly hydrophobic environment. Amphiphilic helices have low overall average hydrophobicity but possess a large hydrophobic moment. They are found at interfaces between hydrophobic and hydrophilic environments, on the surface of proteins, or in peptides such as mellitin, which associate with plasma membranes. The asymmetric distribution of polar and non-polar side chains allows these structures to interact favorably with both polar and non-polar environments when they are situated at the interface.

In practice, with proteins of unknown structure, a hydrophobic moment is calculated over a small window, typically six residues, first assuming an α-helical structure and then assuming a β-sheet structure (Eisenberg et al., 1987). These are both plotted against residue number and compared with hydrophobicity and Chou–Fasman secondary structure calculations. In that way, one can identify probable amphipathic helices and β-structures and probable membrane-spanning helices.

APPLICATION OF SEQUENCE ANALYSIS TO MODELING AND DESIGN

Construction of detailed molecular models of peptide or protein structure based on sequence data is most fruitful when there is an X-ray crystal structure of a related protein available to use as a template. When the relatedness of the template protein to the protein of unknown structure is relatively distant, the use of secondary structure prediction, hydropathy measurements, and a certain amount of intuition also come into play. This convergent approach is illustrated in the following examples.

The complement protein C5a is a 74-residue protein that is released by proteolysis of a larger precursor during the complement activation cascade. It is thought to be involved in inflammatory responses by recruiting monocytes and polymorphonuclear leukocytes by chemotaxis and activating them to release free radicals and degradative enzymes. Inhibitors of C5a activity can have therapeutic application in the treatment of inflammatory diseases such as rheumatoid arthritis.

In an attempt to develop C5a antagonists, it would be very useful to have a model of the C5a structure. There is no X-ray crystal structure of C5a available, but there is a crystal structure of the related complement protein, C3a. Greer (1985) used this structure as a template on which to construct a model of C5a. The initial difficulty was that the sequences of C5a and C3a were only about 36% identical. Greer chose to begin the alignment with the central region of C3a, which was well defined in the crystal structure, and chose an alignment with C5a that would not cause any deletions or insertions in the helical regions. This alignment would not otherwise be considered the optimal one strictly on the basis of sequence. The model built using this alignment showed that the internal residues, which form the core of the structure, are in fact highly conserved, and that the residues on the surface are not conserved. Thus, while the overall similarity between the two proteins is only modest, the residues important for defining the structure are highly conserved. However, they do not appear adjacent to each other in the linear sequence, so sequence homology does not reveal this close relationship.

The second difficulty encountered was that the C3a amino terminus did not appear in the structure, possibly because it was too flexible in the crystal to give rise to substantial electron density. Thus, there was no template for the amino terminus of C5a. Secondary structural analysis suggested that the amino terminus of C5a had a strong potential for helical structure. When that region was folded into a helix, it was found to be strongly amphipathic, so it was reasonable that the hydrophobic face of this helix interacted with an accessible hydrophobic region elsewhere on the C5a molecule. One such hydrophobic patch was located on the surface in a region where the residues were highly conserved in human and porcine C5a and where the amino terminal helix could be docked.

This yielded a working model for the C5a structure where the core structure was identical to that of C3a, but the surface residues had considerable difference, which is consistent with the two proteins' distinct biological activities. The structure was also consistent with known pieces of structure–activity data. Porcine and human C5a will both bind to human C5a receptor, and there is a region on the surface of both molecules that is relatively conserved. This region is spatially near the C-terminal Arg residue, which is important for biological activity but not for binding to the receptor. The three-dimensional solution structure of C5a has recently been determined using 2-D NMR techniques and distance geometry methods (Zuiderweg et al., 1988). The proposed model is in very good agreement with the deduced solution structure with respect to both the core structure, the helical conformation, and placement of the amino terminal sequence. The carboxyl terminus of the protein, however, seems to be highly mobile in solution and does not adopt a fixed conformation.

A similar approach was used to create initial structural models for the human immunodeficiency virus (HIV-1) protease. HIV-1 protease is responsible for the proteolytic maturation of the HIV structural proteins, reverse transcriptase, integrase, and RNase H, which are encoded by the *gag* and *pol* genes. It was initially known that many related retroviruses encode such a protease that processes the *gag* and *pol* translation products and that the HIV *gag* and *pol* proteins were indeed produced as polyprotein precursors. An approximate location in the genome for the protease was determined by protein sequence homology comparisons of the *gag* and *pol* regions of a number of retroviruses, considering especially two highly conserved tripeptide sequences (Asp-Thr-Gly and Ile-Ile-Gly) (Yasunaga et al., 1986). This was confirmed, and the exact location of the protease and its primary sequence were unambiguously determined when the gene was subsequently cloned and expressed (Debouck et al., 1987).

The tripeptide sequence, Asp-Thr-Gly, which is conserved in retroviral proteases, is also a conserved sequence in aspartyl proteases, and in fact, represents the catalytically active aspartic acid residues of the active site. Aspartyl proteases are relatively well-studied enzymes, and a number of crystal structures are available. They consist of two domains, each of which contributes one Asp to the active site. The two domains have a fair degree of sequence homology, and both domains have a similar three-dimensional structure, suggesting that aspartyl proteases arose by gene duplication and fusion (Tang et al., 1978). The retroviral proteases are only about half the size of eukaryotic aspartyl proteases and contain only one Asp-Thr-Gly sequence. Pearl and Taylor (1987) hypothesized that the retroviral protease represents one lobe of an aspartyl protease, and that the protein forms a symmetrical dimer as the active protease. Using a combination of protein sequence comparisons among a large number of aspartyl proteases and secondary structure prediction, they were able to construct a model for the HIV-1 protease as a symmetrical dimer based on the crystal structure of the fungal aspartyl protease endothiapepsin. The dimeric nature of the HIV-1 protease and its identification as an aspartyl protease was later confirmed experimentally (Meek et al., 1989). The model predicted the general structure of the protease as four strands of β-sheet and a central hydrophobic cleft formed at the dimer interface that contains the substrate binding site and the active site aspartyl residues. This overall structure was confirmed when X-ray crystal structures of the HIV-1 protease were obtained (Lapatto et al., 1989; Wlodawer et al., 1989), although some important structural details varied. For example, the model was based on a sequence that incorrectly located the mature amino terminus of the protease and, therefore, failed to predict the intermolecular β-structure between the amino terminus of one monomer and the carboxyl terminus of the other monomer, which forms an important part of the dimer interface. Nevertheless, early recognition that the HIV-1 protease was likely to be a symmetrical dimer with a symmetrical active site led to the design of novel, symmetrical inhibitors of high potency (Erickson et al., 1990).

Peptide Design Based on Sequence

It has been a long-standing goal in peptide chemistry to be able to design short peptides that mimic some important aspect of protein structure or function. The peptides can be designed to be agonists of a biologically active protein by encompassing those residues necessary for activity. Antagonists can be designed using peptides containing only those residues required for binding to a receptor, a substrate, or a required protein subunit. (Fibrinogen receptor antagonists, which fall into this category, will be discussed in more detail later; see Cyclization). When one has antibodies against the protein that neutralize the protein's activity, epitope mapping (or locating the antibody combining sites) can help identify protein sequence portions involved in the biological activity. Synthetic peptides based on the sequences of neutralizing antigens can be used as synthetic antigens in the preparation of antisera and vaccines.

Designing peptide agonists based on protein sequences is a formidable task. With increasing size of the polypeptide chain, it becomes increasingly likely that residues important for biological activity will be in close proximity in the three-dimensional structure but widely separated in the primary sequence. This makes design of a small peptide that can mimic that arrangement problematic, especially when there may not

TABLE 6 Gastrin-Releasing Peptide and Bombesin Analogs

	Sequence	Relative activity
GRP	A-P-V-S-V-G-G-G-T-V-L-A-K-M-Y-P-R-G-N-H-W-A-V-G-H-L-M-NH$_2$	0.28
Bombesin	<E-Q-R-L-G-N-Q-W-A-V-G-H-L-M-NH$_2$	1.00
	<E-Q-R-L-G-N-Q-W-A-V-G-H-L-M-OH	<0.01
	<E-Q-R-L-G-N-Q-W-A-V-G-H-NH$_2$	<0.01
	Ac-G-N-Q-W-A-V-G-H-L-M-NH$_2$	1.00
	Ac-N-H-W-A-V-G-H-L-M-NH$_2$	0.50
	Ac-H-W-A-V-G-H-L-M-NH$_2$	1.00
	Ac-W-A-V-G-H-L-M-NH$_2$	0.04

<E = Pyroglutamic acid.

be any three-dimensional structural information available about the target protein. The case is often easier with peptides of moderate length, since there is often a minimum active fragment that retains a substantial portion of the biological activity. This can be illustrated with the case of gastrin-releasing peptide and bombesin.

Gastrin-releasing peptide (GRP) is a 27-amino-acid peptide originally isolated from porcine gut on the basis of its ability to cause release of gastrin (McDonald et al., 1979). Its C-terminus has a striking homology with bombesin, a 14-residue peptide isolated from frog skin (Anastasi et al., 1971), as well as a number of related peptides of amphibian origin. The sequences of GRP and bombesin are shown in Table 6. Both GRP and bombesin produce a number of biological responses, including hypothermia or lowering of the core body temperature in rats exposed to cold. A series of truncated analogs was prepared and assayed for hypothermia activity in order to determine the minimum sequence necessary for agonist activity (Table 6; Rivier and Brown, 1978; Märki et al., 1981). Deletions from the conserved C-terminus impaired activity. In contrast, the peptides could be shortened considerably from the amino terminus without loss of activity, and full activity was obtained with just the C-terminal octapeptide sequence.

Peptides derived from an internal protein sequence may require blocking groups at the amino and/or carboxyl terminus to mask the presence of charges that are not present in the native protein. For example, the HIV-1 protease mentioned earlier (see Application of Sequence Analysis to Modelling and Design) is an endopeptidase whose substrate is a viral polyprotein. One of the cleavage sites is encompassed by the heptapeptide Ser-Gln-Asn-Tyr-Pro-Val-Val, with cleavage occurring between the Tyr and Pro residues (Moore et al., 1989). We prepared a series of peptides of varying sizes either with N-acetyl and carboxamide blocking groups or free amino or carboxyl termini and compared their substrate activity. The results are shown in Table 7, where K_m is the Michaelis binding constant, k_{cat} is the catalytic rate constant, and k_{cat}/K_m is a measure of catalytic efficiency. The fully blocked heptapeptide 4, the free amino peptide amide 5, and the acetyl peptide free acid 6 all have fairly similar activities. Removal of the C-terminal Val residue (peptide 7) causes a substantial loss in k_{cat} and catalytic efficiency, and removal of both Val residues (peptide 8) results in an inactive peptide. Truncation at the amino terminus causes a slight loss in activity in the hexapeptide 9 and the pentapeptide 11, with a complete loss of activity when the tetrapeptide 13 is reached. The free amine hexapeptide 10 is as active as its acetylated parent, but the free amine pentapeptide 12 again shows a substantial loss in k_{cat} and catalytic efficiency. These data collectively suggest that the active site of the HIV-1 protease will accommodate a hexapeptide, stretching three residues on either side of the cleavage site. Charged groups and the presence of additional residues have little effect

TABLE 7   HIV-1 Protease Substrates

Compound	Sequence	K_m (mM)	k_{cat} (sec^{-1})	k_{cat}/k_m ($mM^{-1} \cdot s^{-1}$)
4	Ac-Ser-Gln-Asn-Tyr-Pro-Val-Val-NH$_2$	5.5	54	9.8
5	H-Ser-Gln-Asn-Tyr-Pro-Val-Val-NH$_2$	1.2	14.0	11.7
6	Ac-Ser-Gln-Asn-Tyr-Pro-Val-Val-OH	5.0	25.7	5.1
7	Ac-Ser-Gln-Asn-Tyr-Pro-Val-NH$_2$	4.9	0.49	.090
8	Ac-Ser-Gln-Asn-Tyr-Pro-NH$_2$	neg.	neg.	neg.
9	Ac-Ala-Asn-Tyr-Pro-Val-Val-NH$_2$	13.5	9.3	0.69
10	H-Ala-Asn-Tyr-Pro-Val-Val-NH$_2$	18	29.7	1.6
11	Ac-Asn-Tyr-Pro-Val-Val-NH$_2$	17.1	37	2.2
12	H-Asn-Tyr-Pro-Val-Val-NH$_2$	12.8	0.88	0.068
13	Ac-Tyr-Pro-Val-Val-NH$_2$	neg.	neg.	neg.
14	H-Tyr-Pro-Val-Val-NH$_2$	neg.	neg.	neg.

on substrate activity since they extend out of the active site. The loss of activity in uncovering the free amino group in the pentapeptide 12 is probably due to the presence of a charged group in the generally hydrophobic active-site cleft. This interpretation of the extent of the enzyme active site is confirmed by X-ray crystal structures of co-crystals of the protease containing substrate-based inhibitors bound in the active site (Miller et al., 1989; Wlodawer et al., 1991).

The addition of a charged moiety, remote enough from the active site that it does not interfere with catalytic activity, can be advantageous in terms of solubility. Succinyl groups are often added to the amino terminus of enzymes' substrates, e.g., the chymotrypsin substrate succinyl-Ala-Ala-Pro-Phe-p-nitroanilide (DelMar et al., 1979). This modification not only improves solubility of hydrophobic peptides but also blocks potential aminopeptidase degradation at the amino terminus.

Epitope Mapping

One of the most common applications of peptides based on protein sequences is epitope mapping. Determination of antibody combining sites on proteins is used to study the mechanism of immunogenicity of proteins as well as to determine protein topology (on the assumption that most epitopes will be located at the protein surface). If there are monoclonal antibodies that block the protein's activity, epitope mapping can help determine the parts of the protein sequence involved in the expression of biological activity. If the protein is derived from a pathogenic organism, peptides based on neutralizing epitopes can be the basis of synthetic vaccines.

Determination of an antigenic epitope relies on preparation of a peptide representing a portion of the protein sequence and demonstrating competition with the protein in antibody binding. There are several approaches to determining which sequences to examine as potential epitopes. The complete approach would require synthesizing overlapping short peptides, generally hexapeptides, that cover the entire protein sequence. Several techniques for multiple solid-phase peptide synthesis make this approach possible (see Chapter 3). Houghten (1985) devised a method in which small amounts of aminoacyl resin are sealed in small polystyrene mesh bags called "T-bags." The bags can be labeled with an indelible marker for identification. The common steps of washing, deprotection, and neutralization can be performed together for a number of these T-bags, which are then separated only for the coupling of the individual amino acids and at the end of the synthesis for HF cleavage. While there is still a fair amount of labor involved, it is possible to turn out 50-mg quantities of a large number of peptides in a relatively short time. Houghten's method is advantageous because the peptides can be

General Rules for Selecting a Stretch of Protein for the Production of an Antibody that will Recognize the Protein

1. Based on the assumption that the surface of the protein being exposed to solvent will be hydrophilic, use a hydropathy profile to locate areas of hydrophilicity in the sequence.

If you have only a short section of sequence to work with, pick areas that contain side chains that will be charged or polar under the conditions you are working with.

In addition to selecting sections that are more likely to be accessible to solvent molecules, these regions will also aid in the solubility of the peptide during manipulation.

2. In general, the immediate amino and carboxyl ends of a protein may tend to be more solvent accessible than the interior sequence. If you have no other criteria for selection, these may be more feasible regions to use.

3. Mimicking the charge state of the ends of the peptide may be helpful in certain instances. For instance, unless the peptide is at the immediate amino or carboxyl end of the protein, its ends will be in the peptide linkage to the rest of the protein and thus will not be protonatable. Synthesis of the peptide with blocked termini (acetyl-amino or carboxyl-amide) may be helpful in mimicking this situation by reducing the presence of potentially undesirable charge effects at the peptide termini. However, note that it is important to make sure the peptide has sufficient additional charge for solubility.

4. Peptides of 10 to 15 residues are usually adequate for the production of antibodies. Longer peptides are more difficult and expensive to make and usually unnecessary. Peptides do not necessarily need to be rigorously purified before conjugation, especially for polyclonal antibody production, since the antisera will be selected on the basis of specificity as well as titer.

5. Including a cysteine residue at either the amino or the carboxyl end of the peptide can be very helpful for cross-linking to carrier protein with heterobifunctional reagents.

6. Antigenicity: In general, small peptides are not sufficiently antigenic in themselves to elicit a good response. This insufficiency is usually overcome by coupling them to a carrier prior to presentation to the host animal. A variety of carriers and coupling methods have been used, such as heterobifunctional reagents referred to above. A more detailed treatment of this subject can be found in Chapters 3 and 5.

isolated, purified, and characterized, but characterization clearly becomes rate-limiting if a sufficiently large number of peptides is being prepared.

An alternative approach has been developed by Geysen et al. (1988). The peptides are synthesized on the tips of polystyrene pins arranged on a support such that the tips can be immersed in the wells of a standard 96-well microtiter plate. As with Houghten's approach, steps such as washing and deprotection are carried out in common while the couplings take place in the microtiter wells. In the original conception, an ELISA-type assay was used so that the results could be read colorimetrically in microtiter plates. The pins were incubated with the probe antibody, washed, incubated with the conjugate antibody (horseradish peroxidase coupled to anti-IgG for the appropriate species in which the probe antibody was raised), washed again, and then immersed in wells containing the enzyme substrate. Color developed only in those wells where the pins bound the original probe antibody. A large amount of labor is involved in this synthesis approach as well, but an extremely large number of peptides can be prepared in a short amount of time. The advantage to this approach is that, in theory, the pins can be reused with another probe antibody if the original peptide–antibody complex can be disrupted. The disadvantage is that the peptides remain attached to the pins and cannot be characterized; there is the possibility that the derivatized polystyrene matrix of the pin could affect the interaction of peptide and antibody. A recent modification to this method involves the attachment of a protected Lys-Pro to the pins and peptide assembly on the ε-amino group of the lysine. The peptide can be released from the pins by mild base treatment, giving the free peptide with a pendant diketopiperazine moiety (Bray et al., 1990).

One approach to limiting the number of peptides that must be prepared for epitope mapping has already been mentioned, namely the use of a hydropathy profile of the protein sequence (Hopp and Woods, 1981; see Hydrophobicity Measurements). This approach is based on the supposition that most epitopes will be found on the protein surface and that relatively hydrophilic stretches of the protein sequence are more likely to be on the surface. One disadvantage of this method is that the hydropathy profile does not give an unambiguous or very detailed indication of protein topology. It is also true that all hydrophilic regions of the protein do not necessarily correspond to epitopes, that all epitopes are not hydrophilic or even on the protein surface, and that the major antigenic regions are not necessarily the most hydrophilic.

The more typical approach to epitope mapping involves using protein biochemistry and molecular genetics techniques to create fragments of the protein, narrowing down the location of the epitope by determining successively smaller fragments that cross-react with the target antibody. With modern techniques and the judicious use of restriction endonucleases, it is often possible to clone and express the protein of interest along with a library of partial sequences. Using sensitive methods of detection such as Western blotting, it is possible to rapidly identify fairly small regions of protein sequence as containing a particular epitope, which can then be refined further through the use of synthetic peptides. A typical case is the identification of the principal neutralizing epitope of the HIV envelope protein.

The HIV envelope protein, gp120, is the viral ligand for the CD4 receptor protein on T4 lymphocytes and is involved both in binding of the virus to target cells and in fusion of the cellular and viral membranes, allowing viral genetic material to enter the cell. Antisera to gp120 have been shown to block both infectivity and fusion (Allen et al., 1985; Veronese et al., 1985). Using protein and peptide fragments produced by recombinant techniques, limited proteolysis and peptide synthesis, the major neutralizing epitope was localized with both polyclonal antisera (Rusche et al., 1988) and a monoclonal antibody (Matsushita et al., 1988) to a 24-residue peptide which occurs in a 36-residue disulfide loop on the gp120 molecule. Synthetic peptides corresponding to portions of this loop were prepared, guided partially by sequence homology comparisons using the sequences of gp120 from different isolates of HIV-1. An octapeptide sequence (Ile-Gln-Arg-Gly-Pro-Gly-Arg-Ala) located at the middle of the loop was shown to be the minimum sequence that could block the effects of neutralizing antibodies and that, when coupled to a carrier protein, could itself generate neutralizing antisera (Javaherian et al., 1989).

OTHER ASPECTS OF PEPTIDE DESIGN

Enzymatic Stability

For simple peptides, the serum half-life and biological activity is often limited by proteolytic hydrolysis of susceptible peptide bonds. To make these peptides pharmacologically or physiologically useful, it is necessary to increase their serum half-life by increasing their resistance to proteolysis. The exact pathway for proteolysis depends on the peptide in question, and determining the metabolic fate of a given peptide can be a difficult and tedious process. However, inspection of the peptide sequence can usually provide insight into the most likely sites for proteolytic degradation.

There are two types of proteolytic enzymes: exopeptidases, which cleave amino acids from either end of a peptide chain, and endopeptidases, which cleave peptide bonds in the middle of a peptide chain. Exopeptidases include aminopeptidases, which cleave the N-terminal amino acid from a peptide chain; carboxypeptidases, which cleave the C-terminal amino acid from a peptide chain; and dipeptidyl carboxypeptidases like angiotensin converting enzyme (ACE), which cleave the C-terminal dipeptide from a peptide chain. Their recognition sites include the charged amino or carboxyl termini of the peptide, and they may show side-chain specificities as well. Endopeptidases generally have specificities for a particular amino acid side chain or a discrete sequence. The most relevant endopeptidase activities to consider are trypsin-like, in which Lys-X or Arg-X bonds are cleaved; chymotrypsin-like, in which Phe-X, Trp-X or bulky aliphatic-X bonds are cleaved; and post-proline cleaving enzyme, in which Pro-X bonds are cleaved.

There are many ways to stabilize a peptide against specific proteolysis. One approach is as simple as acetylating the amino terminus to block aminopeptidase action because aminopeptidases require the presence of a free α-amino group. Another approach involves modifying or deleting an amino acid side chain or inverting its chirality (most proteolytic enzymes have greatly reduced activity against D-amino acid residues). The susceptible peptide bond itself may be altered as well. Introduction of steric bulk along the peptide backbone by substitution of methyl groups for α-hydrogens or amide hydrogens reduces the rate of enzymatic hydrolysis adjacent to those substitutions. Peptide bonds can also be substituted for by a variety of hydrolytically inert isosteric structures such as *trans* olefins. All of these approaches involve some modification to the peptide structure that can have effects on biological activity apart from any effects on proteolytic stabilization. Modifications that reduce susceptibility to proteolytic cleavage can, however, have a profound effect on the in vivo potency of peptides.

Arginine vasopressin (AVP) is a naturally occurring peptide hormone produced in the pituitary and exhibits antidiuretic activity in vivo. The antidiuretic activity from a bolus injection of AVP has a very short duration of action, suggestive of metabolic instability. Indeed, immunoreactive AVP disappears rapidly from the circulation after a bolus injection and is degraded rapidly upon exposure to serum in vitro (Edwards et al., 1973). The structure of arginine vasopressin is given in Figure 14 (compound 15). From the structure, several potential sites for enzymatic degradation can be identified. The peptide has a free amino group and might be susceptible to aminopeptidase action. There is an arginine residue at position 8, which could be a site for tryptic-like cleavage. The amino-terminal hexapeptide is cyclized by means of a disulfide bond. Cyclic peptides are often resistant to proteolytic degradation due to the conformational constraints imposed by cyclization. Disulfides, however, can be opened by reduction or by interchange reactions with sulfhydryl-containing molecules such as glutathione (γ-glutamyl-cysteinyl-glycine). In addition, most small cyclic peptides require the cyclic structure for biological activity since it folds them into conformations that are otherwise relatively unlikely for linear peptides. AVP is no exception, and the linear Ala^{1,6} analog is devoid of antidiuretic activity (Walter et al., 1967).

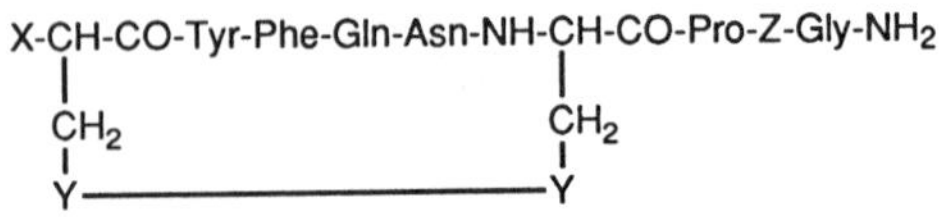

X-CH-CO-Tyr-Phe-Gln-Asn-NH-CH-CO-Pro-Z-Gly-NH$_2$

Number	X	Y	Z	Compound
15	H$_2$N	S	Arg	AVP
16	H	S	Arg	dAVP
17	H$_2$N	S	D-Arg	DAVP
18	H	S	D-Arg	dDAVP
19	H	CH$_2$	Arg	[Asu^{1,6}]-AVP

FIGURE 14   Structure of arginine vasopressin and analogs.

Figure 14 shows some AVP analogs that were modified in ways that would expectedly increase the stability of the peptide. Analog 16 (dAVP) is the desamino version, which would be expected to be resistant to any aminopeptidase activity. Analog 17 (DAVP) is the D-Arg^8 peptide, which should be resistant to tryptic-like cleavages. Analog 18 (dDAVP) contains both modifications together. Analog 19 is a dicarba analog, one in which the disulfide is replaced with a non-reducible ethylene bridge; it is a desamino analog as well. The biological activities of these analogs are given in Table 8. The column *ADH activity* is a measure of antidiuretic potency in vivo in the rat. AVP has 332 International Units of antidiuretic activity per milligram of peptide (Sawyer et al., 1974). The desamino peptide 16 exhibits a fivefold enhanced in vivo potency compared to AVP (Bankowski et al., 1978), suggesting increased metabolic stability. The D-Arg8 analog 17 also retains good in vivo antidiuretic potency (Zaoral, et al., 1967a). The doubly substituted analog 18 shows roughly additive effects for the individual substitutions (Zaoral et al., 1967b). The desamino–dicarba analog 19 also shows enhanced potency in vivo, roughly equivalent to the disulfide analog 16 (Hase et al., 1972).

The peptide dDAVP (compound 18) not only exhibits enhanced in vivo antidiuretic activity but has a fivefold-longer serum half-life after bolus injection and is completely stable to incubation with plasma in vitro (Edwards et al., 1973), establishing the utility of enhancing metabolic stability. What is even more remarkable can be found by comparing the in vitro activity of these analogs. Vasopressin acts in the kidney by binding to specific receptors activating adenylate cyclase. Both the bind-

TABLE 8   Biological Activity of Modified Arginine
Vasopressin Analogs

Number	Compound	ADH (IU/mg)	K_{bind} (nM)	K_a (nM)
15	AVP	332	0.44	0.1
16	dAVP	1745	—	—
17	DAVP	114	1400	2800
18	dDAVP	955	3000	3450
19	[Asu^{1,6}]-AVP	1274	1.4	0.4

ing to rat kidney membrane preparations and the hormone-dependent activation of adenylate cyclase are listed in Table 8. AVP has a sub-nanomolar binding affinity and activation constant (Hechter et al., 1978). In contrast, both DAVP and dDAVP show more than three orders of magnitude decrease in receptor affinity and activation constant (Roy et al., 1975; Butlen et al., 1978). The increased metabolic stability of dDAVP more than makes up for a tremendous loss in intrinsic activity at the receptor, allowing this analog to exhibit enhanced antidiuretic potency in vivo to the extent that it is clinically useful for the treatment of diabetes insipidis. This is a remarkable result and underscores the importance of metabolic stability in the expression of the in vivo biological activity of peptides.

There are more exotic ways to stabilize individual peptide bonds against proteolysis. These approaches involve replacing the amide linkage with a proteolytically resistant linkage. Some peptide bonds, like the *trans* double bond or the thioamide ($C = SNH$), are isosteric with the amide bond and preserve their geometry while others, like the ethylene (CH_2CH_2) or thiomethyl ether linkages (CH_2S), preserve the connectivity and chain length only. All of them alter the polarity and hydrogen-bonding potential of the amide bond as well as imparting resistance to proteolysis.

A number of these substitutions were examined in the endogenous opiate peptide leucine enkephalin, which has the sequence H-Tyr-Gly-Gly-Phe-Leu-OH. Leu-enkephalin is susceptible to a number of enzymatic cleavages including aminopeptidase cleavage of the Tyr-Gly bond (Hambrook et al., 1976), carboxypeptidase cleavage of the Phe-Leu bond (Marks et al., 1977), and dipeptidyl carboxypeptidase cleavage of the Gly-Phe bond (Malfroy et al., 1978). Biological activity requires the free α-amino group and the L-tyrosine side chain at position 1 (Frederickson, 1977), so desamino or D-Tyr substitutions are not useful to decrease aminopeptidase degradation. Substitution of D-Ala for Gly^2 gives an

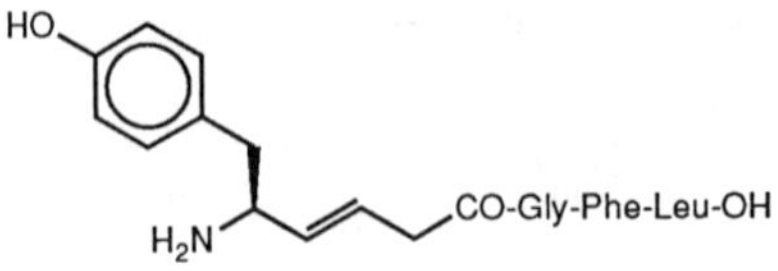

20

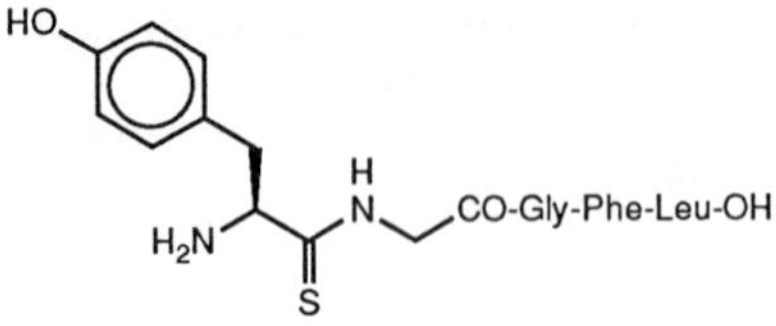

21

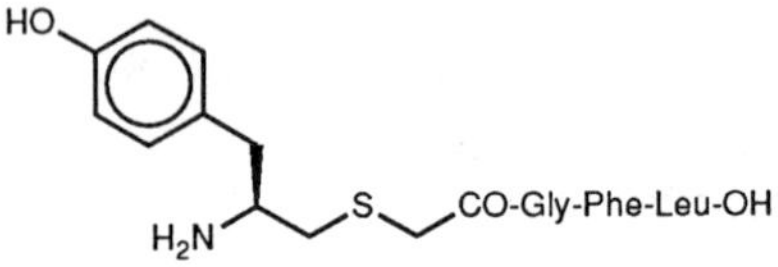

22

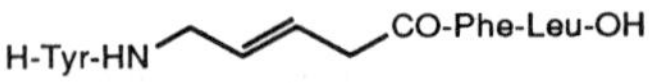

23

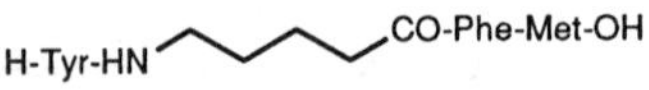

24

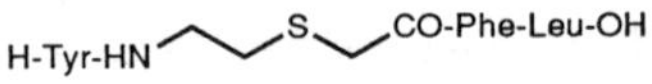

25

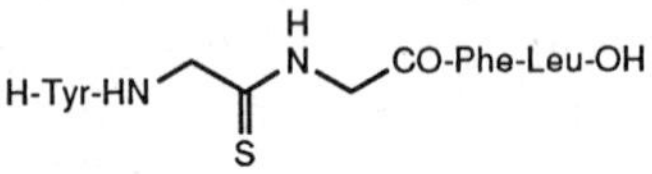

26

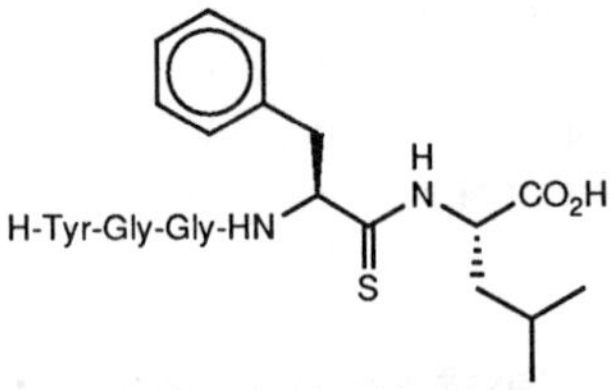

27

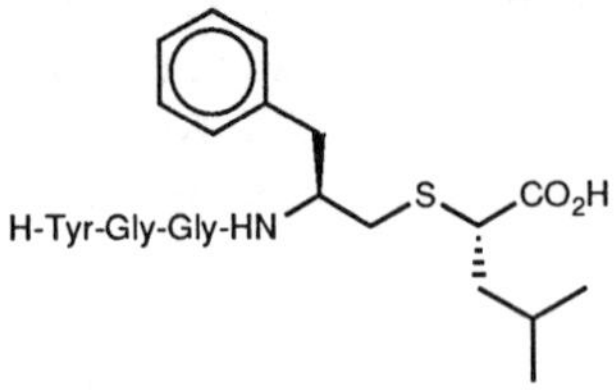

28

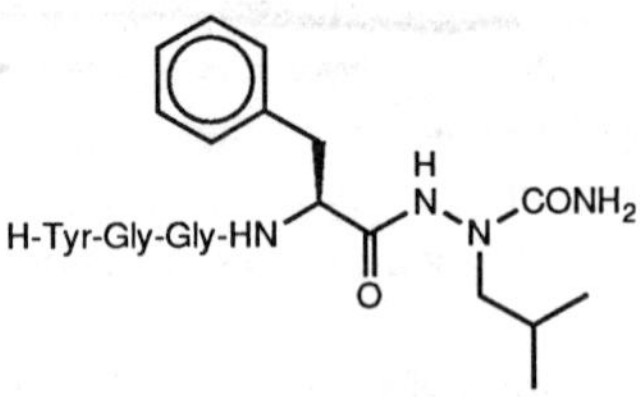

29

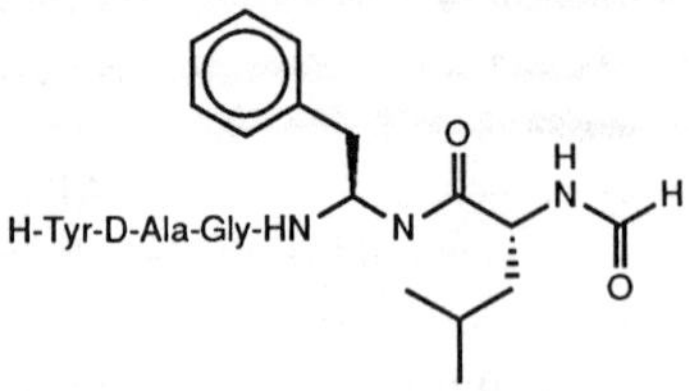

30

FIGURE 15 Enkephalin analogs with modified peptide bonds.

analog with somewhat higher biological activity than the parent compound (Kosterlitz et al., 1980) and improved resistance to proteolysis
(Miller et al., 1977). However, in the quest to further improve metabolic
stability, a number of peptide bond replacements were also examined,
some of which are summarized in Figure 15. The results from amide
bond replacement differ with the particular amide bond being replaced.
Replacement of the Tyr-Gly bond with a *trans* olefin isostere, which
retains the geometry of the amide bond, provides analog 20, which retains full brain receptor binding affinity (Hann et al., 1982). The case
here is not simple because the thioamide-containing analog 21 does not
retain binding affinity despite the thioamide preserving the geometry of
the amide bond (Clausen et al., 1984). Further, the thiomethyl ether
analog 22, which does not preserve the amide geometry, does retain some
binding affinity, but it behaves more like an antagonist rather than an
agonist at low doses in a functional assay (inhibition of electrically-
induced contraction of guinea pig ileum) (Spatola et al., 1986). The
Gly-Gly bond in Leu-enkephalin cannot be replaced by a *trans* olefin
(analog 23) (Cox et al., 1980), ethylene [analog 24, which is an analog of
the other naturally occurring enkephalin sequence, methionine enkephalin (Kawasaki and Maeda, 1982)], or thiomethyl ether moiety (analog
25; Spatola et al., 1986), but substitution of a thioamide (analog 26) gives
a peptide with enhanced binding and biological activity (Clausen et al.,
1984). The most forgiving bond in terms of replacement is the Phe-Leu
bond, which can be replaced with either a thioamide (analog 27; Clausen
et al., 1984) or a thiomethyl ether (analog 28; Spatola et al., 1986) with
retention of binding affinity.

The lack of a requirement of a Phe-Leu amide bond for biological
activity led to some more innovative peptide bond replacements. Analog
29 contains an α-aza-amino acid (aza-leucine), which gives rise to an
acyl hydrazide linkage rather than an amide bond. The α-aza-amino acid
has no chiral center. This analog again retains full biological activity
(Dutta et al., 1977). In the D-Ala^2 enkephalinamide analog 30, amide
bonds are retained, but the sense of the Phe-Leu and Leu-NH$_2$ amide
bonds is reversed, introducing what are called retro-amide bonds. The
phenylalanine carboxyl group has been replaced by an amino group and
the leucine amino group by a carboxyl group to form a NHCO linkage
rather than a CONH peptide bond. The same reversal of functional
groups is made at the Leu-NH$_2$ peptide bond. (Since both amino and
carboxyl groups of the Leu^5 residue have been "exchanged," it is now in
reality D-leucine, but the relative projection of the leucine side chain
from the peptide backbone has been preserved.) The D-Ala substitution
should block aminopeptidase action, and the retro-peptide bonds should
block chymotryptic-like cleavages. The modified peptide exhibits both
enhanced potency in vitro and prolonged activity in vivo compared to
D-Ala^2 enkephalinamide (Chorev et al., 1979).

Conformational Design and Constraint

Conformational considerations in peptide synthesis remain one of the greatest challenges in the design of biologically active peptides. In addition to the nature of the side chains, the three-dimensional shape assumed by a peptide when interacting with its target, be it receptor, antibody, or enzyme, determines its biological activity. Unlike proteins, peptides of under 15 or so residues in length tend not to exhibit a stable or even a preferred solution conformation. There is generally too little hydrophobic character in a short peptide that can be sequestered from the polar environment by folding. The peptide backbone and polar side chains will have a driving force to be solvated, and entropic considerations will generally favor solvation by the bulk solvent. Thus, there will usually be an ensemble of conformational states in solution. If biological activity involves only one discrete conformer, this conformational ensemble essentially represents a dilution of the biologically active species. The problem is most acute for peptides designed to mimic a portion of a protein structure. In their native environment, these peptide sequences can rely on the protein's structural rigidity to hold them in a particular conformation, while as free peptides they have no such constraining influence. Even for analogs of linear peptide hormones, which have evolved to express high biological activity despite conformational mobility, favorable conformational constraints can impart an appreciable increase in biological activity.

Conformational constraints can be divided conceptually into the three following categories: Local constraints involve restricting the conformational mobility of a single residue in a peptide; regional constraints are those which affect a group of residues that form some secondary structural unit, such as a β-turn; global constraints involve the entire peptide structure.

Peptides that are long enough to adopt stable solution conformations can have conformational preferences optimized globally by modifying the peptide sequence in accordance with the empirical rules of secondary structure prediction such as those devised by Chou and Fasman (1978; see Secondary Structure Prediction). This approach does not "freeze" a peptide into a particular secondary structure, but it stabilizes that structure relative to random coil or other structures, increasing the number of peptide molecules that would be found in that conformation at any given time.

Peptides that are too small to adopt stable conformations on their own require covalent modifications to introduce local or regional conformational constraints. These constraints typically involve addition of sterically bulky substituents adjacent to a rotatable bond to restrict its mobility or the incorporation of cyclic structures. It is impossible to cover all the approaches that have been taken to introduce conformational constraints into peptides, but some of the more successful approaches will be discussed below.

Globally Optimized Secondary Structure

It is possible to design peptides of 15 to 20 residues in length that will exhibit a preference for adopting helical structures in solution. The most successful applications involve the design of peptides containing amphiphilic helices, such as the bee venom peptide mellitin. Mellitin is a 26-residue peptide with the sequence Gly-Ile-Gly-Ala-Val-Leu-Lys-Val-Leu-Thr-Thr-Gly-Leu-Pro-Ala-Leu-Ile-Ser-Trp-Ile-Lys-Arg-Lys-Arg-Gln-Gln-NH_2 with the biological activity of binding to and lysing erythrocytes (Habermann, 1972). The lytic activity requires the basic C-terminal hexapeptide, but the amino-terminal 20-residue peptide can still interact with phospholipid bilayers and membranes (Schröder et al., 1971). This suggested to Kaiser and coworkers that the amino-terminal 20 residues might have formed an amphiphilic helix that bound at the membrane interface and held the basic hexapeptide tail responsible for the lytic activity close to the membrane surface where it could interact (DeGrado et al., 1981). A helical wheel projection of mellitin[1-20] is shown in Figure 16(a). There are several problems with this as an ideal amphipathic helix. The Gly residues and especially the Pro[14] residue tend to reduce the helical propensity of the peptide. Also, several of the Gly residues are found on the hydrophobic face. DeGrado et al. (1981) designed an idealized sequence for the 1–20 region in which the hydrophobic residues were replaced with Leu, and the hydrophilic residues were either Ser or Gln; the C-terminal basic hexapeptide was not modified. The resulting idealized helix is shown in Figure 16(b). The idealized peptide had an identical profile of activity compared to mellitin, but it interacted with lipid monolayers more strongly, formed tetramers at a lower concentration, and lysed erythrocytes at a lower concentration than mellitin, which agrees with the formation of a better amphiphilic helix than mellitin.

Local Conformational Constraints

The simplest local constraints that can be placed on a given residue involve the substitution of a methyl group for a hydrogen adjacent to a rotatable bond. For example, replacing the α-hydrogen on alanine with a methyl group gives aminoisobutyric acid (Aib; Figure 17). The greater steric bulk of the methyl group reduced the rotational freedom of the two adjacent peptide backbone angles ϕ and ψ (see Figure 6; see Protein Secondary Structure). In the case of Aib, the low energy backbone torsional angles are $\phi = \psi \approx \pm 60°$, which are the backbone angles associated with helical structure (Marshall and Bosshard, 1972).

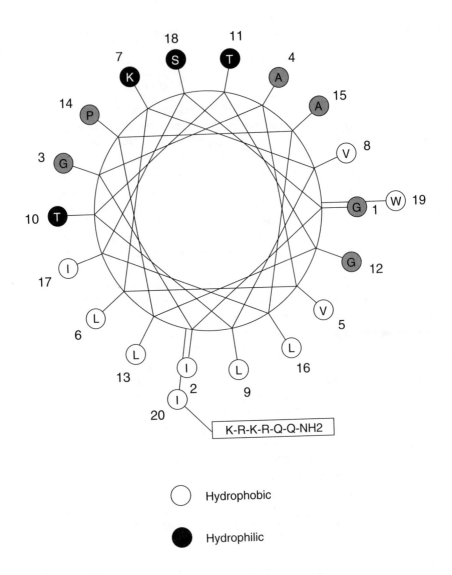

FIGURE 16(a)   Helix wheel projection of a mellitin 1-20 helix. There is some segregation of hydrophilic and hydrophobic residues, but the arrangement is interrupted by Gly and Ala residues and by proline[14], which destabilizes the helix. The basic hexapeptide tail hangs off the helix.

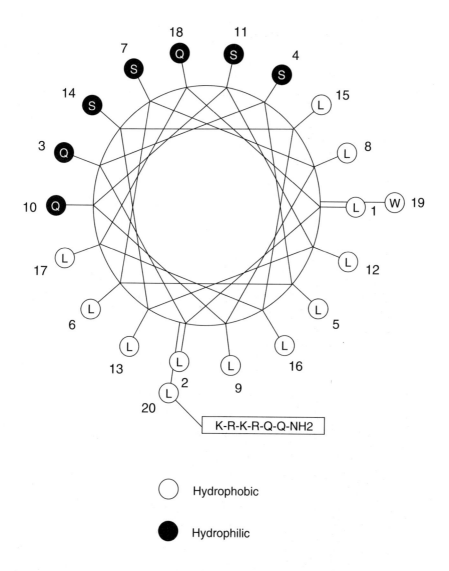

FIGURE 16(b)   Helix wheel projection of the idealized amphipathic lytic peptide based on the mellitin structure. The hydrophobic face is composed solely of leucine residues while the hydrophilic face is composed of serine and glutamine residues. There is little sequence homology in the two helical sequences 1–20, but the length of the amphipathic helix has been preserved while the amphipathicity has been optimized. Tryptophan[19] and the basic hexapeptide tail have been retained.

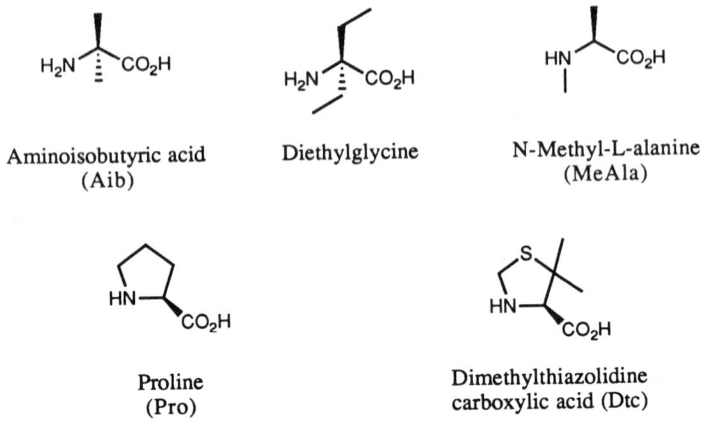

Aminoisobutyric acid
(Aib)

Diethylglycine

N-Methyl-L-alanine
(MeAla)

Proline
(Pro)

Dimethylthiazolidine
carboxylic acid (Dtc)

FIGURE 17   Some constraint-producing amino acids.

Some other local constraint-producing amino acids are shown in Figure 17. The α,α-diethylglycine, with its bulkier substitution at the α-carbon, tends to constrain the backbone angles ϕ and ψ to approximately $180°$, angles typical of a fully extended chain (Benedetti et al., 1988). Substitution of a methyl group for the hydrogen on the amino group, to give N-methyl–alanine for example, has an effect on the *cis-trans* ratio of the amide bond, lowering the relative energy of the *cis* isomer. It has only a modest constraining effect on its own torsional angle ϕ, but it affects the angle ψ of the preceding residue, constraining it to large positive values typical of those found in extended or β-structure (Marshall and Bosshard 1972). Proline is a special case of N-methyl amino acids because its own angle ϕ is constrained to about $-80°$ by being contained in the five-membered pyrrolidine ring. Proline is often found in reverse turn structures, occupying the $i+1$ position in Type II and III' β-turns, the $i+2$ position in Type II' and III β-turns, and in inverse γ-turn conformations (Smith and Pease, 1980). An interesting proline congener is 5,5-dimethylthiazolidine-4-carboxylic acid (Dtc), a β,β-disubstituted thioproline analog. Dtc is conformationally similar to proline, except that the γ-turn is disfavored in Dtc because of steric interaction between the β-methyl groups and the carbonyl carbon. Also, it was suggested as a probe for proline-containing inverse γ-turns (Samanen et al., 1990).

Constraint-producing amino acid substitutions are typically used as probes for the biologically active conformation. If the constrained peptide retains good biological activity, then it can be inferred that the particular backbone constraint is compatible with the biologically active conformation. Negative results are more difficult to interpret because the additional steric bulk producing the constraint may also interfere with the peptide binding to its target molecule regardless of conformational effects. This approach has been applied to a number of peptide hormones including arginine vasopressin (Moore et al., 1985), bradykinin (Turk et al., 1975), and angiotensin (Samanen et al., 1989). In most cases, the constrained analogs do not exhibit any better potency than their unconstrained analogs, in spite of the reduction in conformational mobility. This may be due in part to the fact that these modifications do not eliminate conformational mobility but only reduce it to some extent. The results of incorporation of constraint-producing amino acids can also be somewhat capricious. N-methyl or α-methyl substitutions, for example, have converted angiotensin agonists into antagonist analogs (Peña et al., 1974; Turk et al., 1976). In addition to providing conformational constraints, N-methyl and α-methyl amino acids can impart other properties to peptides, such as increased resistance to proteolysis, and have been successfully used in the design of metabolically stable and orally active renin inhibitors (Pals et al., 1986; Thaisrivongs et al., 1987).

Regional Constraints

Turns are a prominent feature in peptide secondary structure that have often been implicated as important elements for biological activity, as discussed previously (see Turns). Turns or chain reversals must occur in cyclic peptides or in peptides that occur as short loops in proteins, but linear peptides can fold into conformations containing turns as well. There are steric constraints placed on the side chains and backbone torsions of the corner residues in a turn. Gly and Pro residues most readily accommodate these constraints, and their appearance in a linear sequence is suggestive of a potential turn structure. A number of investigators have designed turn mimetics, cyclic moieties designed to replace the internal residue(s) of a turn while maintaining the overall geometry associated with the turn (Nagai and Sato, 1985; Huffman et al., 1988; Kahn et al., 1988). The mimetic most frequently applied to peptide design is the γ-lactam of Freidinger (1981; Figure 18), which induces a Type II' β-turn conformation.

The γ-lactam has successfully been applied to the peptide hormone-luteinizing–hormone-releasing hormone (LHRH), whose sequence is shown in Figure 19. LHRH contains a Gly residue at position 6. When substituted by a D-Ala, the biological activity increased, while the L-Ala analog exhibited low activity (Monahan et al., 1973). This outcome suggests that the Gly residue might be occupying the $i+1$ position of a Type II' β-turn consisting of the sequence Tyr-Gly-Leu-Arg. As discussed previously (see Protein Amino Acids), a D-amino acid at that position would have an equatorial placement of the side chain, while an L-amino acid would have the less favorable axial placement of the side chain. Freidinger et al. (1980) prepared the γ-lactam-containing peptide shown in

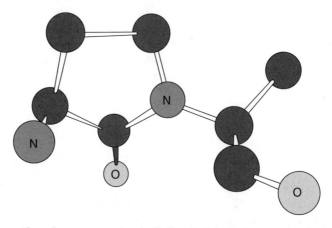

FIGURE 18   γ-Lactam turn mimetic. Ball-and-stick of γ-lactam mimetic showing β-turn conformation; compare with Figure 11(d).

<Glu-His-Trp-Ser-Tyr-Gly-Leu-Arg-Pro-Gly-NH₂

LHRH

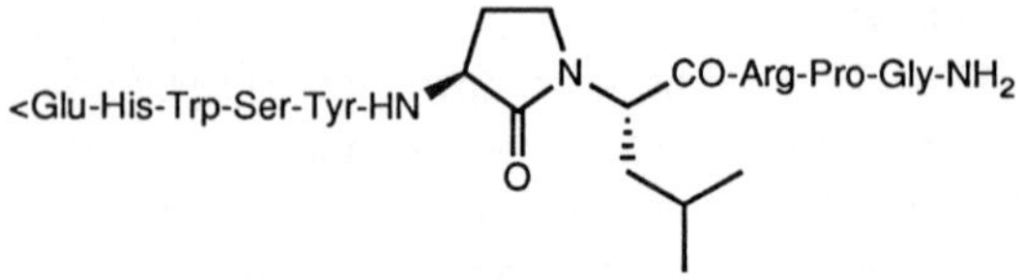

<Glu-His-Trp-Ser-Tyr-HN   CO-Arg-Pro-Gly-NH₂

γ-Lactam analog

FIGURE 19   LHRH and γ-lactam-containing analog.

Figure 19, in which the corner Gly-Leu residues were replaced by the lactam structure. The constrained peptide proved to be more potent than the parent LHRH, both in vivo and in an in vitro assay, suggesting that conformational stabilization is indeed responsible for the improved activity.

Cyclization

The simplest way to introduce a conformational constraint into a linear peptide is by cyclization, either through disulfide bond formation or by coupling the α- or Lys ϵ-amino groups with the carboxyl terminus of an Asp or Glu side chain. Cyclization usually requires peptides of moderate length to adopt some sort of folded conformation consistent with bringing the two ends together, but the exact nature of the conformation(s) induced may not be predictable. As mentioned previously, even a cyclic disulfide hexapeptide still has a great deal of conformational mobility (Hagler et al., 1985). Nevertheless, cyclization greatly reduces the number of accessible conformations compared to the linear peptide and, in favorable cases, can substantially improve or alter peptide biological activity.

One case in point involves the preparation of small peptide fibrinogen receptor antagonists. Platelets express a cell surface glycoprotein, gpIIb/IIIa, which binds to the extracellular matrix protein fibronectin and to the plasma protein fibrinogen. Fibrinogen expresses multiple platelet binding domains and causes aggregation of activated platelets. This adhesive process is important in the formation of blood clots, and inhibitors of this process are potentially useful as antithrombotic agents. Both fibrinogen and fibronectin are extremely large proteins, and their three-dimensional structures are not known. The binding domain on fibronectin was localized to a tetrapeptide sequence Arg-Gly-Asp-Ser by the use of a proteolytic fragment of fibronectin and a series of synthetic peptides based on secondary structure predictions for the sequence of the fragment (Pierschbacher and Ruoslahti, 1984). Small peptides containing this Arg-Gly-Asp-Ser sequence were shown to block binding of platelets both to fibronectin and to fibrinogen, but at concentrations in the 100 mM range (Plow et al., 1985).

In order to try to improve binding affinity, Samanen and coworkers (1991) prepared the series of cyclic analogs shown in Table 9. Both inhibition of aggregation of activated platelets and binding affinity to the gpIIb/IIIa receptor reconstituted in liposomes were measured. The smallest peptide containing the binding sequence is Ac-Arg-Asp-Gly-Ser-NH$_2$ (31), which has about 100 mM antiaggregatory activity and a binding affinity for the reconstituted receptor of 4 mM, 100-fold less potent than fibrinogen itself. Enclosing this sequence in a disulfide loop (32) has only a modest effect on activity, and removing the Ser from the sequence

TABLE 9   Fibrinogen Receptor Binding Peptides

Compound	Sequence	Antiaggregator IC$_{50}$ (μM)	gpIIb/IIIa Binding (μM)
	Fibrinogen		0.043
31	Ac-Arg-Gly-Asp-Ser-NH$_2$	91	4.2
32	Ac-Cys-Arg-Gly-Asp-Ser-Cys-NH$_2$	33	5.3
33	Ac-Cys-Arg-Gly-Asp-Cys-NH$_2$	16	0.78
34	Ac-Cys-Arg-Gly-Asp-Pen-NH$_2$	4.1	N.D.
35	Ac-Cys-MeArg-Gly-Asp-Pen-NH$_2$	0.36	0.027

Pen

(33) enhances activity and binding affinity further. Replacing the C-terminal cysteine with penicillamine (Pen, β,β-dimethyl–cysteine, compound 34), which introduces an additional conformational constraint to the side chain and disulfide, increases the activity still further. The addition of an N-methyl group to the Arg residue (35), which adds yet another conformational constraint, now gives a compound with submicromolar antiaggregatory activity and binding affinity, which is now comparable to that of fibrinogen itself. The end result is a suitably constrained pentapeptide with the binding properties and affinity of a very large protein.

SUMMARY

We have seen examples of peptides designed to mimic protein epitopes or binding sites, and we have seen examples of small peptides that have been modified for enhanced proteolytic stability or for conformational constraints. The exact nature of the design of biologically active peptides depends to a large extent on their intended use. However, several general principles should be kept in mind. Peptides are generally much more flexible and much less soluble than proteins. Smaller peptides do not tend to adopt any preferred solution structure, and longer peptides, which can exhibit strong tendencies to fold into a particular secondary structure, are still in equilibrium with partially folded structures. Peptides are less likely to fold in such a way as to shield hydrophobic regions from solvent exposure, contributing to both the lack of stable solution structures and to insolubility. Conformation, solubility, proteolytic resistance, and other properties are modified by altering the nature of the peptide backbone, the side chains, or both. Even small modification of one residue in a peptide may have a large effect on biological activity, conformation, solubility, or proteolytic susceptibility. Finally, because peptides are synthesized chemically rather than ribosomally, one is not necessarily limited to the use of the 20 naturally occurring amino acids in designing analogs. There are hundreds of synthetically available amino acids with a wide variety of structure, functional groups, charge, polarity, hydrophobicity, or chemical reactivity that can be incorporated to tailor a given peptide to a specific property or use.

REFERENCES

Alderton, G., and Fevold, H. L. 1951. Lanthionine in subtilin. J. Amer. Chem. Soc. 73:463–464.

Allen, J. S., Coligan, J. E., Barin, F., McLane, M. F., Sodroski, J. G., Rosen, C. A., Haseltine, W. A., Lee, T. H., and Essex, M. 1985. Major glycoprotein antigens that induce antibodies in AIDS patients are encoded by HTLV-III. Science 228:1091–1093.

Anastasi, A., Erspamer, V., and Bucci, M. 1971. Isolation and structure of bombesin and alytesin, two analogous active peptides from the skin of the European amphibians *Bombina* and *Alytes*. Experentia 27:166–167.

Baker, E. N. and Hubbard, R. E. 1984. Hydrogen bonding in globular proteins. Prog. Biophys. Mol. Biol. 44:97–179.

Bankowski, K., Manning, M., Haldar, J., and Sawyer, W. H. 1978. Design of potent antagonists of the vasopressor response to arginine-vasopressin. J. Med. Chem. 21:850–853.

Benedetti, E., Barone, V., Bavoso, A., Blasio, B. D., Lelj, F., Pavone, V., Pedone, C., Toniolo, C., Leplawy, M. T., Kaczmarek, K., and Redlinski, A. 1988. Structural versatility of peptides from $C\alpha,\alpha$-dialkylated glycines. I. A conformational energy computation and X-ray diffraction study of homo-peptides from $C^{\alpha,\alpha}$-diethylglycine. Biopolymers 27:357–371.

Berridge, N. J., Newton, G. G. F., and Abraham, E. P. 1952. Purification and nature of the antibiotic nisin. Biochem. J. 52:529–535.

Bierer, B. E., Somers, P. K., Wandless, T. J., Burakoff, S. J., and Schreiber, S. L. 1990. Probing immunosuppressant action with a nonnatural immunophilin ligand. Science 250:556–559.

Bray, A. M., Maeji, N. J., and Geysen, H. M. 1990. The simultaneous multiple production of solution phase peptides; assessment of the Geysen method of simultaneous peptide synthesis. Tetrahedron Lett. 31:5811–5814.

Burley, S. K. and Petsko, G. A. 1985. Aromatic-aromatic interaction: a mechanism of protein structure stabilization. Science 229:23–28.

Burley, S. K. and Petsko, G. A. 1988. Weakly polar interactions in proteins. Adv. Protein Chem. 39:125–189.

Butlen, D., Guillon, G., Rajerison, R. M., Jard, S., Sawyer, W. H., and Manning, M. 1978. Structural requirements for activation of vasopressin-sensitive adenylate cyclase, hormone binding, and antidiuretic actions: effects of highly potent analogues and competitive inhibitors. Mol. Pharmacol. 14:1006–1017.

Chorev, M., Shavitz, R., Goodman, M., Minick, S., and Guillemin, R. 1979. Partially modified retro-inverso enkephalinamides. Science 204:1210–1212.

Chou, P. Y. and Fasman, G. D. Prediction of the secondary structure of proteins from their amino acid sequence. In Advances in Enzymology, Meister, A., ed., John Wiley & Sons, New York, 1978.

Clausen, K., Spatola, A. F., Lemieux, C., Schiller, P. W., and Lawesson, S.-O. 1984. Evidence of a peptide backbone contribution toward selective receptor recognition for leucine enkephalin thioamide analogs. Biochem. Biophys. Res. Commun. 120:305–310.

Cox, M. T., Gormley, J. J., Hayward, C. F., and Peter, N. F. 1980. Incorporation of trans-olefinic dipeptide isosteres into enkephalin and substance P analogs. J. Chem. Soc. Chem. Commun. 800–802.

Dayhoff, M., Schwartz, R. M., and Orcutt, B. C. Atlas of Protein Sequence and Structure. National Biomedical Research Foundation, Silver Spring, Maryland, 1978.

Dayhoff, M. O., Eck, R. V., Chang, M. A., and Sochard, M. R. Atlas of Protein Sequence and Structure 1965. National Biomedical Research Foundation, Silver Spring, Maryland, 1965.

Debouck, C., Gorniak, J. G., Strickler, J. E., Meek, T. D., Metcalf, B. W., and Rosenberg, M. 1987. Human immunodeficiency virus protease expressed in Escherichia coli exhibits autoprocessing and specific maturation of the gag precursor. Proc. Natl. Acad. Sci. USA 84:8903–8906.

DeGrado, W. F., Kézdy, F. J., and Kaiser, E. T. 1981. Design, synthesis and characterization of a cytotoxic peptide with mellitin-like activity. J. Amer. Chem. Soc. 103:679–681.

DelMar, E. G., Largman, C., Brodrick, J. W., and Goekas, M. C. 1979. A sensitive new substrate for chymotrypsin. Anal. Biochem. 99:316–320.

Dutta, A. S., Gormley, J. J., Hayward, C. F., Morley, J. S., Shaw, J. S., Stacey, G. J., and Turnbull, M. T. 1977. Enkephalin analogues eliciting analgesia after intravenous injection. Life Sci. 21:549–562.

Edwards, C. R. W., Kitau, M. J., Chard, T., and Besser, G. M. 1973. Vasopressin analogue DDAVP in diabetes insipidis: clinical and laboratory studies. Br. Med. J. 3:375–378.

Eisenberg, D. 1984. Three-dimensional structure of membrane and surface proteins. Ann. Rev. Biochem. 53:595–623.

Eisenberg, D., Weiss, R. M., and Terwilliger, T. C. 1982. The helical hydrophobic moment: a measure of the amphiphilicity of a helix. Nature 299:371–374.

Eisenberg, D., Wilcox, W., and Eshita, S. Hydrophobic moments as tools for analysis of protein sequences and structures. In Proteins, Structure and Function, L'Italien, J. J., ed., Plenum Press, New York, 1987.

Fischer, G., Wittman-Liebold, B., Lang, K., Kiefhaber, T., and Schmid, F. X.

1989. Cyclophilin and peptidyl-prolyl *cis-trans* isomerase are probably identical proteins. Nature 337:476–478.

Frederickson, R. C. A. 1977. Enkephalin pentapeptides: a review of current evidence for a physiological role in vertebrate neurotransmission. Life Sci. 21:23–42.

Freidinger, R. M. Computer graphics and chemical synthesis in the study of conformation of biologically active peptides. In Peptides, Synthesis, Structure, Function, Proceedings of the Seventh American Peptide Symposium, Rich, D. H. and Gross, E., eds., Pierce Chemical Co., Rockford, Illinois, 1981.

Freidinger, R. M., Veber, D. F., Perlow, D. S., Brooks, J. R., and Saperstein, R. 1980. Bioactive conformation of luteinizing hormone-releasing hormone:evidence from a conformationally constrained analog. Science 210:656–658.

Frömmel, C. and Preissner, R. 1990. Prediction of prolyl residues in cis-conformation in protein structures on the basis of amino acid sequence. FEBS Lett. 277:159–163.

Fukushima, D., Kaiser, E. T., Kézdy, F. J., Kroon, D. J., Kupferberg, J. P., and Yokoyama, S. 1980. Rational design of synthetic models for lipoproteins. Ann. N. Y. Acad. Sci. 348:365–373.

Garnier, J., Osguthorpe, D. J., and Robson, B. 1978. Analysis of the accuracy and implications of simple methods for predicting the secondary structure of globular proteins. J. Biol. Chem. 120:97–120.

Geysen, H. M., Rodda, S. J., and Mason, T. J. A synthetic strategy for epitope mapping. In Peptides, Chemistry and Biology, Proceedings of the Tenth American Peptide Symposium, Marshall, G. R., ed., ESCOM, Leiden, 1988.

Gray, T. M. and Matthews, B. W. 1984. Intrahelical hydrogen bonding of serine, threonine and cysteine residues within α-helices and its relevance to membrane-bound proteins. J. Mol. Biol. 175:75–81.

Greer, J. 1985. Model structure for the inflammatory protein C5a. Science 228:1055–1060.

Habermann, E. 1972. Bee and wasp venoms. Science 177:314–322.

Hagler, A. T., Osguthorpe, D. J., Dauber-Osguthorpe, P., and Hempel, J. C. 1985. Dynamics and conformational energetics of a peptide hormone: vasopressin. Science 227:1309–1315.

Hambrook, J. M., Morgan, B. A., Rance, M. J., and Smith, C. F. C. 1976. Mode of deactivation of the enkephalins by rat and human plasma and rat brain homogenates. Nature 262:782–783.

Hancock, W. S., Prescott, D. T., Marshall, G. R., and Vagelos, P. R. 1972. Acyl carrier protein. XVIII. Chemical synthesis and characterization of a protein with acyl carrier protein activity. J. Biol. Chem. 247:6224–6233.

Hann, M. M., Sammes, P. G., Kennewell, P. D., and Taylor, J. B. 1982. On the double bond isostere of the peptide bond: preparation of an enkephalin analogue. J. Chem. Soc. Perkin Trans I :307–314.

Harding, M. W., Galat, A., Uehling, D. E., and Schreiber, S. C. 1989. A receptor for the immunosuppressant FK506 is a *cis-trans* peptidyl-prolyl isomerase. Nature 341:758–760.

Harrison, R. K. and Stein, R. L. 1990. Substrate specificities of the peptidyl prolyl *cis-trans* isomerase activities of cyclophilin and FK-506 binding protein: evidence for the existence of a family of distinct enzymes. Biochemistry. 29:3813–3816.

Hase, S., Sakakibara, S., Wahrenburg, M., Kirchberger, M., Schwartz, I. L., and Walter, R. 1972. 1,6-Aminosuberic acid analogs of lysine- and arginine-vasopressin and -vasotocin: synthesis and biological properties. J. Amer. Chem. Soc. 94:3590–3600.

Hechter, O., Terada, S., Spitsberg, V., Nakahara, T., Nakagawa, S., and Flouret, G. 1978. Neurohypophyseal hormone-responsive renal adenylate cyclase III. Relationship between affinity and intrinsic activity in neurohypophyseal hormones and structural analogs. J. Biol. Chem. 253:3230–3237.

Hirschmann, R., Nutt, R. F., Veber, D. F., Vitali, R. A., Varga, S. L., Jacob, T. A., Holly, F. W., and Denkewalter, R. G. 1969. Studies on the total synthesis of an enzyme. V. The preparation of enzymatically active material. J. Amer. Chem. Soc. 91:507–508.

Hopp, T. P. and Woods, K. R. 1981. Prediction of protein antigenic determinants from amino acid sequences. Proc. Natl. Acad. Sci. USA 78:3824–3828.

Houghten, R. A. 1985. General method for the rapid solid phase synthesis of large numbers of peptides: specificity of antigen-antibody interaction at the level of individual amino acids. Proc. Natl. Acad. Sci. USA 82:5131–5135.

Howard, J. B. 1981. Reactive site in human α2-macroglobulin: circumstantial evidence for a thiolester. Proc. Natl. Acad. Sci. USA 78:2235–2239.

Huffman, W. F., Callahan, J. F., Eggleston, D. S., Newlander, K. A., Takata, D. T., Codd, E. E., Walker, R. F., Schiller, P. W., Lemieux, C., Wire, W. S., and Burks, T. F. Reverse turn mimics. In Peptides, Chemistry and Biology, Proceedings of the Tenth American Peptide Symposium, Marshall, G. R., ed., ESCOM, Leiden, 1988.

Hunt, S. The non-protein amino acids. In Chemistry and Biochemistry of the Amino Acids, Barrett, G. C, ed., Chapman and Hall, London, 1985.

Ito, N., Phillips, S. E. V., Stevens, C., Ogel, Z. B., McPherson, M. J., Keen, J. N., Yadav, K. D. S., and Knowles, P. F. 1991. Novel thioether bond revealed by a 1.7 Å crystal structure of galactose oxidase. Nature 350:87–90.

IUPAC-IUB Commission on Biological Nomenclature 1970. Abbreviations and symbols for the description of the conformation of polypeptide chains. J. Biol. Chem. 245:6489–6497.

Javaherian, K., Langlois, A. J., McDanal, C., Ross, K. L., Eckler, L. I., Jellis, C. L., Profy, A. T., Rusche, J. R., Bolognesi, D. P., Putney, S. D., and Matthews, T. J. 1989. Principal neutralizing domain of the human immunodeficiency virus type 1 envelope protein. Proc. Natl. Acad. Sci. USA 86:6768–6772.

Kahn, M., Lee, Y.-H., Wilke, S., Chem, B., Fujita, K., and Johnson, M. E. 1988. The design and synthesis of mimetics of peptide β-turns. J. Mol. Recognition 1:75–79.

Kaiser, E. T. and Kézdy, F. J. 1984. Amphiphilic secondary structure: design of peptide hormones. Science 223:249–255.

Kauzmann, W. Some factors in the interpretation of protein denaturation. In Advanced Protein Chemistry, Anfinsen, Jr., Anson, C. B., M. L., Edsall, J. T., and Richards, F. M., eds., Academic Press, New York, 1959.

Kawasaki, K. and Maeda, M. 1982. Amino acids and peptides II. Modification of glycylglycine bond in methionine enkephalin. Biochem. Biophys. Res. Commun. 106:113–116.

Kellis, Jr., J. T., Nyberg, K., Sali, D., and Fersht, A. R. 1988. Contribution of hydrophobic interactions to protein stability. Nature 333:784–786.

Kim, P. S. and Baldwin, R. L. 1982. Specific intermediates in the folding reactions of small proteins and the mechanism of protein folding. Ann. Rev. Biochem. 51:459–489.

Knowles, J. R. 1991. Enzyme catalysis: not different, just better. Nature 350:121–124.

Kosterlitz, H. W., Lord, J. A. H., Paterson, S. J., and Waterfield, A. A. 1980. Effects of changes in the structure of enkephalins and narcotic analgesic drugs on their interactions with μ- and δ-receptors. Br. J. Pharmacol. 68:333–342.

Kyte, J. and Doolittle, R. F. 1982. A simple method for displaying the hydropathic character of a protein. J. Mol. Biol. 157:105–132.

Landry, S. J. and Gierasch, L. M. 1991. Recognition of nascent polypeptide for targeting and folding. TIBS:159–163.

Landschutz, W. H., Johnson, P. F., and McKnight, S. L. 1988. The leucine zipper: a hypothetical structure common to a new class of DNA binding proteins. Science 240:1759–1764.

Lapatto, P., Blundell, T., Hemmings, A., Overington, J., Wilderspin, A., Wood, S., Merson, J. R., Whittle, P. J., Danley, D. E., Geoghegan, K. F., Hawrylik, S. J., Lee, S. E., Scheld, K. G., and Hobart, P. M. 1989. X-ray analysis of HIV-1 proteinase at 2.7 Å resolution confirms structural homology among retroviral enzymes. Nature 342:299-302.

Lipman, D. J. and Pearson, W. R. 1985. Rapid and sensitive protein similarity searches. Science 227:1435–1441.

Malfroy, B., Swerts, J. P., Guyon, A., Roques, B. P., and Schwartz, J. C. 1978. High affinity enkephalin-degrading peptidase in brain is increased after morphine. Nature 276:523–526.

Märki, W., Brown, M., and Rivier, J. E. 1981. Bombesin analogs: effects on thermoregulation and glucose metabolism. Peptides 2, Suppl. 2:169–177.

Marks, N., Grynbaum, A., and Neidle, A. 1977. On the degradation of enkephalins and endorphines by rat and mouse brain extracts. Biochem. Biophys. Res. Commun. 74:1552–1559.

Marshall, G. R. and Bosshard, H. E. 1972. Angiotensin II. Biologically active conformation. Circ. Res., Suppl. II 30–31: II-143-150.

Matsushita, S., Robert-Guroff, M., Rusche, J., Koito, A., Hattori, T., Hoshino, H., Javaherian, K., Takatsuki, K., and Putney, S. 1988. Characterization of a human immunodeficiency virus neutralizing monoclonal antibody and mapping of the neutralizing epitope. J. Virol. 62:2107–2114.

McDonald, T. J., Jornvall, H., Nilsson, G., Vagne, M., Ghatei, M., Bloom, S. R., and Mutt, V. 1979. Characterization of a gastrin releasing peptide from porcine non-antral gastric tissue. Biochem. Biophys. Res. Comm. 90:227.

Meek, T. D., Dayton, B. D., Metcalf, B. W., Dreyer, G. B., Strickler, J. E., Gorniak, J. G., Rosenberg, M., Moore, M. L., Magaard, V. W., and Debouck, C. 1989. Human immunodeficiency virus 1 protease expressed in *Escherichia coli* behaves as a dimeric aspartic protease. Proc. Natl. Acad. Sci. USA 86:1841–1845.

Merrifield, B. 1963. Solid phase peptide synthesis I. The synthesis of a tetrapeptide. J. Am. Chem. Soc. 85:2149–2154

Miller, M., Schneider, J., Sathyanarayana, B. K., Toth, M. V., Marshall, G. R., Lawson, L., Selk, L., Kent, S. B. H., and Wlodawer, A. 1989. Structure of complex of synthetic HIV-1 protease with a substrate-based inhibitor at 2.3 Å resolution. Science 246:1149–1152.

Miller, R. J., Chang, K.-J., Cuatrecasas, P., and Wilkinson, S. 1977. The metabolic stability of enkephalins. Biochem. Biophys. Res. Commun. 74:1311–1317.

Monahan, M. W., Amoss, M. S., Anderson, H. A., and Vale, W. 1973. Synthetic analogs of the hypothalamic luteinizing hormone releasing factor with increased agonist or antagonist properties. Biochemistry 12:4616–4620.

Moore, M. L., Bryan, W. M., Fakhoury, S. A., Magaard, V. W., Huffman, W. F., Dayton, B. D., Meek, T. D., Hyland, L., Dreyer, G. B., Metcalf, B. W., Strickler, J. E., Gorniak, J. G., and Debouck, C. 1989. Peptide substrates and inhibitors of the HIV-1 protease. Biochem. Biophys. Res. Commun. 159:420–425.

Moore, M. L., Huffman, W. F., Bryan, W. M., Silvestri, J., Chang, H.-L., Marshall, G. R., Stassen, F., Stefankiewicz, J., Sulat, L., Schmidt, D., Kinter, L., McDonald, J., and Ashton-Shue, D. Vasopressin antagonist analogs containing α-methyl amino acids at position 4. In Peptides: Structure and Function, Proceedings of the Ninth American Peptide Symposium, Deber, C. M., Hruby V. J., and Kopple, K. D., eds., Pierce Chemical Co., Rockford, Illinois, 1985.

Morgan, B. P., Scholtz, J. M., Ballinger, M. D., Sipkin, I. D., and Bartlett, P. A. 1991. Differential binding energy: a detailed evaluation of the influence of hydrogen bonding and hydrophobic groups on the inhibition of thermolysin by phosphorus-containing inhibitors. J. Amer. Chem. Soc. 113:297–307.

Morishima, H., Takita, T., Aoyagi, T., Takeuchi, T., and Umezawa, H. 1970. The structure of pepstatin. J. Antibiot. 23:263–265.

Nagai, U. and Sato, K. 1985. Synthesis of a bicyclic dipeptide with the shape of a β-turn central part. Tetrahedron Lett. 26:647–650.

Needleman, S. and Wunsch, C. 1970. A general method applicable to the search for similarities in the amino acid sequence of two proteins. J. Mol. Biol. 48:443-453.

Nutt, R. F., Brady, S. F., Darke, P. L., Ciccarone, T. M., Colton, C. D., Nutt, E. M., Rodkey, J. A., Bennett, C. D., Waxman, L. H., Sigal, I. S., Anderson, P. S., and Veber, D. F. 1988. Chemical synthesis and enzymatic activity of a 99-residue peptide with a sequence proposed for the human immunodeficiency virus protease. Proc. Natl. Acad. Sci. USA 85:7129–7133.

O'Shea, E. K., Rutkowski, R., and Kim, P. S. 1989. Evidence that the leucine zipper is a coiled coil. Science 243:538–542.

Pals, D. T., Thaisrivongs, S., Lawson, J. A., Kati, W. M., Turner, S. R., De-Graaf, G. L., Harris, D. W., and Johnson, G. A. 1986. An orally active inhibitor of renin. Hypertension 8:1105–1112.

Pandey, R. C., Cook, Jr., J. C., and Rinehart, Jr., K. L. 1977. High resolution and field desorption mass spectrometry studies and revised structures of alamethicins I and II. J. Amer. Chem. Soc. 99:8469–8483.

Payne, J. W., Jakes, R., and Hartley, B. S. 1970. The primary structure of alamethicin. Biochem. J. 117:757–766.

Pearl, L. H. and Taylor, W. R. 1987. A structural model for the retroviral proteases. Nature 329:351–354.

Pearson, W. R. and Lipman, D. J. 1988. Improved tools for biological sequence comparison. Proc. Natl. Acad. Sci. USA 85:2444–2448.

Peña, C., Stewart, J. M., and Goodfriend, T. C. 1974. A new class of angiotensin inhibitors:N-methylphenylalanine analogs. Life Sci. 14:1331.

Pierschbacher, M. D. and Ruoslahti, E. 1984. Cell attachment of fibronectin can be duplicated by small synthetic fragments of the molecule. Nature 309:30–33.

Plow, E. F., Pierschbacher, M. D., Ruoslahti, E., Marguerie, G. A., and Ginsberg, M. H. 1985. The effect of Arg-Gly-Asp-containing peptides on fibrinogen and von Willebrand factor binding to platelets. Proc. Natl. Acad. Sci. USA 82:8057–8061.

Polácek, I., Krejcí, I., Nesvada, H., and Rudinger, J. 1970. Action of [1,6-di-alanine]-oxytocin and [1,6-di-serine]-oxytocin on the rat uterus and mammary gland in vitro. Eur. J. Pharmacol. 9:239–248.

Rinehart, Jr., K. L. , Gloer, J. B., Cook, Jr., J. C., Miszak, S. A., and Scahill, T. A. 1981. Structures of the didemnins, antiviral and cytotoxic depsipeptides from a Caribbean tunicate. J. Amer. Chem. Soc. 103:1857–1859.

Rivier, J., Brown, M., and Vale, W. 1976. D-Trp8 somatostatin: an analog of somatostatin more potent than the native molecule. Biochem. Biophys. Res. Comm. 65:746–751.

Rivier, J. E. and Brown, M. R. 1978. Bombesin, bombesin analogues, and related peptides: effects on thermoregulation. Biochem. 17:1766–1771.

Rose, G. D. and Roy, S. 1980. Hydrophobic basis of packing in globular proteins. Proc. Natl. Acad. Sci. USA 77:4643–4647.

Roy, C., Barth, T., and Jard, S. 1975. Vasopressin-sensitive kidney adenylate cyclase. J. Biol. Chem. 250:3149–3156.

Rüegger, A., Kuhn, M., Lichti, H., Loosli, H.-R., Huguenin, R., Quiquerez, C., and Wartburg, A. V. 1976. Cyclosporin A, ein immunosuppressiv wirksamer Peptidmetabolit aus Trichoderma polysporem (Link ex Pers.) *Rifai.* Helv. Chim. Acta 59:1075–1092.

Rusche, J. R., Javaherian, K., McDanal, C., Petro, J., Lynn, D. L., Grimaila, R., Langlois, A., Gallo, R. C., Arthur, L. O., Fischinger, P. J., Bolognesi, D. P., Putney, S. D., and Matthews, T. J. 1988. Antibodies that inhibit fusion of human immunodeficiency virus-infected cells bind a 24-amino acid sequence of the viral envelope, gp120. Proc. Natl. Acad. Sci. USA 85:3198–3202.

Samanen, J., Ali, F. E., Romoff, T., Calvo, R., Sorenson, E., Bennett, D., Berry, D., Kostler, P., Vasko, J., Powers, D., Stadel, J., and Nichols, A. SK&F 106760: Reinstatement of high receptor affinity on a peptide fragment (RGDS) of a large glycoprotein (fibrinogen) through conformational constraints. In Peptides 1990, Proceedings of the 21st European Peptide Symposium, Giralt, E. and Andreu, D., eds., ESCOM, Leiden, 1991.

Samanen, J., Narindray, D., Cash, T., Brandeis, E., Adams, J., W., Yellin, T., Eggleston, D., Debrosse, C., and Regoli, D. 1989. Potent angiotensin II antagonists with non-β-branched amino acids in position 5. J. Med. Chem. 32:466–472.

Samanen, J., Zuber, G., Bean, J., Eggleston, D., Romoff, T., Kopple, K., Saunders, M., and Regoli, D. 1990. 5,5-Dimethylthiazolidine-4-carboxylic acid (Dtc) as a proline analog with restricted conformation. Int. J. Pept. Protein Res. 35:501–509.

Sarantakis, D., McKinley, W. A., and Grant, N. H. 1973. The synthesis and biological activity of Ala^3,14-somatostatin. Biochem. Biophys. Res. Commun. 55:538-542.

Sawyer, W. H., Acosta, M., Balaspiri, L., Judd, J., and Manning, M. 1974. Structural changes in the arginine vasopressin molecule that enhance antidiuretic activity and specificity. Endocrinology 94:1106–1115.

Schröder, E., Lübke, K., Lehmann, M., and Beetz, I. 1971. Peptide syntheses. XLVII. Mellitin. 3. Hemolytic activity and action on surface tension of aqueous solutions on synthetic mellitins and their derivatives. Experientia 27:764–765.

Siekierka, J. J., Hung, S. H. Y., Poe, M., Lin, C. S., and Sigal, N. H. 1989. A cytosolic binding protein for the immunosuppressant FK506 has the peptidyl-prolyl isomerase activity but is distinct from cyclophilin. Nature 341:755–757.

Sigal, N. H., Dumont, F., Durette, P., Siekierka, J. J., Peterson, L., Rich, D. H., Dunlap, B. E., Staruch, M. J., Melino, M. R., Koprak, S. L., Williams, D., Witzel, B., and Pisaon, J. M. 1991. Is cyclophilin involved in the immunosuppressive and nephrotoxic mechanism of action of cyclosporin A? J. Exp. Med. 173:619–628.

Smith, J. A. and Pease, L. G. Reverse turns in peptides and proteins. In CRC Critical Reviews in Biochemistry, Fasman, G. D., ed., CRC Press, Boca Raton, Florida, 1980.

Sothrup-Jensen, L., Petersen, T. E., and Magnuson, S. 1980. A thiol-ester in a2-macroglobulin is cleaved during proteinase complex formation. FEBS Lett. 121:275–279.

Spatola, A. F., Saneii, H., Edwards, J. V., Bettag, A. L., Anwer, M. K., Rowell, P., Browne, B., Lahti, R., and von Voightlander, P. 1986. Structure-activity relationships of enkephalins containing serially replaced thiomethylene amide bond surrogates. Life. Sci. 38:1243–1249.

Steinmann, B., Bruckner, P., and Superti-Furga, A. 1991. Cyclosporin A slows collagen triple-helix formation *in vivo*: indirect evidence for a physiologic role of peptidyl-prolyl *cis-trans* isomerase. J. Biol. Chem. 266:1299–1303.

Tack, B. F., Harrison, R. A., Janatova, J., Thomas, M. L., and Prahl, J. W. 1980. Evidence for the presence of an internal thioester bond in third component of human complement. Proc. Natl. Acad. Sci. USA 77:5764–5768.

Takahashi, N., Hayano, T., and Suzuki, M. 1989. Peptidyl-prolyl *cis-trans* isomerase is the cyclosporin A-binding protein cyclophilin. Nature 337:473–475.

Tanford, C. The Hydrophobic Effect. John Wiley & Sons, Inc., New York, 1973.

Tang, J., James, M. N. G., Hsu, I. N., Jenkins, J. A., and Blundell, T. L. 1978. Structural evidence for gene duplication in the evolution of the acid proteases. Nature 271:618–621.

Thaisrivongs, S., Pals, D. T., Lawson, J. A., Turner, S. R., and Harris, D. W. 1987. α-Methylproline-containing renin inhibitory peptides: *in vivo* evaluation in an anesthetized, ganglion-blocked, hog renin infused rat model. J. Med. Chem. 30:536–541.

Thanki, N., Thornton, J. M., and Goodfellow, J. M. 1990. Influence of secondary structure on the hydration of serine, threonine and tyrosine residues in proteins. Protein Eng. 3:495–508.

Turk, J., Needleman, P., and Marshall, G. R. 1975. Analogs of bradykinin with restricted conformational freedom. J. Med. Chem. 18:1139–1142.

Turk, J., Needleman, P., and Marshall, G. R. 1976. Analogs of angiotensin II with restricted conformational freedom including a new antagonist. Mol. Pharmacol. 12:217–224.

Veronese, F. D., DeVico, A. L., Copeland, T. D., Oroszlan, S., Gallo, R. C., and Sarngadharan, M. G. 1985. Characterization of gp41 as the transmembrane protein coded by the HTLV-III/LAV envelope gene. Science 229:1402–1405.

Walter, R., Rudinger, J., and Schwartz, I. L. 1967. Chemistry and structure-activity relations of the antidiuretic hormones. Amer. J. Med. 42:653-677.

Walter, R., Smith, C. W., Mehta, P. K., Boonjarern, S., Arruda, J. A. L., and Kurtzman, N. A. Conformational considerations of vasopressin as a guide to development of biological probes and therapeutic agents. In Disturbances in Body Fluid Osmolality, Andreoli, T. E., Grantham, J. J., and Rector, F. C., Jr., eds., American Physiological Society, Bethesda, Maryland, 1977.

Wickner, W. T. and Lodish, H. F. 1985. Multiple mechanisms of protein insertion into and across membranes. Science 230:400–407.

Wieland, T. 1968. Poisonous principles of mushrooms of the genus Amanita. Science 159:946–952.

Wlodawer, A., Miller, M., Jaskólski, M., Sathyanarayana, B. K., Baldwin, E., Weber, I. T., Selk, L. M., Clawson, L., Schneider, J., and Kent, S. B. H. 1989. Conserved folding in retroviral proteases: crystal structure of a synthetic HIV-1 protease. Science 245:616–621.

Wlodawer, A., Miller, M., Swain, A. L., and Jaskólski, M. Structure of three inhibitor complexes of HIV-1 protease. In Methods in Protein Sequence Analysis, Jörnvall, Höög, and Gustavsson, eds., Birkhauser Verlag, Basel, 1991.

Yajima, H. and Fujii, N. 1981. Studies on peptides. 103. Chemical synthesis of a crystalline protein with the full enzymic activity of ribonuclease A. J. Amer. Chem. Soc. 103:5867–5871.

Yasunaga, T., Sagata, N., and Ikawa, Y. 1986. Protease gene structure and env gene variability of the AIDS virus. FEBS Lett. 199:145–150.

Zaoral, M., Kolc, J., and Sorm, F. 1967a. Amino acids and peptides. LXX. Synthesis of D-Arg[8]- and D-Lys8 vasopressin. Coll. Czech. Chem. Comm. 32:1242–1248.

Zaoral, M., Kolc, J., and Sorm, F. 1967b. Amino acids and peptides. LXXI. Synthesis of 1-deamino-8-D-g-aminobutyrine-vasopressin, 1-deamino-8-D-lysine-vasopressin and 1-deamino-8-D-arginine-vasopressin. Coll. Czech. Chem. Comm. 32:1250–1257.

Zuiderweg, E. R. P., Henkin, J., Mollison, K. W., Carter, G. W., and Greer, J. 1988. Comparison of model and nuclear magnetic resonance structures for the human inflammatory protein C5a. Proteins: Struct. Funct. Genet. 3:139–145.

3

Principles and Practice of Solid-Phase Peptide Synthesis

Gregg B. Fields, Zhenping Tian,
and George Barany

Peptides play key structural and functional roles in biochemistry, pharmacology, and neurobiology. Naturally occurring and designed peptides are also important probes for research in enzymology, immunology, and molecular biology. The amino acid building blocks can be among the 20 genetically encoded L residues or they may be unnatural. Further, the sequences can be linear, cyclic, or branched. It follows that rapid, efficient, and reliable methodology for the chemical synthesis of these molecules is of utmost interest. A number of synthetic peptides are significant commercial or pharmaceutical products, ranging from the sweet dipeptide L-Asp-L-Phe-OMe (aspartame) to clinically used hormones such as oxytocin, adrenocorticotropic hormone, and calcitonin. Synthesis can lead to potent and selective new drugs by judicious substitutions that change functional groups, conformations, or both. These include introduction of *N*- or *C*-alkyl substituents, unnatural or D-amino acids, side-chain modifications, including sulfate groups, phosphate groups, or carbohydrate moieties, and constraints such as disulfide bridges between half-cystines or side-chain lactams between Lys and Asp or Glu (see Chapter 2). Most of the biologically or medicinally important peptides that are the targets for useful structure-function studies by chemical synthesis comprise fewer than 50 amino acid residues, but occasionally a synthetic approach can lead to important conclusions about small proteins in the 100-residue size range.

Methods for synthesizing peptides are divided conveniently into two categories: solution (classical) and solid phase (SPPS). The classical methods have evolved since the beginning of the twentieth century, and they are described amply in several reviews and books (Wünsch, 1974; Finn and Hofmann, 1976; Bodanszky and Bodanszky, 1984). The solid-phase alternative was conceived and elaborated by R.B. Merrifield beginning in 1959, and it has also been covered comprehensively (Erickson and Merrifield, 1976; Birr, 1978; Barany and Merrifield, 1979; Stewart and Young, 1984; Merrifield, 1986; Barany et al., 1987; Barany et al., 1988; Kent, 1988; Clark-Lewis and Kent, 1989; Atherton and Sheppard, 1989; Fields and Noble, 1990; Barany and Albericio, 1991b). Solution synthesis retains value in large-scale manufacturing and for specialized laboratory applications. However, the need to optimize reaction conditions, yields, and purification procedures for essentially every intermediate (each of which has unpredictable solubility and crystallization characteristics) renders classical methods time-consuming and labor-intensive. Consequently, most workers now requiring peptides for their research opt for the more accessible solid-phase approach.

In this chapter, we discuss critically the scope and limitations of the best available procedures for solid-phase synthesis of peptides. At the same time, we mention briefly important new developments and trends in this field. Literature citations are weighted toward detailed reports with full experimental descriptions, with some bias toward those describing procedures with which we and our collaborators have laboratory experience.

OVERVIEW OF SOLID-PHASE STRATEGY

The concept of SPPS (Figure 1) is to retain chemistry proved in solution (protection scheme, reagents), but adding a covalent attachment step (anchoring) that links the nascent peptide chain to an insoluble *polymeric support*. Subsequently, the anchored peptide is extended by a series of addition (deprotection/coupling) cycles, which are required to proceed with exquisitely high yields and fidelities. It is the essence of the solid-

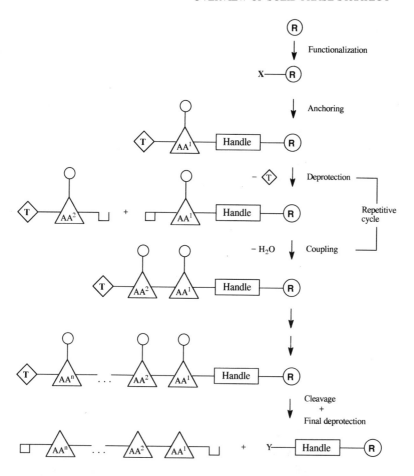

FIGURE 1   Stepwise solid-phase synthesis of linear peptides. **R**, insoluble polymeric support; AA^1, AA^2, ..., AA^n, amino acid residues numbered starting from C-terminus; **T**, "temporary protection; **O**, "permanent" protection; ⌐⌐, free carboxyl; ⌐⌐, free amino group. See text for further details. Figure adapted from Barany et al., 1988.

phase approach that reactions are driven to completion by the use of *excess* soluble reagents, which can be removed by simple *filtration* and *washing* without manipulative losses. Because of the speed and simplicity of the repetitive steps, which are carried out in a single reaction vessel at ambient temperature, the major portion of the solid-phase procedure is readily amenable to *automation*. Once chain elaboration has been accomplished, it is necessary to *release* protected residues and to release (cleave) the crude peptide from the support under conditions that are minimally destructive toward sensitive residues in the sequence. Finally, there must follow prudent *purification* and appropriate *characterization* of the synthetic peptide product to verify that the desired structure is indeed the one obtained.

An appropriate polymeric support (resin) must be chosen that has adequate mechanical stability as well as desirable physicochemical properties that facilitate solid-phase synthesis (see Polymeric Support). In practice, such supports include those that exhibit significant levels of swelling in useful reaction/wash solvents. Swollen resin beads are reacted and washed batchwise with agitation and filtered either with suction or under positive nitrogen pressure. Alternatively, solid-phase synthesis may be carried out in a continuous-flow mode, by pumping reagents and solvents through resins that are packed into columns. The usual batchwise resins often lack the rigidity and strength necessary for column procedures. More appropriate supports, which are usually, but not always, lower in terms of functional capacity, are obtained when mobile polymer chains are chemically grafted onto, or physically embedded within, an inert matrix.

Regardless of the structure and nature of the polymeric support chosen, it must contain appropriate functional groups onto which the first amino acid can be anchored. In early schemes that still have considerable popularity, chloromethyl groups are introduced onto a polystyrene resin by a direct Friedel-Crafts reaction, following which an N^α-protected amino acid, as its triethylammonium or cesium salt, is added to provide a polymer-bound benzyl ester. More recently, it has been recognized that greater control and generality is possible by use of "handles," which are defined as bifunctional spacers that, on one end, incorporate features of a smoothly cleavable protecting group. The other end of the handle contains a functional group, often a carboxyl, that can be activated to allow coupling to functionalized supports, for example ones containing aminomethyl groups. Particularly advantageous, though more involved to prepare, are "preformed" handles, which serve to link the first amino acid to the resin in two discrete steps, and thereby provide maximal control over this essential step of the synthesis (see Attachment to Support).

The next stage of solid-phase synthesis is the systematic elaboration of the growing peptide chain. In the vast majority of solid-phase syntheses, suitably N^α- and side-chain protected amino acids are added stepwise in the $C \rightarrow N$ direction. A particular merit of this strategy is that the best practical realizations have been shown experimentally to proceed with only negligible levels of racemization. A "temporary" protecting group is removed quantitatively at each step to liberate the N^α-amine of the peptide resin, following which the next incoming protected amino acid is introduced with its carboxyl group suitably activated (see Formation of Peptide Bond). It is frequently worthwhile to verify that the coupling has gone to completion by some monitoring technique (see Monitoring).

Once the desired linear sequence has been assembled satisfactorily on the polymeric support, the anchoring linkage must be cleaved. Depending on the chemistry of the original handle and on the cleavage reagent selected, the product from this step can be a C-terminal peptide acid, amide, or other functionality. The cleavage can be conducted so as to retain "permanent" side-chain protecting groups and thus yield protected segments that, once purified, are suitable for further condensation. Alternatively, selected "permanent" groups can be retained on sensitive residues for later deblocking in solution. However, the approach that is most widely used involves final deprotection which is carried out essentially concurrent with cleavage; in this way, the released product is directly the free peptide.

PROTECTION SCHEMES

The preceding section outlined the key steps of the solid-phase procedure but dealt only tangentially with combinations of "temporary" and "permanent" protecting groups and the corresponding methods for their removal. The choice and optimization of protection chemistry is perhaps the key factor in the success of any synthetic endeavor. Even when a residue has been incorporated safely into the growing resin-bound polypeptide chain, it may still undergo irreversible structural modification or rearrangement during subsequent synthetic steps. The vulnerability to damage is particularly pronounced at the final deprotection/cleavage step, since these are usually the harshest conditions. At least two levels of protecting-group stability are required, insofar as the "permanent" groups used to prevent branching or other problems on the side chains must withstand repeated applications of the conditions for quantitative removal of the "temporary" N^{α}-amino protecting group. On the other hand, structures of "permanent" groups must be such that conditions can be found to remove them with minimal levels of side reactions that affect the integrity of the desired product. The necessary stability is often approached by kinetic "fine tuning," which is a reliance on quantitative rate differences whenever the same chemical mechanism (usually acidolysis) serves to remove both classes of protecting groups. An often limiting consequence of such schemes, based on graduated lability, is that they force adoption of relatively severe final deprotection conditions. Alternatively, *orthogonal* protection schemes can be used. These schemes involve two or more classes of groups that are removed by differing chemical mechanisms, and therefore can be removed in any order and in the presence of the other classes. Orthogonal schemes offer the possibility of substantially milder overall conditions, because selectivity can be attained on the basis of differences in chemistry rather than in reaction rates.

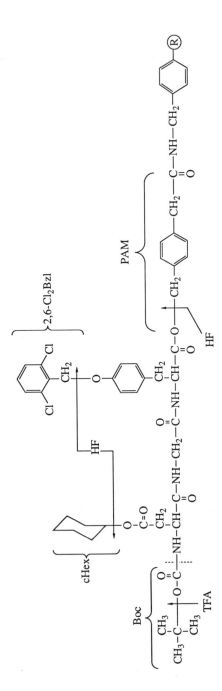

FIGURE 2 "Merrifield" protection scheme for solid-phase synthesis, based on graduated acidolysis. Temporary N^α-amino protection is provided by the Boc group, removed at each step by the moderately strong acid TFA. Permanent Bzl-based and cHex side-chain protecting groups, and the PAM linkage, are then cleaved simultaneously by HF or other strong acids, with a free peptide acid being formed in high yield.

"Temporary" Protection of N^α-Amino Groups

Boc Chemistry

The so-called "standard Merrifield" system is based on graduated acid lability (Figure 2, in a modern, improved version). The acidolyzable "temporary" N^α-tert-butyloxycarbonyl (Boc) group is introduced onto amino acids with either di-*tert*-butyl dicarbonate or 2-*tert*-butyloxycarbonyloximino-2-phenylacetonitrile (Boc-ON) in aqueous 1,4-dioxane containing NaOH or triethylamine (Et₃N) (Bodanszky and Bodanszky, 1984). The Boc group is stable to alkali and nucleophiles and removed rapidly by inorganic and organic acids (Barany and Merrifield, 1979). Boc removal is usually carried out with trifluoroacetic acid (TFA) (20 to 50%) in dichloromethane (DCM) for 20 to 30 minutes, and, for special situations, HCl (4 N) in 1,4-dioxane for 35 minutes. Deprotection with neat (100%) TFA, which offers enhanced peptide resin solvation compared to TFA-DCM mixtures, proceeds in as little as 4 minutes (Kent and Parker, 1988; Wallace et al., 1989). Following acidolysis, a rapid diffusion-controlled neutralization step with a tertiary amine, usually 5 to 10% Et₃N or *N,N*-diisopropylethylamine (DIEA) in DCM for 3 to 5 minutes, is interpolated to release the free N^α-amine. Alternatively, Boc amino acids may be coupled without prior neutralization by using "in situ" neutralization, i.e., coupling in the presence of DIEA or NMM (Suzuki et al., 1975; Schnolzer et al., 1992). "Permanent" side-chain protecting groups are ether, ester, and urethane derivatives based on benzyl alcohol, suitably "fine tuned" with electron-donating methoxy or methyl groups or electron-withdrawing halogens for the proper level of acid stability/lability. Alternatively, ether and ester derivatives based on cyclopentyl or cyclohexyl alcohol are sometimes applied, because their use moderates certain side reactions. These "permanent" groups are sufficiently stable to repeated cycles of Boc removal, yet they are cleaved cleanly in the presence of appropriate scavengers by the use of liquid anhydrous hydrogen fluoride (HF) at 0 °C or trifluoromethanesulfonic acid (TFMSA) at 25 °C (see Cleavage). The 4-(hydroxymethyl)phenylacetic acid (PAM) or 4-methylbenzhydrylamine (MBHA) anchoring linkages are similarly "fine tuned" to be cleaved at the same time (see Attachment to Support).

Fmoc Chemistry

A mild orthogonal alternative is constructed using Carpino's base-labile "temporary" N^α-9-fluorenylmethyloxycarbonyl (Fmoc) group (Figure 3). The optimal reagent for preparation of Fmoc amino acids is fluorenylmethyl succinimidyl carbonate (Fmoc-OSu), applied in a partially aqueous/organic mixture in the presence of base; the alternative procedure involving derivatization by Fmoc chloride is accompanied by unaccep-

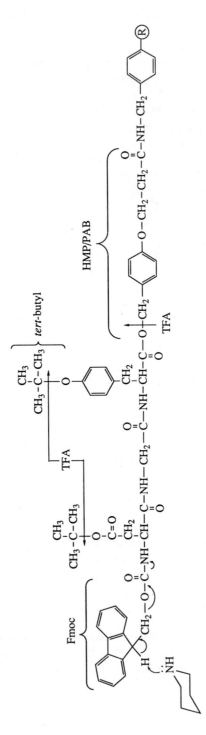

FIGURE 3   A mild two-dimensionsl orthogonal protection scheme for solid-phase synthesis. Temporary N^α-amino protection is provided by the Fmoc group, removed by the indicated base-catalyzed β-elimination mechanism. Permanent tBu-based side-chain protecting groups and the HMP/PAB ester linkage are both cleaved by treatment with TFA to yield the free peptide acid. A third dimension of orthogonality may be added with an acid-stable, photolabile anchoring linkage (details in text).

A problem that occurs during preparation of Fmoc amino acids is the precipitation of either the Fmoc-OSu reagent or the base (NaHCO$_3$ or Na$_2$CO$_3$) upon mixing of the organic (1,4-dioxane, acetone, DMF, and acetonitrile have been proposed) and aqueous cosolvents. This problem is best overcome by use of aqueous dimethoxyethane with Na$_2$CO$_3$ (Fields et al., 1989; Netzel-Arnett et al., 1991). A solution of Fmoc-OSu (3.0 mmol) in dimethoxyethane (10 mL) is added slowly to the amino acid (2.0 mmol) dissolved in 10% aqueous Na$_2$CO$_3$ (10 mL); final yields of Fmoc amino acids after workup are in a range of 75 to 95%.

The Fmoc group has been shown to be completely stable to treatment with TFA, HBr in acetic acid (HOAc), or HBr in nitromethane for 1 to 2 days (Carpino and Han, 1972). Somewhat less stability was found in dipolar aprotic solvents (Atherton et al., 1979). Fmoc-Gly was deprotected after 7 days in dimethylacetamide (DMA), *N,N*-dimethylformamide (DMF), and NMP to the extent of 1, 5, and 14%, respectively. Although these low levels of decomposition are considered to be relatively insignificant, it is nevertheless prudent to purify the aforementioned solvents just before use. For NMP, Fmoc group removal is attributed directly to the presence of methylamine as an impurity (Otteson et al., 1989). The addition of HOBt (0.01 to 0.1 M) greatly reduces the detrimental effect of methylamine (Albericio and Barany, 1987a); Fmoc-Gly-HMP resin was less than 0.05% deprotected after 12 hours in NMP containing 0.01 M HOBt (Otteson et al., 1989). Fmoc amino acids can be stored in purified or synthesis-quality NMP with little decomposition for 6 to 8 weeks, in the dark at 25 °C (Kent et. al., 1991).

table levels (2 to 20%) of Fmoc dipeptide formation (Pacquet, 1982; Sigler et al., 1983; Lapatsanis et al., 1983; Tesser et al., 1983; Ten Kortenaar et al., 1986; Milton et al., 1987; Fields et al., 1989). The Fmoc group may also be added via [4-(9-fluorenylmethyloxycarbonyloxy)-phenyl]dimethylsulfonium methyl sulfate (Fmoc-ODSP) in H$_2$O with Na$_2$CO$_3$ or Et$_3$N (Azuse et al., 1989). Removal of the Fmoc group is achieved usually with 20 to 55% piperidine in DMF or N-methylpyrrolidone (NMP) for 10 to 18 minutes (Atherton et al., 1978a; Atherton et al., 1978b; Chang et al., 1980a; Albericio et al., 1990a; Fields and Fields, 1991); piperidine in DCM is not recommended, because an amine salt precipitates after relatively brief standing. The base abstracts the acidic proton at the 9 position of the fluorene ring system; β-elimination follows to give a highly reactive dibenzofulvene intermediate, which is trapped by excess secondary amine to form a stable, harmless adduct

(Carpino and Han, 1972). The Fmoc group may also be removed by 2% 1,8-diazabicyclo[5.4.0]undec-7-ene (DBU) in DMF; however, this reagent is recommended for continuous-flow syntheses only, because the dibenzofulvene intermediate does not form an adduct with DBU and thus must be washed rapidly from the peptide resin (Wade et al., 1991). After Fmoc removal, the liberated N^α-amine of the peptide resin is free and ready for immediate acylation without an intervening neutralization step (compare to the previous paragraph on Boc chemistry). "Permanent" protection compatible with N^α-Fmoc protection is provided primarily by ether, ester, and urethane derivatives based on tert-butanol. These derivatives are cleaved at the same time as appropriate anchoring linkages by use of TFA at 25 °C. Scavengers must be added to the TFA to trap the reactive carbocations that form under the acidolytic cleavage conditions.

"Permanent" Protecting Groups for Reactive Amino Acid Side Chains

Once the means for N^α-amino protection has been selected, compatible protection for the side chains of trifunctional amino acids must be specified. These choices are made in the context of potential side reactions, which should be minimized. Problems may be anticipated either during the coupling steps or at the final cleavage/deprotection step. For certain residues (e.g., Cys, Asp, Glu, and Lys), side-chain protection is absolutely essential, whereas for others, an informed decision should be made depending upon the length of the synthetic target and other considerations. Most solid-phase syntheses follow maximal rather than minimal protection strategies. Almost all of the useful N^α-Boc and N^α-Fmoc protected derivatives can be manufactured in bulk, and they are found in the catalogues of the major suppliers of peptide synthesis chemicals. The most widely used "permanent" protecting groups for the trifunctional amino acids have been listed (Table 1), together with information on how derivatives are prepared, conditions for their intentional deblocking, and conditions under which the indicated side-chain protection is either entirely stable or cleaved prematurely by reagents used for peptide synthesis.

The side-chain carboxyls of Asp and Glu are protected as benzyl (OBzl) esters for Boc chemistry and as tert-butyl (OtBu) esters for Fmoc chemistry. A sometimes serious side reaction with protected Asp residues involves an intramolecular elimination to form an aspartimide, which can then partition in water to the desired α-peptide and the undesired by-product with the chain growing from the β-carboxyl (Bodanszky and Kwei, 1978; Barany and Merrifield, 1979; Tam et al., 1988). Aspartimide formation is sequence dependent, with Asp(OBzl)-Gly, -Ser, -Thr, -Asn, and -Gln sequences showing the greatest tendency to cyclize under basic conditions (Bodanszky et al., 1978; Bodanszky and Kwei, 1978; Nicolás et al., 1989); the same sequences are also quite susceptible in strong acid

(Barany and Merrifield, 1979; Fujino et al., 1981; Tam et al., 1988). For models containing Asp(OBzl)-Gly, the rate and extent of aspartimide formation was substantial both in base (100% after 10 minutes treatment with 20% piperidine in DMF, 50% after 1 to 3 hours treatment with Et$_3$N or DIEA) and in strong acid (a typical value is 36% after 1 hour treatment with HF at 25°C). Sequences containing Asp(OtBu)-Gly are somewhat susceptible to base-catalyzed aspartimide formation (11% after 4 hours treatment with 20% piperidine in DMF) (Nicolás et al., 1989), but they do not rearrange at all in acid (Kenner and Seely, 1972).

To minimize the imide/$\alpha \rightarrow \beta$ rearrangement side reaction, Fmoc-Asp may be protected with the 1-adamantyl (O-1-Ada) group (Okada and Iguchi, 1988) and Boc-Asp with either the 2-adamantyl (O-2-Ada) (Okada and Iguchi, 1988) or cyclohexyl (OcHex) (Tam et al., 1988) groups. The base-labile 9-fluorenylmethyl (OFm) group offers orthogonal side-chain protection for Boc-Asp/Glu (Bolin et al., 1989; Albericio et al., 1990c; Al-Obeidi et al., 1990), while the palladium-sensitive allyl (OAl) group (Belshaw et al., 1990; Lyttle and Hudson, 1992) offers orthogonal side-chain protection for both Boc- and Fmoc-Asp/Glu; neither of the esters mentioned in this sentence are as yet available commercially.

The side-chain hydroxyls of Ser, Thr, and Tyr are protected as Bzl and tBu ethers for Boc and Fmoc SPPS, respectively. In strong acid, the benzyl (Bzl) protecting group blocking the Tyr phenol can migrate to the 3-position of the ring (Erickson and Merrifield, 1973a). This side reaction is decreased greatly when Tyr is protected by the 2,6-dichlorobenzyl (2,6-Cl$_2$Bzl) (Erickson and Merrifield, 1973a) or 2-bromobenzyloxycarbonyl (2-BrZ) (Yamashiro and Li, 1973) group; consequently, the latter two derivatives are much preferred for Boc SPPS. No corresponding C-alkylation occurs in Fmoc chemistry.

The ε-amino group of Lys is best protected by the 2-chlorobenzyloxycarbonyl (2-ClZ) or Fmoc group for Boc chemistry and, reciprocally, by the Boc group for Fmoc chemistry. The 2-ClZ group offers the desired acid stability for Boc chemistry, by comparison to the benzyloxycarbonyl (Z) and other ring-chlorinated Z groups. Branching due to premature side-chain deprotection by TFA is avoided, but the 2-ClZ group is still readily removable by strong acids (Erickson and Merrifield, 1973b). Orthogonal side-chain protection for both Boc- and Fmoc-Lys is provided by the palladium-sensitive allyloxycarbonyl (Aloc) group (Lyttle and Hudson, 1992).

The highly basic trifunctional guanidino side-chain group of Arg may be protected or unprotected (i.e., protonated). Appropriate benzenesulfonyl derivatives are the 4-toluenesulfonyl (Tos) or mesitylene-2-sulfonyl (Mts) groups in conjunction with Boc chemistry, and either 4-methoxy-2,3,6-trimethylbenzenesulfonyl (Mtr) or 2,2,5,7,8-pentamethylchroman-6-sulfonyl (Pmc) with Fmoc chemistry. These groups most likely block the

Principles and Practice of Solid-Phase Peptide Synthesis

Table 1 Amino Acid Side-Chain Protection for Solid-Phase Peptide Synthesis[a]

Side-Chain Protecting Group	Protected Amino Acid Derivative	Stability	Removal[a]	Reagent(s) for Introduction	Reference(s)[a]
Compatible with Boc chemistry					
Benzyl (–CH₂–)	Asp/Glu(OBzl)	TFA	Strong Acid	Bzl-OH + H⁺	Barany and Merrifield, 1979#; Bodanszky and Bodanszky, 1984*[b]; Tam and Merrifield, 1987#
	Ser/Thr/Tyr(Bzl)	TFA Base	Strong Acid	Bzl-Br + base	Yajima et al., 1988#; Kiso et al., 1989#
2-Adamantyl	Asp(O-2-Ada)	TFA Piperidine 1 M HCl	Strong Acid	Ada-2-OH + DCC	Okada and Iguchi, 1988*#
Cyclohexyl	Asp(OcHex)	TFA	Strong Acid	cHex-OH + carbodiimide + DMAP or cHex-OH + H⁺	Tam and Merrifield, 1987#; Tam et al., 1988*#; Yajima et al., 1988#; Penke and Tóth, 1989*; Kiso et al., 1989#
2,6-Dichlorobenzyl	Tyr(2,6-Cl₂Bzl)	TFA	Strong Acid	2,6-Cl₂Bzl-Br + base	Erickson and Merrifield, 1973a*#; Tam and Merrifield, 1987#; Yajima et al., 1988#; Kiso et al., 1989#

Group	Abbreviation			Introduction	References
2-Bromobenzyloxycarbonyl	Tyr(2-BrZ)	TFA	Strong Acid	2-BrZ-ONp + base	Yamashiro and Li, 1973*#; Tam and Merrifield, 1987#
2-Chlorobenzyloxycarbonyl	Lys(2-ClZ)	TFA	Strong Acid	2-ClZ-OSu + base	Erickson and Merrifield, 1973b*#; Bodanszky and Bodanszky, 1984*; Tam and Merrifield, 1987#; Kiso et al., 1989#
9-Fluorenylmethyloxycarbonyl	Lys(Fmoc)	TFA	Piperidine TBAF	Fmoc-N_3 + MgO	Albericio et al., 1990c*#
2,4-Dinitrophenyl	His(Dnp)	Acids	Thiophenol	1-fluoro-2,4-dinitrobenzene + base	Chillemi and Merrifield, 1969*#; Stewart et al., 1972#; Tam and Merrifield, 1987#; Applied Biosystems, Inc., 1989a#

Table 1 (continued)

Side-Chain Protecting Group	Protected Amino Acid Derivative	Stability	Removal[a]	Reagent(s) for Introduction	Reference(s)[a]
4-Toluenesulfonyl	His(Tos)	TFA	Strong acid Ac$_2$O-pyridine HOBt[@]	Tos-Cl + base	Stewart et al., 1972[#] Barany and Merrifield, 1979[#] van der Eijk et al., 1980[#] Bodanszky and Bodanszky, 1984*
	Arg(Tos)	HBr-TFA TFA	HF	Tos-Cl + base	Tam and Merrifield, 1987[#]
Benzyloxymethyl	His(Bom)	TFA	Strong acid	ClCH$_2$OBzl	Brown et al., 1982*[#] Tam and Merrifield, 1987[#] Kiso et al., 1989[#]
Mesitylene-2-sulfonyl	Arg(Mts)	TFA	Strong acid	Mts-Cl + base	Yajima et al., 1978*[#] Tam and Merrifield, 1987[#] Yajima et al., 1988[#]
4-Methylbenzyl	Cys(Meb)	TFA	Strong acid Tl(Tfa)$_3$	Meb-Br + base	Erickson and Merrifield, 1973a*[#] Barany and Merrifield, 1979[#] Tam and Merrifield, 1987[#] Fujii et al., 1987[#]

Protecting group	Derivative	Cleavage	Cleavage	Formation	References
3-Nitro-2-pyridinesulfenyl	Cys(Npys)	TFA HF[c]	Thiols HF[c] HOBt[@] Piperidine[@]	Npys-Cl	Matsueda and Walter, 1980* Albericio et al., 1989b[#] Rosen et al., 1990[#]
9-Fluorenylmethyl	Cys(Fm)	TFA HF	Piperidine[d]	Fm-OTos + base	Albericio et al., 1990c*[#]
	Asp/Glu(OFm)	TFA HF	Piperidine TBAF	Fm-OH + DCC/DMAP	Albericio et al., 1990c*[#]
9-Xanthenyl	Asn/Gln(Xan)	Base	Strong acid TFA[@]	Xan-OH + HOAc	Dorman et al., 1972*[#] Stewart and Young, 1984[#] Tam and Merrifield, 1987[#]
Formyl (on N^{in})	Trp(CHO)	TFA HF[c]	TFMSA piperidine HF[c]	HCOOH	Bodanszky and Bodanszky, 1984* Tam and Merrifield, 1987[#] Yajima et al., 1988[#]

Table 1 (continued)

Side-Chain Protecting Group	Protected Amino Acid Derivative	Stability	Removal[a]	Reagent(s) for Introduction	Reference(s)[a]
Sulfoxide (on thioether)	Met(O)	TFA Base HF^c	MMA DMF-SO$_3$ HF^c NH$_4$I TMSBr + thioanisole	H$_2$O$_2$	Houghton and Li, 1979#; Bodanszky and Bodanszky, 1984*; Tam and Merrifield, 1987#; Fujii et al., 1987#; Futaki et al., 1990b#
Compatible with Fmoc chemistry					
tert-Butyl CH$_3$—C—CH$_3$ / CH$_3$	Asp/Glu(OtBu)	Base	TFA	C$_4$H$_8$/H$^+$ or Boc-F	Chang et al., 1980b*; Bodanszky and Bodanszky, 1984*; Meienhofer, 1985#; Loffet et al., 1989*; Lajoie et al., 1990*e; Greene, 1991#
	Ser/Thr/Tyr(tBu)	Base	TFA	C$_4$H$_8$/H$^+$	
1-Adamantyl	Asp(O-1-Ada)	Base	TFA	Ada-1-OH + DCC	Okada and Iguchi, 1988*#

Protecting group	Residue			Reference
tert-Butyloxycarbonyl	Lys(Boc)	Base	Boc$_2$O *or* Boc-ON	Bodanszky and Bodanszky, 1984*; Meienhofer, 1985#
	His(Boc)	Piperidine	Boc$_2$O^f	Atherton and Sheppard, 1989*#
Triphenylmethyl	His(Trt)	Piperidine / 1 N HCl	Trt-Cl + Me$_2$SiCl$_2$	Barlos et al., 1982*; Sieber and Riniker, 1987#
	Cys(Trt)	Piperidine	HOAc / TFA / Hg(II) / I$_2$ — Trt-OH BF$_3$•Et$_2$O	Photaki et al., 1970#; Bodanszky and Bodanszky, 1984*; Meienhofer, 1985#; Greene, 1991#
	Asn/Gln(Trt)	Piperidine / 1 N HCl	Trt-OH Ac$_2$O/H$^+$	Sieber & Riniker, 1991*#
tert-Butoxymethyl	His(Bum)	Piperidine	ClCH$_2$OtBu	Colombo et al., 1981*#

Note: TFA is listed as the acidic cleavage condition for Lys(Boc), His(Boc), His(Trt), Asn/Gln(Trt), and His(Bum).

Table 1 (continued)

Side-Chain Protecting Group	Protected Amino Acid Derivative	Stability	Removal[a]	Reagent(s) for Introduction	Reference(s)[a]
4-Methoxy-2,3,6-trimethyl-benzenesulfonyl	Arg(Mtr)	Piperidine	TFA[g]	2,3,5-Trimethylanisole + ClSO$_3$H	Fujino et al., 1981*#; Atherton and Sheppard, 1989#
2,2,5,7,8-Pentamethyl-chroman-6-sulfonyl	Arg(Pmc)	Piperidine	TFA	2,2,5,7,8-Pentamethyl-chroman + ClSO$_3$H	Ramage and Green, 1987*#
2,4,6-Trimethoxybenzyl	Asn/Gln(Tmob)	Piperidine	TFA	Tmob-NH$_2$[h] + DCC/HOSu	Weygand et al., 1968a*; Hudson, 1988b#
	Cys(Tmob)	Piperidine	TFA	Tmob-OH + TFA	Munson et al., 1992*#

Compatible with both Boc and Fmoc chemistries

Name	Structure	Amino acid	Stable to	Cleaved by	Reagent	References
Allyl	$CH_2{=}CH{-}CH_2{-}$	Asp/Glu(OAl)	Acids, Base	Pd(0)	Al-OH + TMS-Cl or H_2SO_4	Belshaw et al., 1990*; Greene, 1991*#; Lyttle and Hudson, 1992*#
Allyloxycarbonyl	$CH_2{=}CH{-}CH_2{-}O{-}\overset{O}{\overset{\|}{C}}{-}$	Lys(Aloc)	Acids, Base	Pd(0)	Aloc-Cl	Lyttle and Hudson, 1992*#
Acetamidomethyl	$CH_3{-}\overset{O}{\overset{\|}{C}}{-}NH{-}CH_2{-}$	Cys(Acm)	Piperidine, Acids	Hg(II), I_2, Tl(Tfa)$_3$	Acm-OH/TFA	Veber et al., 1972*#; Bodanszky and Bodanszky, 1984*; Atherton et al., 1985a#; Albericio et al., 1987a*e; Tam and Merrifield, 1987#; Albericio et al., 1987*g; Fujii et al., 1987#; Brady et al., 1988#; McCurdy, 1989#
Trimethylacetamidomethyl	$CH_3{-}\overset{CH_3}{\underset{CH_3}{C}}{-}\overset{O}{\overset{\|}{C}}{-}NH{-}CH_2{-}$	Cys(Tacm)	Piperidine, Acids	Hg(II), I_2, AgBF$_4$	Tacm-OH/TFA	Yoshida et al., 1990#; Kiso et al., 1990*#

Table 1 (concluded)

Side-Chain Protecting Group	Protected Amino Acid Derivative	Stability	Removal[a]	Reagent(s) for Introduction	Reference(s)[a]
tert-Butylsulfenyl	Cys(StBu)	Piperidine Acids	Thiols Phosphines	tBu-SH tBu-SCN	Wünsch and Spangenberg, 1971* Atherton et al., 1985a[#] Romani et al., 1987[#]

[a] The protecting group structure drawn in the left column does not include the functional group that is being protected through a point of attachment at the far right of the structure. The order corresponds, by category, to first mention in the text. Abbreviations are provided in the second column. Conditions or reagents listed under Removal are intended for quantitative deprotection (see Cleavage), unless marked by @, in which case the removal is an unacceptable side reaction that indicates an incompatibility with the protecting group. Strong acid means HF, TFMSA, or equivalent reagents (see Cleavage). TFA cleavages are best carried out in the presence of appropriate scavengers (see Cleavage). References marked by * refer to preparation of the side-chain derivative; references marked by # refer to stability and lability of each derivative.

[b] Boc-Thr(Bzl)-OH is best prepared by the procedure of Chen et al., 1989.

[c] Stable to "high" (90%) HF, but removed by "low" HF-scavengers; see Cleavage.

[d] Complete deprotection requires 4-hour treatment by piperidine-DMF (1:1).

[e] One-pot synthesis of the Fmoc-amino acid side-chain protected derivative.

[f] Fmoc-His(Boc)-OH is prepared by reacting (Boc)2O with commercially available Fmoc-His(Fmoc)-OH.

[g] Arg(Mtr) deprotection by TFA can be slow, especially in multiple Arg(Mtr)-containing peptides. See Cleavage.

[h] Tmob-amine is reacted with a carboxyl protected Asp or Glu to produce Asn(Tmob) or Gln(Tmob), respectively.

ω-nitrogen of Arg, and their relative acid lability is Pmc > Mtr >> Mts > Tos (Fujino et al., 1981; Green et al., 1988). The Tos and Mts groups are removed by the same strong acids that cleave Bzl-type groups. The Mtr group may require extended TFA-thioanisole treatment (2 to 8 hours) for removal, while Arg(Pmc) is deprotected readily by 50% TFA (< 2 hours). A number of other Arg protecting groups have been proposed, particularly N^{ω}-mono or $N^{\delta,\omega}$-bis-urethane derivatives. However, based on current information, the aforementioned benzenesulfonyl derivatives seem to offer the best prospects of clean incorporation of Arg without contaminating ornithine (Orn) (Rink et al., 1984).

Due to the high pK_a of the guanidino group (~12.5), it can be protected selectively by protonation with HCl or HBr. Successful SPPS using protonated Arg requires a proton source (i.e., HOBt) for all subsequent coupling steps. This protocol is recommended to suppress intermolecular acylation of the guanidino group, which would lead to Orn formation (Atherton et al., 1984; Atherton and Sheppard, 1989).

A common side reaction of most Boc/Fmoc-Arg derivatives is δ-lactam formation (Barany and Merrifield, 1979; Rzeszotarska and Masiukiewicz, 1988). During carbodiimide activation, δ-lactam formation (intramolecular aminolysis) competes with peptide bond formation (intermolecular aminolysis). Acylations in the presence of HOBt are commonly used to inhibit δ-lactam formation. N^{α}-protected Arg derivatives may also be coupled as preformed esters, from which δ-lactam side products have been separated from the desired ester prior to use (Atherton et al., 1988b). However, preformed esters will undergo conversion to the δ-lactam relatively shortly after dissolving in DMF (D. Hudson, unpublished results).

Activated His derivatives are uniquely prone to racemization during stepwise SPPS due to an intramolecular abstraction of the proton on the optically active α-carbon by the imidazole π-nitrogen (Jones et al., 1980). Racemization could be suppressed either by reducing the basicity of the imidazole ring or by blocking the base directly (Riniker and Sieber, 1988). Consequently, His side-chain protecting groups can be categorized depending on whether the τ- or π-imidazole nitrogen is blocked. The Tos group blocks the N^{τ} of Boc-His and is removed by strong acids. However, the Tos group is also lost prematurely during SPPS steps involving HOBt; this allows acylation or acetylation (during capping) of the imidazole group, followed by chain termination due to $N^{im} \rightarrow N^{\alpha}$-amino transfer of the acyl or acetyl group (Ishiguro and Eguchi, 1989; Kusunoki et al., 1990). Therefore, HOBt should never be used during couplings of amino acids once a His(Tos) residue has been incorporated into the peptide resin. An HF-stable, orthogonally removable, N^{τ}-protecting group for Boc strategies is the 2,4-dinitrophenyl (Dnp) function. Final Dnp deblocking is best carried out at the peptide resin level prior to the HF cleavage step by use of thiophenol in DMF

Dnp removal from His(Dnp)-containing peptide resins is achieved by use of 20 mmol thiophenol per millimole His(Dnp) residue. Peptide resin is suspended in DMF (5 mL per gram of resin), thiophenol is added, and the reaction proceeds for 1 hour at 25 °C. After thorough washing of the Boc-peptide resin with DMF, H2O, ethanol, and DCM, the N^α-Boc group is removed and the peptide is cleaved with HF or TFMSA. If Dnp groups still remain, the peptide is dissolved in 6 M guanidine-HCl, 50 mM Tris acetate, pH 8.5 (10 to 20 mg peptide per mL solution), then deprotected by adding 2-mercaptoethanol to 20% (v/v) and treating for 2 hours at 37 °C. The peptide should then be purified immediately by gel filtration or HPLC (Applied Biosystems, Inc., 1989a).

(see box). The τ-nitrogen of Fmoc-His can be protected by the Boc and triphenylmethyl (Trt) groups. The commercially available N^τ-protected Fmoc-His(Fmoc) derivative is not recommended because it is poorly soluble; furthermore, immediate side-chain deprotection at the first exposure to piperidine allows a variety of side reactions (Bodanszky et al., 1977; Barany and Merrifield, 1979; Riniker and Sieber, 1988). When His is N^τ-protected by the Boc group, the basicity of the imidazole ring is reduced sufficiently so that acylation by the preformed symmetrical anhydride (PSA) method proceeds with little racemization (Atherton and Sheppard, 1989). Contrary to earlier reports, His(Boc) now appears to be reasonably stable to repetitive base treatment (Atherton and Sheppard, 1989). His(Trt) is completely stable to piperidine, but it is removed with TFA (Sieber and Riniker, 1987). The Trt group reduces the basicity of the imidazole ring (the pK_a decreases from 6.2 to 4.7), although racemization by the PSA method is not eliminated completely (Sieber and Riniker, 1987). Since Dnp and Trt N^τ-protection do not allow PSA coupling with low racemization, it is recommended that the appropriate derivatives be coupled as preformed esters or in situ with carbodiimide in the presence of HOBt (Sieber and Riniker, 1987; Riniker and Sieber, 1988). Boc-His (Tos) is coupled efficiently using benzotriazolyl N-oxytrisdimethyl-aminophosphonium hexafluorophosphate (BOP) (3 equiv) in the presence of DIEA (3 equiv); these conditions minimize racemization and avoid premature side-chain deprotection by HOBt (Forest and Fournier, 1990).

Blocking of the π-nitrogen of the imidazole ring has been shown to be effective in reducing His racemization (Fletcher et al., 1979). The N^π of His is protected by the benzyloxymethyl (Bom) and tert-butoxymethyl (Bum) groups for Boc and Fmoc chemistry, respectively. Couplings using N^π-protected Boc-His(Bom) or Fmoc-His(Bum) PSAs result in racemization-free incorporation of His (Brown et al., 1982; Colombo et al., 1984). HF deprotection of His(Bom) and TFA deprotection of

Boc- and Fmoc-His are prepared by the general procedures described earlier for Boc and Fmoc amino acids. Crude Boc-His is dissolved in methanol (MeOH)-0.1 M pyridinium acetate buffer (pH 3.8) (1:1) and purified by ion-exchange chromatography over Dowex 50 (H^+ form) using a pH gradient from 3.8 to 5.8 (Kawasaki et al., 1989). Crude Fmoc-His is purified by washing with H2O and hot MeOH (Kawasaki et al., 1989). N^τ-protected Boc- and Fmoc-His derivatives are prepared by straightforward reactions of appropriate side-chain derivatizing reagents with either N^α-protected or free His (see references in Table 1). N^π-protected derivatives are prepared by protecting His-OMe at the N^α (by the Z or Boc group) and N^τ (by the Boc group), derivatizing N^π with the appropriate chloromethyl ether (simultaneously removing the N^τ–Boc group), and saponifying the methyl ester with NaOH (Brown et al., 1982; Colombo et al., 1984). The N^α-Z group is removed by hydrogenolysis and replaced by the Fmoc group (Colombo et al., 1984).

His(Bum) liberates formaldehyde, which can modify susceptible side chains (see Cleavage).

The carboxamide side chains of Asn and Gln are often left un-protected in SPPS, but this approach leaves open the danger of dehydra-tion to form nitriles upon activation with in situ reagents. On the other hand, acylations by activated esters result in minimal side-chain dehydra-tion (Barany and Merrifield, 1979; Mojsov et al., 1980; Gausepohl et al., 1989b) (see Formation of Peptide Bond). Nitrile formation is also in-hibited during in situ carbodiimide acylations when HOBt is added (Moj-sov et al., 1980; Gausepohl et al., 1989b) (see Formation of Peptide Bond). However, the presence of HOBt does not effectively inhibit N^α-protected Asn dehydration during BOP in situ acylations (Gausepohl et al., 1989b).

At the point where an N^α-amino protecting group is removed from Gln, the possibility exists for an acid-catalyzed intramolecular amin-olysis, which displaces ammonia and leads to pyroglutamate formation (Barany and Merrifield, 1973; Barany and Merrifield, 1979; DiMarchi et al., 1982; Orlowska et al., 1987). Cyclization occurs primarily during couplings; N^α-protected amino acids and HOBt promote this side reac-tion (DiMarchi et al., 1982). Consequently, it is recommended that the incoming residue that is to be incorporated onto Gln be activated as a non-acidic species, e.g., PSA or a preformed ester (see Formation of Peptide Bond).

Although conditions are available for the safe incorporation of Asn and Gln with free side chains during SPPS, there are compelling reasons for their protection. Side-chain protecting groups, such as 9-xanthenyl (Xan), 2,4,6-trimethoxybenzyl (Tmob), and Trt minimize the occurrence

In peptides where several Asn, Gln, His, and Cys residues are close in sequence, it may be worthwhile to limit the global use of Trt side-chain protection. Interspersing side-chain unprotected Asn and/or Gln residues in such congested sequences should limit difficult couplings due to steric hindrance. In addition, Asn or Gln adjacent to Trp should be left unprotected, since the Tmob, Trt, and Xan side-chain protecting groups can modify Trp during TFA deprotection/cleavage (Southard, 1971; see also Cleavage for additional references).

of dehydration (Mojsov et al., 1980; Hudson, 1988b; Gausepohl et al., 1989b; Sieber and Riniker, 1990) and pyroglutamate formation (Barany and Merrifield, 1979), and they may also inhibit hydrogen bonding that otherwise leads to secondary structures that substantially reduce coupling rates. Unprotected Fmoc-Asn and -Gln have poor solubility in DCM and DMF; solubility is improved considerably by Tmob or Trt side-chain protection. The Xan group has been used in Boc chemistry, but it does not entirely survive the TFA deprotection conditions (Dorman et al., 1972; Stewart and Young, 1984).

The highly sensitive side chains of Trp and Met generally survive cycles of Fmoc chemistry, but their protection during Boc chemistry is often advisable. For these purposes, the base-labile N^{in}-formyl (CHO) and reducible sulfoxide functions are applied respectively. Trp(CHO) is best deprotected at the peptide resin level by treatment with piperidine-DMF (9:91), 0 °C, 2 hours, *prior* to HF cleavage; the formyl group also is removed by 20 to 25% HF in the presence of dimethylsulfide and 4-thiocresol (see Cleavage). Smooth deblocking of Met(O) occurs in 20 to 25% HF in the presence of dimethylsulfide (see Cleavage), or by N-methylmercaptoacetamide (MMA) (10%) in 10% aqueous HOAc at 37 °C for 24 to 36 hours (Houghten and Li, 1979), NH$_4$I-dimethylsulfide (20 equiv each) in TFA at 0 °C for 1 hour (Fujii et al., 1987), or DMF·SO$_3$-EDT (5 equiv each) in 20% pyridine/DMF at 20 °C for 1 hour (Futaki et al., 1990b). DMF·SO$_3$-EDT treatment of Met(O) can be carried out only while hydroxyl residues are side-chain protected, because free hydroxyls will be sulfated (Futaki et al., 1990a). Unprotected Trp may be incorporated by Boc chemistry when 2.5% anisole plus either 2% dimethyl phosphite or indole are added to the TFA deprotection solution (Stewart and Young, 1984; Hudson et al., 1986).

The most challenging residue to manage in peptide synthesis is Cys, which is required for some applications in the free sulfhydryl form and, for others, as a contributor to a disulfide linkage. Another issue is the selective formation of multiple disulfides by the concurrent use of two or more classes of Cys protecting groups (see Post-Translational Modifications and Unnatural Structures). Compatible with Boc chemistry are the

4-methylbenzyl (Meb), acetamidomethyl (Acm), trimethylacetamido-methyl (Tacm), *tert*-butylsulfenyl (S*t*Bu), 3-nitro-2-pyridinesulfenyl (Npys), and Fm β-thiol protecting groups; compatible with Fmoc chemistry are the Acm, Tacm, S*t*Bu, Trt, and Tmob groups. The Trt and Tmob groups are labile in TFA; due to the tendency of the resultant stable carbonium ions to realkylate Cys (Photaki et al., 1970), effective scav-engers are needed (see Cleavage). The Meb group is optimized for re-moval by strong acid (Erickson and Merrifield, 1973a); Cys(Meb) resi-dues may also be directly converted to the oxidized (cystine) form by thallium (III) trifluoroacetate [Tl(Tfa)$_3$], although some cysteic acid forms at the same time. Cys(Npys) and Cys(Fm) are stable to acid and cleaved, respectively, by thiols and base. The Acm and Tacm groups are acid- and base-stable and removed by mercuric (II) acetate or silver tetrafluoroborate, followed by treatment with H$_2$S or excess mercaptans to free the β-thiol. In multiple Cys(Acm)-containing peptides, mercuric (II) acetate may not be a completely effective removal reagent (Kenner et al., 1979). Alternatively, Cys(Acm) and Cys(Tacm) residues are con-verted directly to disulfides by treatment with I$_2$, Tl(Tfa)$_3$, or methyl-trichlorosilane in the presence of diphenylsulphoxide. Finally, the acid-stable S*t*Bu group is removed by reduction with thiols or phosphines.

   C-terminal esterified (but not amidated) Cys residues are racemized by repeated piperidine deprotection treatments during Fmoc SPPS. Fol-lowing 4 hours exposure to piperidine-DMF (1:4), the extent of racem-ization found was 36% D-Cys from Cys(S*t*Bu), 12% D-Cys from Cys(Trt), and 9% D-Cys from Cys(Acm) (Atherton et al., 1991). At least two examples have been provided where the use of Cys(S*t*Bu) as the *C*-terminal residue, esterified to an HMP/PAB-type resin, was entirely incompatible with formation of the desired peptide. Instead, TFA cleav-age gave reduction-resistant by-products (structures not fully deter-mined), retaining the *t*Bu group but evidently missing two molecules of water based on mass spectrometric evidence (Eritja et al., 1987). *N*-ter-minal Cys residues are modified covalently by formaldehyde, liberated during HF deprotection of His(Bom) residues (see Cleavage). Additional difficulties, often poorly understood, have arisen with a range of protec-ted Cys derivatives in a variety of applications (Barany and Merrifield, 1979; Atherton et al., 1985b).

   As is clear from the preceding discussion, the Boc and Fmoc groups have risen to the fore as the most widely used and commercially viable N^α-amino protecting groups for SPPS. A plethora of other N^α-amino protecting groups, some illustrating remarkably creative organic chemistry, have been proposed over the years (Figure 4). Among these, the 2-(4-biphenyl)propyl[2]oxycarbonyl (Bpoc) (Wang and Merrifield 1969; Kemp et al., 1988), 2-(3,5-dimethoxyphenyl)propyl[2]oxycar-bonyl (Ddz) (Birr et al., 1972; Voss and Birr, 1981), 1-(1-adamantyl)-1-methylethoxycarbonyl (Adpoc) (Voelter et al., 1987; Shao et al., 1991),

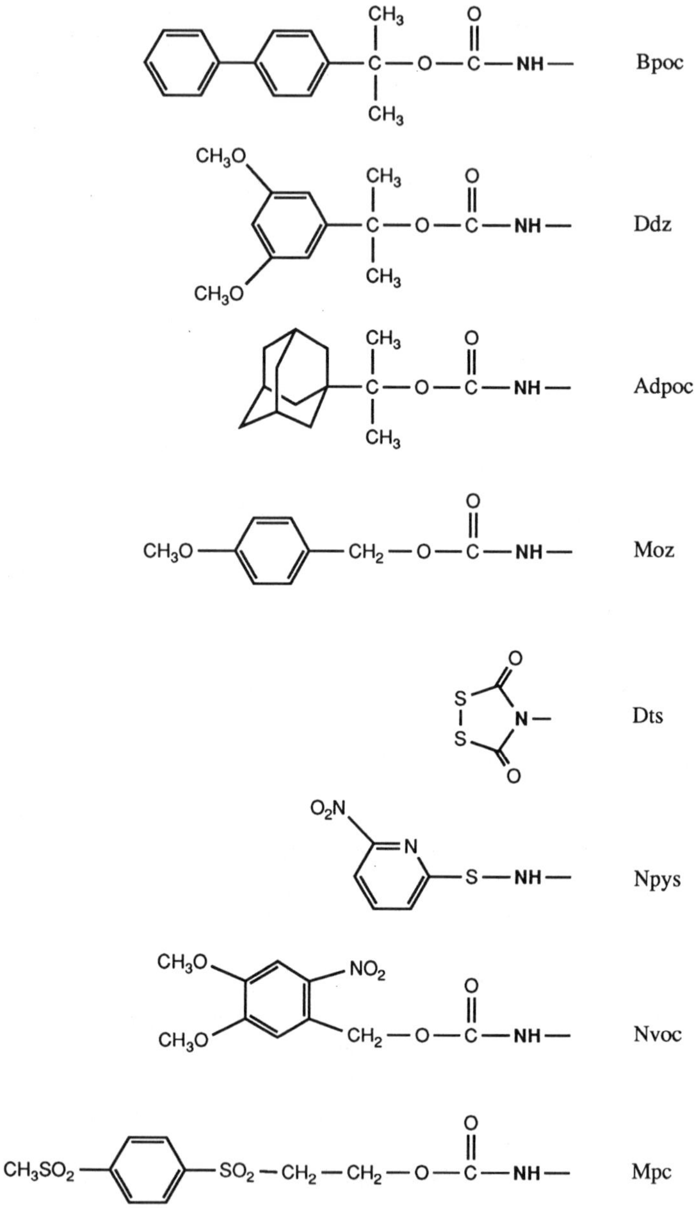

FIGURE 4   Alternative N^{α}-amino protecting groups for SPPS. The protected nitrogen that is part of the amino acid is shown in boldface.

and 4-methoxybenzyloxycarbonyl (Moz) (Wang et al., 1987; Chen et al., 1987) groups are removed in dilute TFA, the dithiasuccinoyl (Dts) (Barany and Merrifield, 1977; Barany and Albericio, 1985; Albericio and Barany, 1987a; Zalipsky et al., 1987), and 3-nitro-2-pyridinesulfenyl (Npys) (Matsueda and Walter, 1980; Wang et al., 1982; Ikeda et al., 1986; Hahn et al., 1990) groups are removed by thiolysis, the 6-nitroveratryl-oxycarbonyl (Nvoc) group is removed by photolysis (Patchornik et al., 1970; Fodor et al., 1991), and the 2-[4-(methylsulfonyl)phenylsul-fonyl]ethoxycarbonyl (Mpc) is base-labile (Schielen et al., 1991). Chemistries relying on these protecting groups are beyond the scope of this chapter.

POLYMERIC SUPPORT

The term *solid phase* often conjures misleading images among the uninitiated. Supports that lead to successful results for macromolecule synthesis are far from static, and because of the need for reasonable capacities, it is rare for solid-phase chemistry to take place exclusively on surfaces. The resin support is quite often a polystyrene suspension polymer cross-linked with 1% of 1,3-divinylbenzene; the level of functionalization is typically 0.2 to 1.0 mmol/g. Dry polystyrene beads have an average diameter of about 50 μm, but with the commonly used solvents for peptide synthesis, namely DCM and DMF, they swell 2.5- to 6.2-fold in volume (Sarin et al., 1980). Thus, the chemistry of solid-phase synthesis takes place within a well-solvated gel containing mobile and reagent-accessible chains (Sarin et al., 1980; Live and Kent, 1982). Polymer supports have also been developed based on the concept that the insoluble support and peptide backbone should have comparable polarities (Atherton and Sheppard, 1989). A resin of copolymerized dimethyl-acrylamide, N,N'-bisacryloylethylenediamine, and acryloylsarcosine methyl ester (typical loading 0.3 mmol/g), commercially known as polyamide or Pepsyn, has been synthesized to satisfy this criterion (Arshady et al., 1981). Under the best solvation conditions for both polystyrene and polyamide supports, reaction rates approach, but generally do not reach, those attainable in solution. It has been shown for a polystyrene carrier that macroscopic dimensions of both dry and solvated beads change dramatically once an appreciable level of peptide has been built up (Sarin et al., 1980). Thus, for this specific case, reactions continued to occur efficiently throughout the interior of a peptide resin that was fourfold the weight of the starting support.

A fertile area of inquiry has been the testing of supports with macroscopic physical properties and possibly other characteristics differing from 1% cross-linked polystyrene and polyamide gel beads. These include membranes (Bernatowicz et al., 1990), cotton and other appropriate carbohydrates (Frank and Döring, 1988; Lebl and Eichler, 1989;

Eichler et al., 1989), controlled-pore silica glass (Büttner et al., 1988), and linear polystyrene chains grafted covalently onto dense Kel-F particles (Tregear, 1972; Kent and Merrifield, 1978; Albericio et al., 1989a) or polyethylene sheets (Berg et al., 1989). Supports developed specifically to withstand the back pressures that arise during continuous-flow procedures have been low-density, highly permeable inorganic matrices with polyamide embedded within. These embedded matrices include polyamide-kieselguhr (known commercially as Pepsyn K) (Atherton et al., 1981b) and polyamide-Polyhipe (Small and Sherrington, 1989). Pepsyn K has a typical loading of 0.1 mmol/g, while Polyhipe loadings range from 0.3 to 1.8 mmol/g. An exciting recent development involves the use of polyethylene glycol-polystyrene graft supports (0.1 to 0.4 mmol/g), which swell in a range of solvents and have excellent physical and mechanical properties for both batchwise and continuous-flow SPPS (Hellermann et al., 1983; Zalipsky et al., 1985; Bayer and Rapp, 1986; Bayer et al., 1990; Barany et al., 1992). Poly N-[2-(4-hydroxyphenyl)ethyl]-acrylamide (core Q) is also suitable for both batchwise and continous-flow SPPS with high loading capacities (5 mmol/g) (Epton et al., 1987; Baker et al., 1990).

ATTACHMENT TO SUPPORT

Almost all syntheses by the solid-phase method are carried out in the $C \rightarrow N$ direction and, therefore, generally start with the intended C-terminal residue of the desired peptide being linked to the support either directly or by means of a suitable handle. Anchoring linkages have been designed so that eventual cleavage provides either a free acid or amide at the C-terminus, although, in specialized cases, other useful end groups can be obtained. The discussion that follows focuses on linkers that are either commercially available, readily prepared, and/or of special interest (Table 2); more complete listings are available (Barany et al., 1987; Fields and Noble, 1990).

Note that several handles (Table 2) have a free or activated carboxyl group that is intended to attach to the polymeric support. Such handles are most frequently coupled onto supports that have been functionalized with amino groups. Aminomethyl-polystyrene resin is optimally prepared essentially as described by Mitchell et al. (1978), except that methanesulfonic acid (0.75 g per 1 g polystyrene) is preferred as the catalyst instead of the originally described TFMSA (S.B.H. Kent and K.M. Otteson, unpublished results). Amino groups are introduced onto a variety of polyamide supports by treatment with ethylenediamine to displace carboxylate derivatives (Atherton and Sheppard, 1989). All else being equal, there are significant advantages to those anchoring methods in which the key step is amide bond formation by reaction of an activated handle carboxyl with an amino support, since such reactions can be made to go readily to completion. This approach allows control of loading levels and obviates difficulties that may arise due to extraneous or unreacted functionalized groups. As indicated earlier, the best control is achieved by coupling "preformed handles," which are protected amino acid derivatives that have been synthesized and purified in solution prior to the solid-phase anchoring step.

Peptide Acids

For Boc chemistry, the most common approach to peptide acids uses substituted benzyl esters that are cleaved in strong acid at the same time that other benzyl-type protecting groups are removed (Figure 2). The classical procedures starting with chloromethyl resin are still favored by many (Gutte and Merrifield, 1971; Gisin, 1973; Stewart and Young, 1984), although preformed handle approaches with 4-(hydroxymethyl)phenylacetic acid (PAM) (Mitchell et al., 1978; Tam et al., 1979; Clark-Lewis and Kent, 1989) are preferable for a number of applications. Other resins and linkers proposed for Boc chemistry are cleaved by orthogonal modes, allowing their use for the preparation of partially protected peptide segments (see Auxiliary Issues). The 4-(2-hydroxyethyl)-3-nitrobenzoic acid (NPE) (Eritja et al., 1991; Albericio et al., 1991b) and 9-(hydroxymethyl)-2-fluoreneacetic acid (HMFA) (Liu et al., 1990) linkers are cleaved by bases (see Cleavage). The HMFA linker is also cleaved by free N^{α}-amino groups from the peptide resin; the addition of HOBt during SPPS inhibits premature cleavage from this source (Liu et al., 1990). A 4-nitrobenzophenone oxime resin (DeGrado and Kaiser, 1980; DeGrado and Kaiser, 1982; Findeis and Kaiser, 1989; Scarr and Findeis, 1990) yields a peptide acid upon cleavage by either N-hydroxypiperidine (HOPip) or amino acid tetra-n-butylammonium salts [AA⁻ ⁺N(nBu)₄] (Findeis and Kaiser, 1989; Lansbury et al., 1989; Sasaki and Kaiser, 1990). Note that in the latter mode of oxime resin cleavage, the penultimate residue of the desired peptide is the one initially attached to the support. Finally, the acid-stable 2-bromopropionyl (α-methylphenacyl ester) linker (Wang, 1976) is of interest because it can be cleaved by photolysis (350 nm).

For Fmoc chemistry, peptide acids have been generated traditionally using the 4-alkoxybenzyl alcohol resin/4-hydroxymethylphenoxy (HMP/PAB) linker (Wang, 1973; Lu et al., 1981; Sheppard and Williams, 1982; Colombo et al., 1983; Albericio and Barany, 1985; Bernatowicz et al., 1990), which is cleaved in 1 to 2 hours at 25 °C with 50 to 100% TFA. The precise lability of the resultant 4-alkoxybenzyl esters depends on the spacer between the phenoxy group and the support. The HMP/PAB moiety can be established directly on the resin, or it can be introduced as a handle; preformed handles are best coupled to amino-functionalized resins as their 2,4,5-trichlorophenyl- or 2,4-dichlorophenyl-activated

Table 2   Resin Linkers and Handles[a]

Linker/Handle/Resin	Cleavage Conditions	Resulting C-Terminus	Reference(s)
4-Chloromethyl resin	Strong acid	Acid	Gutte and Merrifield, 1971 Stewart and Young, 1984
4-Hydroxymethylphenylacetic acid (PAM)	Strong acid	Acid	Mitchell et al., 1978 Tam et al., 1979
3-Nitro-4-(2-hydroxyethyl)benzoic acid (NPE)	Piperidine DBU	Acid	Eritja et al., 1991 Albericio et al., 1991b
9-(Hydroxymethyl)-2-fluoreneacetic acid (HMFA)	Piperidine	Acid	Mutter and Bellof, 1984c Liu et al., 1990

Name / Structure	Cleavage reagent	Product	Reference
4-Nitrobenzophenone oxime resin	HOPip AA·⁺N(nBu)$_4$ AA-NH$_2$	Acid[b] Amide	DeGrado and Kaiser, 1982[c] Findeis and Kaiser, 1989 Scarr and Findeis, 1990
α-Bromophenacyl	hv (350 nm)	Acid	Wang, 1976
4-Alkoxybenzyl alcohol resin	TFA	Acid	Wang, 1973[c] Lu et al., 1981
4-Hydroxymethylphenoxyacetic acid (HMPA/PAB)	TFA	Acid	Sheppard and Williams, 1982

Table 2 (continued)

Linker/Handle/Resin	Cleavage Conditions	Resulting C-Terminus	Reference(s)
3-(4-Hydroxymethylphenoxy)propionic acid (PAB) HOCH₂—⟨ ⟩—O(CH₂)₂CO₂H	TFA	Acid	Albericio and Barany, 1985
3-Methoxy-4-hydroxymethylphenoxyacetic acid CH₃O—, HOCH₂—⟨ ⟩—CH₂CO₂H	Dilute TFA	Acid	Sheppard and Williams, 1982
4-(2′,4′-Dimethoxyphenylhydroxymethyl) phenoxymethyl resin HO—CH—⟨ ⟩—OCH₃, CH₃O—, —O—Ⓡ	Dilute TFA	Acid	Rink, 1987

2-Methoxy-4-alkoxybenzyl alcohol resin (SASRIN™)	Dilute TFA	Acid	Mergler et al., 1988a
2-Chlorotrityl chloride resin	TFA, HOAc	Acid	Barlos et al., 1989 Barlos et al., 1991a
4-(4-Hydroxymethyl-3-methoxyphenoxy)butyric acid (HMPB)	Dilute TFA	Acid	Flörsheimer and Riniker, 1991

Table 2 (continued)

Linker/Handle/Resin	Cleavage Conditions	Resulting C-Terminus	Reference(s)
5-(4-Hydroxymethyl-3,5-dimethoxyphenoxy)valeric acid (HAL)	Dilute TFA	Acid	Albericio and Barany, 1991
Hydroxy-crotonyl-aminomethyl resin (HYCRAM™)	Pd(0) + NMM *or* dimedone	Acid	Kunz and Dombo, 1988 Guibé et al., 1989 Lloyd-Williams et al., 1991b
3-Nitro-4-hydroxymethylbenzoic acid (ONb)	hv (350 nm)	Acid	Rich and Gurwara, 1975c Giralt et al., 1982 Barany and Albericio, 1985 Kneib-Cordonier et al., 1990

4-Methylbenzhydrylamine resin (MBHA)	Strong acid	Amide	Matsueda and Stewart, 1981 Gaehde and Matsueda, 1981
4-(2',4'-Dimethoxyphenylaminomethyl)phenoxymethyl resin	Dilute TFA	Amide	Rink, 1987
4-(4'-Methoxybenzhydryl)phenoxyacetic acid (Dod)	TFA	Amide	Stüber et al., 1989

Table 2 (concluded)

Linker/Handle/Resin	Cleavage Conditions	Resulting C-Terminus	Reference(s)
3-(Amino-4-methoxybenzyl)-4,6-dimethoxyphenyl-propionic acid	Dilute TFA	Amide	Breipohl et al., 1989
4-Succinylamino-2,2′,4′-trimethoxybenzhydrylamine resin (SAMBHA)	TFA	Amide	Penke et al., 1988

5-(4-Aminomethyl-3,5-dimethoxyphenoxy)valeric acid (PAL)	TFA	Amide	Albericio and Barany, 1987b; Albericio et al., 1990a
5-(9-Aminoxanthen-2-oxy)valeric acid (XAL)	Dilute TFA	Amide	Sieber, 1987c; Barany and Albericio, 1991a; Bontems et al., 1992
3-Nitro-4-aminomethylbenzoic acid (Nonb)	hv (350 nm)	Amide	Hammer et al., 1990

PAL structure: CH_3O-substituted aminomethylphenoxy ring with $O(CH_2)_4CO_2H$ side chain; H_2NCH_2 and two CH_3O groups.

XAL structure: 9-amino-xanthene (H_2N) with $O(CH_2)_4CO_2H$ side chain.

Nonb structure: $H_2N(CH_2)_2$ and O_2N substituents on a benzene ring bearing CO_2H.

[a] Structural diagrams are oriented so that the resin or point of attachment to support is on the far right, and the site for anchoring the C-terminal amino acid residue is on the far left. Benzyl ester type linkages may also be cleaved by a range of nucleophiles to give acids, esters, and other derivatives. See text discussion under Cleavage, and consult Barany and Merrifield (1979) and Barany et al. (1987) for further examples.

[b] Cleavage by aminolysis results in 0.6 to 2% racemization (DeGrado and Kaiser, 1980).

[c] Reference for historical reasons; preparation of linker/resin has been improved in later references.

esters, sometimes in the presence of HOBt (Albericio and Barany, 1985; Bernatowicz et al., 1990; Albericio and Barany, 1991). A number of supports and linkers are available that can be cleaved in dilute acid; under optimal circumstances, these can be used to prepare protected peptide segments retaining side-chain *tert*-butyl protection. These include 3-methoxy-4-hydroxymethylphenoxyacetic acid (Sheppard and Williams, 1982), 4-(2',4'-dimethoxyphenyl-hydroxymethyl)phenoxy resin (Rink acid) (Rink, 1987; Rink and Ernst, 1991), 2-methoxy-4-alkoxybenzyl alcohol (SASRIN^{TM}) (Mergler et al., 1988a), 2-chlorotrityl-chloride resin (Barlos et al., 1989; Barlos et al., 1991a), 4-(4-hydroxymethyl-3-methoxyphenoxy)butyric acid (HMPB) (Flörsheimer and Riniker, 1991), and 5-(4-hydroxymethyl-3,5-dimethoxyphenoxy)-valeric acid (HAL) (Albericio and Barany, 1991). Because of its acute acid lability, and in order to prevent premature loss of peptide chains, the Rink acid linker is used in conjunction with N^{α}-protected amino acid preformed HOBt esters in the presence of excess DIEA (3 equiv) (Rink and Ernst, 1991). The HMP/PAB and SASRIN^{TM} linkers are available as the corresponding chlorides or bromides (Colombo et al., 1983; Mergler et al., 1989a; Bernatowicz et al., 1990).

Linkers for preparing peptide acids that are compatible with both Boc and Fmoc chemistries include hydroxy-crotonyl (HYCRAM^{TM}) (Kunz and Dombo, 1988; Kunz, 1990; Lloyd-Williams et al., 1991b), which is cleaved by Pd(0) catalyzed transfer of the allyl linker to a weak nucleophile, and 3-nitro-4-hydroxymethylbenzoic acid (ONb) (Rich and Gurwara, 1975; Giralt et al., 1982; Barany and Albericio, 1985; Kneib-Cordonier et al., 1990), which is cleaved photolytically at 350 nm.

All of the anchoring linkages that ultimately provide peptide acids are esters; rates and yields of reactions for ester bond formation (Table 3) are lower than those for corresponding methods for amide bond formation (see Formation of Peptide Bond). Consequently, compromises are needed to achieve reasonable loading reaction times and substitution levels, while ensuring that the extent of racemization remains acceptably low. As a point of departure, esterification of N^{α}-protected amino acid PSA catalyzed by 1 equiv of 4-dimethylaminopyridine (DMAP) in DMA results in significant (1.5 to 20%) racemization (Atherton et al., 1981a). In general, racemization levels can be reduced to acceptable levels (0.2 to 1.2%) when catalytic (0.06 equiv) amounts of DMAP are used and loadings are performed with carbodiimides in situ (Mergler et al., 1988a), sometimes in the presence of N-methylmorpholine (NMM) (0.9 equiv) (D. Hudson, personal communication). Alternatively, in situ carbodiimide loading with HOBt (2 equiv) and DMAP (1 equiv) at low temperature (0 to 3 °C) provides a good compromise of minimized racemization (0.1 to 0.3%) and reasonable loading times (16 hours) (van Nispen et al., 1985). No racemization was detected (0.05%) when in situ loadings were carried out at 25 °C with only N,N'-dicyclohexylcarbodiimide (DCC) (4 equiv) and HOBt (3 equiv) and no DMAP (Grandas et al., 1989a). Esterifications of Fmoc amino acids mediated by N,N-dimethylformamide dineopentyl acetal (Albericio and Barany, 1984; Albericio and Barany, 1985), 2,6-dichlorobenzoyl chloride (DCBC) (Sieber, 1987a), diethyl azodicarboxylate (DEAD) (Sieber, 1987a; Stanley et al., 1991), or 2,4,6-mesitylene-sulfonyl-3-nitro-1,2,4-triazolide (MSNT) (Blankemeyer-Menge et al., 1990) have all been reported to suppress racemization, as is the case with preformed Fmoc amino acid 2,5-diphenyl-2,3-dihydro-3-oxo-4-hydroxythiophene dioxide (OTDO) esters (Kirstgen et al., 1987; Kirstgen et al., 1988; Kirstgen and Steglich, 1989), Fmoc amino acid chlorides (Akaji et al., 1990a), or Fmoc amino acid N-carboxyanhydrides (NCA) (Fuller et al., 1990). The well-established cesium salt method (Gisin, 1973) also allows loading of N^α-protected amino acids to chloromethyl linkers and resins with low levels of racemization (Colombo et al., 1983; Mergler et al., 1989b) while effectively preventing alkylation of susceptible residues (Cys, His, Met) (Gisin, 1973). Bromomethylated linkers may be loaded directly by Boc amino acids in the presence of KF (Tam et al., 1979) or by Fmoc amino acids in the presence of DIEA (Bernatowicz et al., 1990), in each case with little racemization. While premature removal of the Fmoc group during loading can result in dipeptide formation (Pedroso et al., 1983), the efficient ester bond formation methods described in this paragraph minimize this side reaction. However, care should be taken when preparing cesium salts of Fmoc amino acids, because Cs_2CO_3 may promote partial removal of the Fmoc group. Of the ester bond formation methods discussed here, the most generally applicable are esterification of N^α-protected amino acids in situ by carbodiimide (DCC or DIPCDI) in the presence of catalytic amounts of DMAP (0.06 to 0.1 equiv), or by N,N-dimethylformamide dineopentyl acetal. For esterifications in the presence of DMAP, time and temperature should be carefully mediated, and HOBt (1 to 2 equiv) may be included.

   Fmoc-His and Fmoc-Cys derivatives are particularly difficult to load efficiently while suppressing racemization. Low racemization loadings have been documented using the cesium salts of Fmoc-His(Trt) (0.4% D-His) and Fmoc-Cys(Acm) (0.5% D-Cys) (Mergler et al., 1989b) and for Fmoc-His(Bum) (0.3% D-His) esterification by MSNT (Blankemeyer-Menge, 1990). Since the last-mentioned result is undoubtedly due to the fact that the Bum group blocks N^π of His, it may be noted that a more efficient loading procedure involves Fmoc-His(Bum) (2 equiv) esterified in situ by N,N'-diisopropylcarbodiimide (DIPCDI) (2 equiv) and DMAP (0.16 equiv) in DCM-DMF (1:3) for 1 hour (Fields and Fields, 1990). The best reported results for loading Fmoc-Cys(Trt) and Fmoc-Cys(StBu) are 2.1% D-Cys during Fmoc-Cys(StBu) esterification by MSNT (Blankemeyer-Menge, 1990) and 2.0 to 2.5% D-Cys during Fmoc-Cys(Trt) cesium salt loading (Mergler et al., 1989b) or in situ

Table 3. Formation of Ester Bonds to Attach N^α-Protected Amino Acids to Linkers[a]

Reagent(s)	Linker[a]	Stoichiometries[b]	Conditions[c,d,e]	Racemization (%)[e]	Loading (%)[e]	Reference
Boc-AA-OH: DCHA	Br-PAM*	2:2:1	DMF (4 h, 50 °C + 14 h, 25 °C)	<0.1	86	Mitchell et al., 1978
Boc-AA-OH: KF	Br-PAM*	2.2:1.1:1	CH3CN (18-48 h)	N.R.	~100	Tam et al., 1979
Fmoc-AA-OH: DMF-dineopentyl acetal	HO-HMP*,#	1.7:1.7:1	DMF (72 h)	<0.05	67 - 75	Albericio and Barany, 1984; Albericio and Barany, 1985
Fmoc-AA-OH: DMAP: HOBt: DCC	HO-HMP#	2:2:4:2:1	DMF (18 h, 0 °C)	0.1 - 0.3	47	van Nispen et al., 1985
Fmoc-AA-OH: DCBC: pyridine	HO-HMP#	2:2:3.3:1	DMF (15-20 h)	0.1 - 0.7 Arg(Mtr) <1.0 His(Trt) 27.0 His(Bum) 2.2 Cys(Acm) 1.2 Cys(Trt) 4.0	55 - 66 His(Bum) 16	Sieber, 1987a
Fmoc-AA-OH: DEAD: Ph3P	HO-HMP#	3:3:3:1	THF (16 h, 0 °C)	0.3 Cys(Acm) 1.6	53 - 61	Sieber, 1987a; Stanley et al., 1991
Fmoc-AA-OH: DCC: DMAP	HO-SASRIN#	1.5:1.2: 0.01-0.1	DMF-DCM (1:3) (20 h, 0 °C)	0.2 - 1.0 Ile 1.2 Cys(Acm) 4.0 Cys(Trt) 18.3 His(Trt) 26.0	~80	Mergler et al., 1988a; Mergler et al., 1989b

Reagents	Linker	Stoichiometry	Solvent (time)	% Racemization	Yield (%)	Reference
Fmoc-AA-OH : DCC : HOBt	HO-HMP#	4 : 4 : 4 : 3 : 1	DMA (17 h)	<0.05	60 - 94, Arg 6, Asn 31, Gln 29, Pro 54	Grandas et al., 1989a
Fmoc-AA-OH : MSNT : MeIm	HO-HMP#	2 : 2 : 1.5 : 1	DCMf (0.5 h, twice) His(Bum) CHCl3	0 - 0.6, Cys(StBu) 2.1, Asp(OtBu) 1.4	72 - 100	Blankemeyer-Menge et al., 1990
Fmoc-AA-OH : DIEA	Br-HMP*	1.1 : 1 : 1	DMF (2-3 h)	<0.1	61 - 99	Bernatowicz et al., 1990
Fmoc-AA-OH : DIEA	Cl-Trt#	0.3-1.0 : 2.5 : 1.6	DCE (0.5 h)	<0.05	61 - 99, Trp 51	Barlos et al., 1991a
Fmoc-AA-O⁻ Cs⁺	Cl-CH₂-®#	1 : 1	DMF (16 h, 50 °C) N.R.	N.R.	72 - 95	Gisin, 1973
Fmoc-AA-O⁻ Cs⁺	Cl-HMP#	2 : 1	DMA (15-24 h, 50 °C)	0.01 - 0.07	89 - 98	Colombo et al., 1983
Fmoc-AA-O⁻ Cs⁺ : NaI	Cl-SASRIN#	1.5-3.0 : 1 : 1	DMA (24 h)	0.1 - 0.7, Cys(Trt) 2.5	N.R.	Mergler et al., 1989b
Fmoc-AA-OTDO : DIEA	HO-HMP#	40 mM : 1 : 1	DCMf (1-2 h), Ile 10 h	<0.2, Cys(Acm) 1.0	81 - 97	Kirstgen et al., 1987; Kirstgen and Steglich, 1989
Fmoc-AA-Cl	HO-HMP#	5 : 1	pyridine-DCM (2:3), (1 h)	<0.5, Met 1.7, Ala 0.7	~100	Akaji et al., 1990a
Fmoc-NCA : NMM	Rink acid#	3 : 0.02 : 1	toluenef (0.5-1 h)	<0.1	N.R.	Fuller et al., 1990

a The column entitled Linker distinguishes between ester bond formation *first* to provide a preformed handle (*) (often the preferred route, as discussed further in the text) and ester bond formation *directly* to the linker resin (#).
b Stoichiometries (equivalents) are stated in the same order as the reagents are listed, followed last by the linker or resin.
c When publications state overnight reactions, this table indicates 18 h.
d Reactions are at room temperature (25 °C) unless otherwise stated.
e Values are representative for incorporation of most amino acids. Individual amino acids falling outside of the general range are also listed.
f The solvent must be dry, or yields decrease dramatically.

Fmoc-Cys(Trt) loading with DCC and HOBt (Atherton et al., 1991). Loading of Fmoc-Cys(Acm), Fmoc-Cys(Trt), and Fmoc-His(Boc) to bromomethylated-linkers in the presence of DIEA has been reported to result in less than 0.1% D-isomers (Bernatowicz et al., 1990).

Peptide Amides

Most anchoring linkages that ultimately provide C-terminal peptide amides in a useful and general manner are benzhydrylamide derivatives. The attachment step is a direct coupling of an N^α-protected amino acid by means of its carboxyl to an appropriate benzhydrylamine resin, with eventual cleavage at a different locus providing the desired carboxamide. The 4-methylbenzhydrylamine (MBHA) linkage has been fine tuned with an electron-donating 4-methyl group (Matsueda and Stewart, 1981; Gaehde and Matsueda, 1981) to cleave in strong acid with good yields, yet it is completely stable to the conditions of Boc chemistry. The benzhydrylamide system also has been fine tuned with electron-donating methoxy groups to create the TFA-sensitive 4-(2′,4′-dimethoxyphenyl-aminomethyl)phenoxy (Rink amide) (Rink, 1987), 4-(4′-methoxy-benzhydryl)phenoxyacetic acid (Dod) (Stüber et al., 1989), 3-(amino-4-methoxybenzyl)-4,6-dimethoxyphenylpropionic acid (Breipohl amide) (Breipohl et al., 1989), and 4-succinylamino-2,2′,4′-trimethoxy-benzhydrylamine (SAMBHA) (Penke et al., 1988) linkers for use in Fmoc chemistry. Other structural themes are compatible with Fmoc chemistry and provide anchoring linkages that cleave in TFA to give peptides amides. These include the 5-(4-aminomethyl-3,5-dimethoxy-phenoxy)valeric acid (PAL) (Albericio and Barany, 1987b; Albericio et al., 1990a) and 5-(9-aminoxanthen-2-oxy)valeric acid (XAL) (Sieber, 1987c; Barany and Albericio, 1991a; Bontems et al., 1992) handles, both of which have highly desirable features by direct comparison to alternative procedures. Also, the photolabile 3-nitro-4-aminomethylbenzoic acid (Nonb) handle is an option with both Boc and Fmoc chemistries (Hammer et al., 1990). Dod, Breipohl amide, XAL, PAL, and Nonb handles in their N-protected (usually Fmoc) forms are attached to the appropriate amino-functionalized supports in situ with DIPCDI or BOP/DIEA in the presence of Dhbt-OH or HOBt (Stüber et al., 1989; Breipohl et al., 1989; Albericio et al., 1990a; Hammer et al., 1990).

Esterification of N^α-protected Asn and Gln can be sluggish (Barany and Merrifield, 1979; Wu et al., 1988; Fields and Fields, 1990). As an alternative, Boc-Glu(OH)-OBzl has been coupled (by means of an un-protected γ-carboxyl side chain) to benzhydrylamine resin, with HF cleavage yielding a peptide containing C-terminal Gln (Li et al., 1976). In parallel fashion, Fmoc-Asp(OH)-OtBu or Fmoc-Glu(OH)-OtBu have been coupled (by means of an unprotected β- or γ-carboxyl side-chain) to PAL, Rink amide, or Breipohl amide, with TFA cleavage yielding pep-

The substitution level of Fmoc amino acid resins is determined by quantitative spectrophotometric monitoring following piperidine deblocking. Fmoc amino acyl resins (4 to 8 mg) are shaken or stirred in piperidine-DMF (3:7) (0.5 mL) for 30 minutes, following which MeOH (6.5 mL) is added, and the resin is allowed to settle. The resultant fulvene-piperidine adduct has UV absorption maxima at 267 nm ($\varepsilon = 17,500$ $M^{-1}cm^{-1}$), 290 nm ($\varepsilon = 5800$ $M^{-1}cm^{-1}$), and 301 nm ($\varepsilon = 7800$ $M^{-1}cm^{-1}$). For reference, a piperidine-DMF-MeOH solution (0.3:0.7:39) is prepared. Spectrophotometric analysis is typically carried out at 301 nm, with comparison to a free Fmoc amino acid (i.e., Fmoc-Ala) of known concentration treated under identical conditions. The substitution level (mmol/g) = ($A_{301} \times 106$ µmol/mol $\times 0.007$ L/7800 $M^{-1}cm^{-1} \times 1$ cm $\times$ mg of resin) (Meienhofer et al., 1979; D. Hudson, unpublished results).

tides containing C-terminal Asn or Gln (Albericio et al., 1990b; Breipohl et al., 1990; Fields and Fields, 1990).

FORMATION OF PEPTIDE BOND

There are currently four major kinds of coupling techniques that serve well for the *stepwise* introduction of N^{α}-protected amino acids for solid-phase synthesis. In the solid-phase mode, coupling reagents are used in excess to ensure that reactions reach completion. The ensuing discussion will skirt some of the rather complicated mechanistic issues and focus on practical details. Recommendations for coupling methods are included in Tables 4 and 5.

In situ Reagents

The classical example of an in situ coupling reagent is N,N'-dicyclohexylcarbodiimide (DCC) (Rich and Singh, 1979; Merrifield et al., 1988) (see Figure 5). The related N,N-diisopropylcarbodiimide (DIPCDI) is more convenient to use under some circumstances, because the resultant urea coproduct is more soluble in DCM. The generality of carbodiimide-mediated couplings is extended significantly by the use of 1-hydroxybenzotriazole (HOBt) as an additive, which accelerates carbodiimide-mediated couplings, suppresses racemization, and inhibits dehydration of the carboxamide side chains of Asn and Gln to the corresponding nitriles (König and Geiger, 1970a; König and Geiger, 1973; Mojsov et al., 1980). Recently, protocols involving benzotriazol-1-yloxy-tris(dimethylamino)phosphonium hexafluorophosphate (BOP), 2-(1H-benzotriazol-1-yl)-1,1,3,3-tetramethyluronium hexafluorophosphate (HBTU), 2-(1H-benzotriazol-1-yl)-1,1,3,3-tetramethyluronium tetrafluoroborate (TBTU), and 2-[2-oxo-1(2H)-pyridyl]-1,1,3,3-bispenta-

Table 4   Typical Protocols for Automated Boc Chemistry SPPS[a]

Cycle	Function	Time
1	DCM wash	3×1 min
2	TFA-DCM (1:1) deprotection	$2 + 20$ min
3	DCM wash	3×1 min
4	DIEA-DCM (1:9) neutralization	2×2 min
5	DCM wash	3×1 min
6	DMF or NMP wash	3×1 min
7a	Boc-amino acid (3 equiv) in DMF	5 min
7b	DIPCDI (3 equiv) in DMF[b]	60 min
or		
7a	Boc-amino acid (3 equiv) in DMF	5 min
7b	BOP (3 equiv):DIEA (5.3 equiv) in DMF[c]	45 min
or		
7	Boc-amino acid PSA (2 equiv) in DMF or NMP[c]	60 min
or		
7	Boc-amino acid preformed ester (3 equiv) in DMF or NMP	60 min
8	DMF or NMP wash	3×1 min

[a] Refer to original research papers for additional specifications and variations.

[b] HOBt (3 equiv) is added for Boc-Asn, -Gln, -Arg(Tos), -Arg(Mts), and -His(Dnp).

[c] Boc-Asn and -Gln require side-chain protection in this variation.

methyleneuronium tetrafluoroborate (TOPPipU) have deservedly achieved popularity. BOP, HBTU, TBTU, and TOPPipU require a tertiary amine, such as NMM or DIEA, for optimal efficiency (Dourtoglou et al., 1984; Fournier et al., 1988; Ambrosius et al., 1989; Gausepohl et al., 1989a; Seyer et al., 1990; Fields et al., 1991; Knorr et al., 1991). HOBt has been reported to accelerate further the rates of BOP- and HBTU-mediated couplings (Hudson, 1988a; Fields et al., 1991). In situ activations by excess HBTU or TBTU can cap free amino groups (Gausepohl et al., 1992); it is not known whether HOBt can suppress this side reaction. Acylations using BOP result in the liberation of the carcinogen hexamethylphosphoramide, which might limit its use in large-scale work. The modified BOP reagent benzotriazole-1-yl-oxy-tris-pyrrolidinophosphonium hexafluorophosphate (PyBOP) liberates potentially less toxic by-products (Coste et al., 1990). Protocols have been reported for the use of BOP to incorporate side-chain unprotected Thr and Tyr (Fournier et al., 1988; Fournier et al., 1989).

Table 5   Typical Protocols for Automated Fmoc Chemistry SPPS[a]

Cycle	Function	Time
1	DMF or NMP wash	3 × 1 min
2	Piperidine-DMF or NMP (1:4) deprotection	3 + 17 min
3	DMF or NMP wash	3 × 1 min
4a	Fmoc-amino acid (4 equiv) in DMF or NMP	5 min
4b	DIPCDI (4 equiv):HOBt (4 equiv) in DMF	60 min
or		
4a	Fmoc amino acid (4 equiv) in DMF or NMP	5 min
4b	BOP (3 equiv):NMM (4.5 equiv):HOBt (3 equiv) in DMF[b]	45 min
or		
4a	Fmoc amino acid (4 equiv): HBTU (3.8 equiv): HOBt (4 equiv) in DMF[b]	5 min
4b	DIEA (7.8 equiv)	45 min
or		
4	Fmoc-amino acid preformed ester (4 equiv) in DMF or NMP	60 min
5	DMF or NMP wash	3 × 1 min

[a] Refer to original research papers for additional specifications and variations.

[b] Fmoc-Asn and -Gln require side-chain protection in this variation.

Active Esters

A long-known but steadfast coupling method involves the use of active esters. The classical 2- and 4-nitrophenyl esters (ONo and ONp, respectively), used in DMF, allow relatively slow but dehydration-free introduction of Asn and Gln (Mojsov et al., 1980) (see Figure 6). ONo and ONp esters of Boc- and Fmoc amino acids are prepared from DCC and either 2- or 4-nitrophenol, and the undesired nitrile contaminant is separated easily (Bodanszky et al., 1973; Bodanszky et al., 1980). *N*-hydroxysuccinimide (OSu) esters of Fmoc amino acids have been used successfully in SPPS (Fields et al., 1988), but they are not recommended for general use due to the formation of succinimidoxycarbonyl-β-alanine-*N*-hydroxysuccinimide ester (Gross and Bilk, 1968; Weygand et al., 1968b).

More recently, workers have concentrated on pentafluorophenyl (OPfp), HOBt, 3-hydroxy-2,3-dihydro-4-oxo-benzotriazine (ODhbt), and substituted 1-phenylpyrazolinone enol esters. Boc and Fmoc amino acid OPfp esters are prepared from DCC and pentafluorophenol (Kisfaludy et al., 1973; Penke et al., 1974; Kisfaludy and Schön, 1983) or pentafluorophenyl trifluoroacetate (Green and Berman, 1990). Although

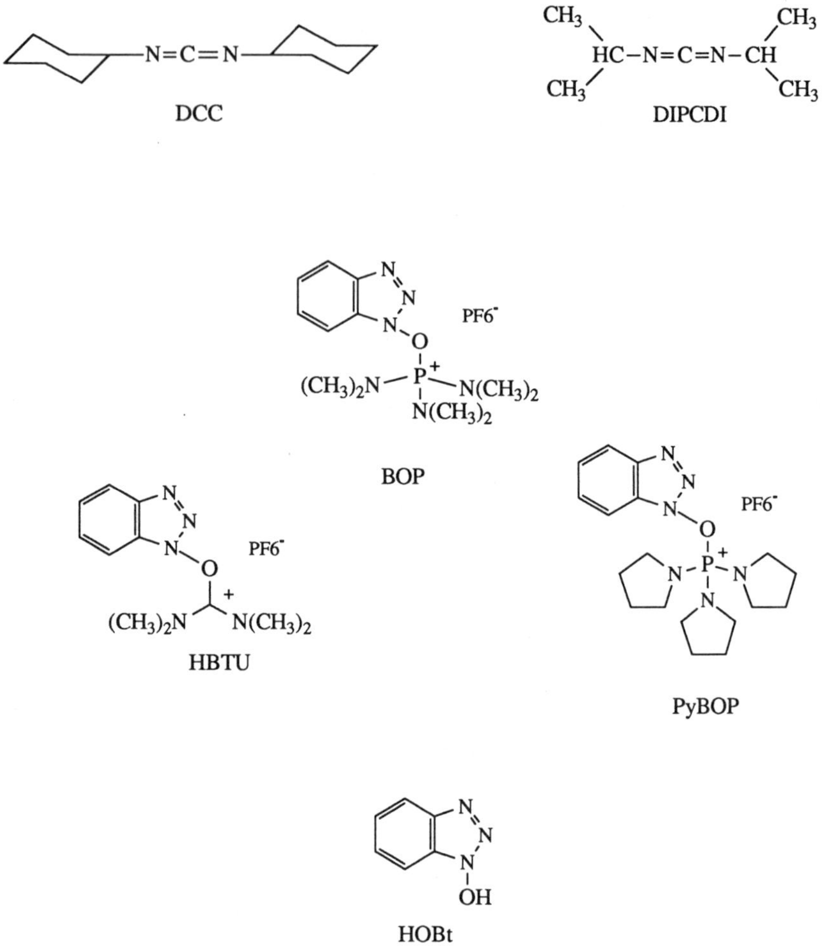

FIGURE 5   In situ coupling reagents and additives for SPPS.

OPfp esters alone couple slowly, the addition of HOBt (1 to 2 equiv) increases the reaction rate (Atherton et al., 1988a; Hudson, 1990b). Fmoc-Asn-OPfp allows for efficient incorporation of Asn with little side-chain dehydration (Gausepohl et al., 1989b). HOBt esters of Fmoc amino acids are formed rapidly (with DIPCDI) and highly reactive (Harrison et al., 1989; Fields et al., 1989), as are Boc amino acid HOBt esters (Geiser et al., 1988). N^{α}-protected amino acid ODhbt esters suppress racemization and are highly reactive, in similar fashion to HOBt esters (König and Geiger, 1970b). Preparation of ODhbt esters (from Dhbt-OH and DCC) is accompanied by the formation of the by-product 3-(2-azidobenzoyloxy)-4-oxo-3,4-dihydro-1,2,3-benzotriazine (König and Geiger,

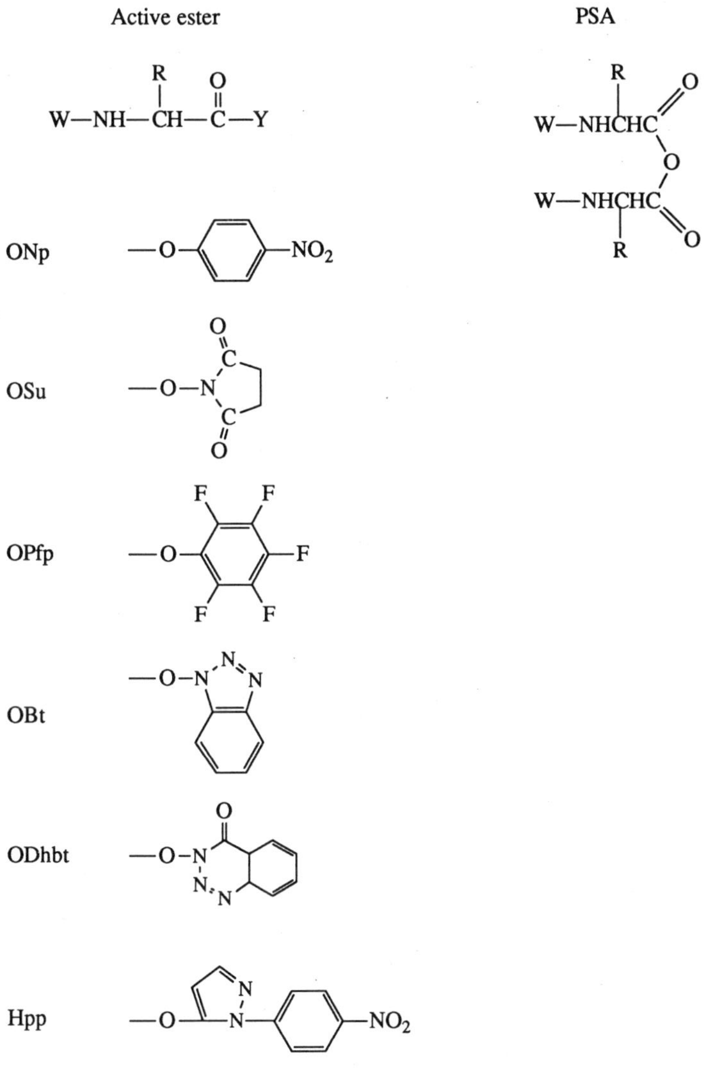

Active ester   PSA

$$W-NH-CH-C-Y$$

with R above CH and O double bonded to C

ONp   —O—⟨benzene ring⟩—NO$_2$

OSu   —O—N⟨succinimide ring with two C=O⟩

OPfp   —O—⟨pentafluorophenyl ring: F F F F F⟩

OBt   —O—N⟨benzotriazole ring, N=N-N⟩

ODhbt   —O—N⟨3,4-dihydro-benzotriazin-4-one ring with O, N·N⟩

Hpp   —O—⟨pyrazolinone ring, N-N⟩—⟨benzene ring⟩—NO$_2$

FIGURE 6   Activated N^α-protected amino acids. W is either Boc or Fmoc, Y is the structure specified next to the abbreviation of the active ester derivative.

1970c). Fmoc amino acid ODhbt esters are far more stable than HOBt esters and, therefore, can be isolated from the side product before use (Atherton et al., 1988b). Aminolysis by Fmoc amino acid esters of 1-(4-nitrophenyl)-2-pyrazolin-5-one (Hpp), 3-phenyl-1-(4-nitrophenyl)-2-pyrazolin-5-one (Pnp), and 3-methyl-1-(4-nitrophenyl)-2-pyrazolin-5-

one (Npp) also proceeds rapidly (Hudson, 1990a; Johnson et al., 1992). Competition experiments have shown ester reactivity usually to be Pnp > Hpp ~ Npp > ODhbt > OPfp > OSu > ONp > ONo (Hudson, 1990a; Hudson, 1990b; Johnson et al., 1991), although Hpp esters were found to be superior to Pnp and Npp esters for "difficult" couplings (Johnson et al., 1991). Both Fmoc-Tyr and Fmoc-Ser have been incorporated successfully as preformed active esters without side-chain protection (Fields et al., 1989; Otvös et al., 1989a).

Preformed Symmetrical Anhydrides

Preformed symmetrical anhydrides (PSAs) are favored by some workers because of their high reactivity (see Figure 6). They are generated in situ from the corresponding N^α-protected amino acid (2 or 4 equiv) plus DCC (1 or 2 equiv) in DCM; following removal of the urea by filtration, the solvent is exchanged to DMF for optimal couplings. Detailed synthetic protocols based on PSAs have been described for Boc (Merrifield et al., 1982; Yamashiro, 1987; Geiser et al., 1988; Kent and Parker, 1988; Wallace et al., 1989) and Fmoc (Chang et al., 1980a; Heimer et al., 1981; Atherton and Sheppard, 1989) chemistries.

The use of the PSA procedure to introduce Boc/Fmoc-Gly or -Ala occasionally results in inadvertent coupling of a diglycyl or dialanyl unit (Merrifield et al., 1974; Benoiton and Chen, 1987; Merrifield et al.,

Efficient couplings using Boc amino acid PSAs are critically dependent on the concentration of the activated species in solution. It has been recommended that Boc amino acid PSA couplings proceed at a concentration of 0.15 M and that double couplings be standard practice for syntheses of >50 residues (Kent and Parker, 1988).

Fmoc amino acids have variable solubility properties in relatively nonpolar solvents, such as DCM. Fmoc-Asp(OtBu), -Glu(OtBu), -Ile, -Leu, -Lys(Boc), -Ser(tBu), -Thr(tBu), and -Val are soluble in DCM, while Fmoc-Ala, -Gly, -Met, -Trp, and -Tyr(tBu) require the presence of a more polar solvent (i.e., DMF) for solubilization. Fmoc-Asn, -Gln, -His(Bum), and -Phe require at least 60% DMF to remain in solution. Conversion of Fmoc amino acids to the corresponding PSAs results in poorer solubilities in nonpolar solvents. Thus, whether Fmoc amino acids are coupled in situ or as preactivated species, relatively polar solvents (DMF or NMP) should be used to ensure that all reactants are in solution.

1988). Also, side-chain unprotected Asn and Gln, all Arg derivatives, and N^τ-protected His should not be used as PSAs due to the potential side reactions discussed previously (see Protection Schemes). The solubilities of some Fmoc amino acids make PSAs a less-than-optimum activated species. Not all Fmoc amino acids are readily soluble in DCM, thus requiring significant DMF for solubilization. Optimum activation conditions, which require neat DCM (Rich and Singh, 1979), cannot be obtained. In addition, the resulting Fmoc amino acid PSAs are even less soluble than the parent Fmoc amino acid (Harrison et al., 1989).

Acid Halides

N^α-protected amino acid chlorides have a long history of use in solution synthesis. Their use in solid-phase synthesis has been limited, because the Boc group is not completely stable to reagents used in the preparation of acid chlorides. The Fmoc group, on the other hand, survives acid chloride preparation with thionyl chloride (Carpino et al., 1986), while both the Fmoc and Boc groups are stable to acid fluoride preparation with cyanuric fluoride (Carpino et al., 1990; Bertho et al., 1991; Carpino et al., 1991a). For derivatives with tBu side-chain protection, only the acid fluoride procedure can be used (Carpino et al., 1990). Fmoc amino acid chlorides and fluorides react rapidly under SPPS conditions in the presence of HOBt/DIEA and DIEA, respectively, with very low levels of racemization (Carpino et al., 1990; Carpino et al., 1991b).

MONITORING

A crucial issue for stepwise solid-phase peptide synthesis is the repetitive yield per deprotection/coupling cycle. There are a number of ways of monitoring these steps, including some with a possibility for "real-time" feedback based on the kinetics of appearance or disappearance of appropriate soluble chromophores measured in a flow-through system. Most accurate and meaningful are qualitative and quantitative tests for the presence of unreacted amines after an acylation step. Such tests should ideally be negative before proceeding further in the chain assembly. For certain active ester methods, the leaving group has "self-indicating" properties insofar as a colored complex is noted for as long as unreacted amino groups remain on the support. These various techniques reveal that high efficiencies can, indeed, be achieved in stepwise synthesis.

The best known qualitative monitoring methods are the ninhydrin (Kaiser et al., 1970) and isatin (Kaiser et al., 1980) tests for free N^α-amino and -imino groups, respectively, where a positive colorimetric response to an aliquot of peptide resin indicates the presence of unreacted N^α-amino/imino groups. These tests are easy, reliable, and require only a

Free amino groups are quantitated based on their reaction with ninhydrin to produce Ruhemann's purple. Three reagent solutions are required: solution 1 is phenol-ethanol (7:3), solution 2 is 0.2 mM KCN in pyridine, and solution 3 is 0.28 M ninhydrin in ethanol. A sample of Boc-peptide resin (2 to 5 mg) is incubated with 75 μL of solution 1, 100 μL of solution 2, and 75 μL of solution 3 for 7 minutes at 100 °C (Sarin et al., 1981; Applied Biosystems, Inc., 1989c). For Fmoc-peptide resins, premature removal of the Fmoc group (by pyridine) is minimized by adding 2 to 3 drops (20 to 40 μL) of glacial HOAc to each resin sample and heating the reaction mixture for 5 minutes instead of 7 minutes (Applied Biosystems, Inc., 1989b). Immediately following the designated incubation time, 60% aqueous ethanol (4.8 mL) is added to each sample with vigorous mixing. Once the peptide resin has settled, the absorbance of each sample solution is read at 570 nm; 60% ethanol is used as a reference. The concentration of free amino groups (mmol/g) = $(A_{570} \times 10^6 \ \mu\text{mol/mol} \times 0.005$ L)/$(15000 \ \text{M}^{-1}\text{cm}^{-1} \times 1 \ \text{cm} \times \text{mg of resin})$.

few minutes to perform, allowing the chemist to make a quick decision concerning how to proceed. A highly accurate quantitative modification of the ninhydrin procedure has been developed (see Box).

Other monitoring techniques exist that are generally nondestructive (noninvasive) and, therefore, can be carried out on the total batch. Resin-bound N^α-amino groups can be titrated with picric acid, 4,4′-dimethoxytrityl chloride, bromphenol blue dye, and quinoline yellow dye. Picric acid is removed from resin-bound amines with 5% DIEA in DCM, and the resulting chromophore is quantitated at 362 nm (Hodges and Merrifield, 1975; Arad and Houghten, 1990). For trityl monitoring, 4,4′-dimethoxytritylchloride and tetra-n-butylammonium perchlorate are reacted with the resin, released with 2% dichloroacetic acid in DCM, and quantitated at 498 nm (Horn and Novak, 1987; Reddy and Voelker, 1988). The effect of the dilute acid on Fmoc amino acid side-chain protecting groups and linkers has not been reported. For bromphenol blue and quinoline yellow monitoring, the dye is bound to free amino groups following deprotection, then displaced as acylation proceeds. Quantitative monitoring can be carried out at 600 and 495 nm for bromphenol blue and quinoline yellow, respectively (Krchnák et al., 1988; Flegel and Sheppard, 1990; Young et al., 1990). Gel-phase nuclear magnetic resonance (NMR) spectroscopy has been proposed for direct examination of resin-bound reactants (Epton et al., 1980; Giralt et al., 1984; Butwell et al., 1988), but that procedure would appear to suffer from the problems of sensitivity, expense, and time needed to accumulate data.

As an alternative to quantitating resin-bound species, soluble reactants or coproducts can be analyzed. Continuous measurement of electrical conductivity can be used to evaluate coupling and Fmoc deprotection efficiencies (Nielson et al., 1989; Fox et al., 1990; McFerran et al., 1991). The progress of Fmoc chemistry can be evaluated by observing at 300 to 312 nm, the decrease of absorbance when Fmoc amino acids are taken up during coupling and by the increase in absorbance when the Fmoc group is released with piperidine (Chang et al., 1980a; Atherton and Sheppard, 1985; Frank and Gausepohl, 1988; Atherton and Sheppard, 1989). Monitoring a decrease in Fmoc amino acid concentration at 300 nm can be complicated when OPfp esters are utilized (Atherton et al., 1988b). More straightforward acylation monitoring is possible when Fmoc amino acid ODhbt, Hpp, Pnp, and Npp esters are used. During the coupling of Fmoc amino acid ODhbt esters, the liberated HO-Dhbt component binds to free N^α-amino groups, producing a bright yellow color, which diminishes as acylation proceeds (Cameron et al., 1988). Real-time spectrophotometric monitoring proceeds at 440 nm (Cameron et al., 1988). In a similar way, ionization of the leaving group from Fmoc amino acid Npp esters by free N^α-amino groups results in a blood-red color (Hudson, 1990a). As coupling proceeds, the color change (to gold) can be monitored at 488 nm.

Unfortunately, continuous-flow monitoring is inherently insensitive for direct judgment of reaction endpoints. Assuming an initial twofold excess of activated incoming amino acid, absorbance will drop from 2.00 units to 1.05 units or 1.01 units, respectively, with 95 percent or 99 percent coupling. It is difficult to distinguish accurately between 1.05 and 1.01. In contrast, if unreacted resin-bound components are titrated for the same two cases, the fivefold difference between 0.05 and 0.01 is easily noted. The sensitivities of techniques monitoring resin-bound components is limited by nonspecific binding or irreversible reactions of the titrant with the protected peptide or resin, which contribute to background readings despite complete reactions.

Invasive monitoring of both synthetic efficiency and amino acid composition of peptide resins can be achieved by a powerful quantitative variation of the Edman sequential degradation, called preview analysis (Tregear et al., 1977; Matsueda et al., 1981; Kent et al., 1982). Crude peptide resins are sequenced directly; each Edman degradation cycle serves to identify a primary amino acid residue and preview the next amino acid in the sequence. Because preview is cumulative, quantitation of peaks after a number of cycles indicates the average level of deletion peptides and thus the overall synthetic efficiency. Since the linkers as well as most side-chain protecting groups used in Boc chemistry are stable to the conditions of Edman degradation, sequencing is a true solid-phase process; moreover, identification of amino acid phenylthiohydantoins requires a set of protected standards (Simmons and Schlesinger, 1980; Steiman et al., 1985). Preview sequence analysis of peptide resins made by Fmoc chemistry requires initial TFA cleavage, followed by immobilization (covalent or noncovalent) of the crude peptide on a suitable support (Kochersperger et al., 1989). Most side-chain protecting groups used in conjunction with Fmoc chemistry are not stable to the conditions of Edman degradation; hence, the usual free side-chain phenylthiohydantoin standards can be used.

The technique of "internal reference amino acids" (IRAA) is often very useful to accurately measure yields of chain assembly and retention of chains on the support during synthesis and after cleavage (Matsueda and Haber, 1980; Atherton and Sheppard, 1989; Albericio et al., 1990a). In addition, amino acid analysis of peptide resins may be used to monitor synthetic efficiency; the advent of microwave hydrolysis technology may permit rapid analysis (Yu et al., 1988).

AUTOMATION OF SOLID-PHASE SYNTHESIS

A significant advantage of solid-phase methods lies in the ready automation of the repetitive steps (see Tables 4 and 5). The first instrument for synthesis of peptides was built by Merrifield, Stewart, and Jernberg (1966) and is currently on display at the Smithsonian Institution. Numerous models for both batchwise and continuous-flow operation at various scales of operation are now commercially available. Some of these instruments include features to facilitate monitoring (compare to previous section). Supported procedures have also been introduced for the generation of large numbers of (usually related) peptide sequences in a reasonably short time, although with some sacrifices in the usual standards for purity and characterization.

Automated Synthesizers, 1-3 Simultaneous Syntheses

The Applied Biosystems, Inc., Models 430A (Kent et al., 1984) and 431A utilize either Boc or Fmoc chemistries, with reaction mixing by vigorous vortex or gas bubbling. The automated 430A and 431A cycles for Boc amino acids (PSA in DMF, HOBt ester in NMP) and Fmoc amino acids (HOBt ester in either DMF or NMP) have been described in detail (Geiser et al., 1988; Fields et al., 1989; Fields et al., 1990). The Boc amino acid PSA cycles feature solvent exchange, so that activation occurs in DCM and coupling in DMF. The 430A and 431A also use fully automated HBTU + HOBt in situ cycles (Fmoc chemistry in NMP), called *FastMoc*[TM] (Fields et al., 1991). The synthesis scale is from 0.1 to 0.25 mmol, with microprocessor control by means of an internal computer. The MilliGen/Biosearch 9600 also utilizes either Boc or Fmoc chemistries, with reaction mixing by nitrogen bubbling. Coupling for both Boc and Fmoc amino acids can be in situ (In-Reservoir Activation[TM]) with BOP/HOBt, or by solution sampling and preactivation using DIPCDI or DIPCDI/HOBt. Cycle control is by Sequence

Driven^{TM} "Expert System" software utilized with an IBM PC/AT Synthesis Workstation. Advanced ChemTech Models 100, 200, and 400 use either Boc or Fmoc chemistries, with reaction mixing by nitrogen bubbling or oscillation. Coupling for the Models 200 and 400 is by preformed mixed or symmetrical anhydrides (with solvent exchange) or by preformed HOBt esters. The Model 100 does not preactivate, and thus it must use preformed or in situ species. Cycles for all Advanced Chem-Tech instruments are controlled by means of an IBM PS/2. Pharmaceutical considerations are fulfilled by the Model 400 (Birr, 1990a; Birr, 1990b), because it can utilize 100 g or more of resin. The Eppendorf Synostat^{TM} P is compatible with either Boc or Fmoc chemistries, utilizing in situ HBTU or BOP activation at scales from 0.1 to 5.0 mmol with adjustable vortex mixing. The Rainin/Protein Technologies PS3 features coupling by in situ BOP with Fmoc amino acids only (all prepackaged) at scales from 0.1 to 0.5 mmol with nitrogen bubbling for reaction mixing. The MilliGen/Biosearch EXCELL is also an Fmoc only instrument, using in situ DIPCDI (in DMF-DCM) or BOP/HOBt for coupling on a 0.1 mmol scale. The Biotech Instruments BT 7600 is designed for Fmoc chemistry and operates at scales from 0.05 to 1.0 mmol, using preformed OPfp esters with continuous conductivity monitoring.

The just-mentioned instruments are designed for batchwise syntheses. Continuous-flow instruments (Fmoc chemistry only) include the MilliGen/Biosearch 9050, the NovaSyn^{TM} Crystal, and the NovaSyn^{TM} Gem. The MilliGen/Biosearch 9050 (Kearney and Giles, 1989) is a 3-column automated instrument with a synthesis scale of 0.1 to 1.0 mmol and flow rates typically from 5 to 15 mL/min. Coupling proceeds by means of in situ BOP/HOBt or DIPCDI, or preformed OPfp/HOBt esters with continuous spectrophotometric monitoring of both coupling and deprotection at 365 nm. An NEC APC IV computer controls the Milli-Gen Express-Peptide^{TM} software. The NovaSyn^{TM} Crystal (AminoTech, 1991) is, in similar fashion to the MilliGen/Biosearch 9050, a 3-column automated instrument with a synthesis scale of 0.05 to 0.7 mmol. Acylation methods include pre–formed OPfp and ODhbt esters, in situ PyBOP/HOBt, or PSA with continuous spectrophotometric analysis of both coupling (by counterion distribution monitoring) and deprotection. The software is controlled by an IBM compatible 1120/S. The NovaSyn^{TM} Gem is a semiautomated, 2-column instrument with similar acylation and monitoring features as the NovaSyn^{TM} Crystal (Amino-Tech, 1991).

Automated and Semiautomated Multiple Peptide Synthesizers

The Zinsser Analytic SMPS 350/Advanced ChemTech 350 utilizes two independent robotic arms, controlled through an IBM PS/2, to synthesize up to 96 peptides simultaneously. Only Fmoc chemistry is used, with coupling by PSAs or HOBt esters or in situ DIPCDI or TBTU, on a 0.5 mmol scale (Schnorrenberg and Gerhardt, 1989; Groginsky, 1990). The ABIMED Model AMS 422 (Gausepohl et al., 1990; Gausepohl et al., 1991) uses Fmoc chemistry to synthesize up to 48 peptides on a scale from 5 to 50 μmol with activation by in situ PyBOP. A single robotic arm dispenses reagents, while resins are contained in fritted polypropylene tubes. The manual Multiple Peptide Synthesis Tea Bag method can synthesize up to 120 peptides simultaneously, using either Boc or Fmoc chemistry and coupling by PSAs or in situ DCC/DIPCDI (Houghten et al., 1986; Beck-Sickinger et al., 1991). The Tea Bag method has been semiautomated (Beck-Sickinger et al., 1991) and commercialized by Labostec and Biotech Instruments (BT 7500). Biotech Instruments also supplies a PepSeal heat sealer for Tea Bag preparation. Cambridge Research Biochemicals markets the Pepscan/PIN method (Geysen et al., 1984; Hoeprich, et al., 1989; Arendt and Hargrave, 1991), which uses either Boc or Fmoc chemistry and coupling in situ with DCC or BOP/HOBt to synthesize up to 96 peptides (10 to 100 nmol) simultaneously. The semiautomated Dupont RaMPs^{TM} (Wolfe and Wilk, 1989) synthesizes up to 25 peptides simultaneously by either Boc or Fmoc chemistry using PSAs or OPfp/HOBt esters. Finally, the just-described Affymax Parallel Chemical Synthesis system (VLSIPS, for "very large-scale immobilized polymer synthesis"), using the photolabile Nvoc N^α-protecting group and preformed HOBt esters, can be used for simultaneous synthesis of an extraordinarily high number of related peptides (e.g., $1024 = 2^{10}$ by 10 stages requiring 2.5 hours each) (Fodor et al., 1991).

CLEAVAGE

Boc SPPS is designed primarily for simultaneous cleavage of the peptide anchoring linkage and side-chain protecting groups with strong acid (HF or equivalent), while Fmoc SPPS is designed primarily to accomplish the same cleavages with moderate strength acid (TFA or equivalent). In each case, careful attention to cleavage conditions (reagents, scavengers, temperature, and times) is necessary in order to minimize a variety of side reactions. Considerations for separate removal of acid-stable side-chain protecting groups have been covered earlier (see Protection Schemes). Nonacidolytic methods for cleavage of the anchoring linkage, each with certain advantages as well as limitations, may also be used in conjunction with either Boc or Fmoc chemistries.

Hydrogen Fluoride (HF)

Treatment with HF simultaneously cleaves PAM and MBHA linkages and removes the side-chain protecting groups commonly applied in Boc

HF cleavage procedures require a special all-fluorocarbon apparatus. The standard method uses HF-anisole (9:1) (1 mL per 20 μmol peptide) at 0 °C for 1 hour, with the addition of 2-mercaptopyridine (10 equiv) for Met-containing peptides. For Cys- and Trp-containing peptides, HF-anisole-dimethylsulfide-4-thiocresol (10:1:1:0.2) is recommended. Following cleavage, HF is evaporated carefully under aspirator suction with ice-bath cooling, and most of the anisole is removed by vacuum from an oil pump. The vessel is triturated with ether to remove benzylated scavenger adducts; at the same time, the ether facilitates transfer of the cleaved resin (with trapped peptide) to a fritted glass funnel. Next, 30% aqueous HOAc (twice, 1.5 mL per 20 μmol peptide) is used to rinse the cleavage vessel and extract the resin on the fritted glass filter. The combined filtrates are diluted with H_2O to bring the HOAc concentration to <10%, and the peptide is usually recovered by lyophilization (Stewart and Young, 1984; Tam and Merrifield, 1987; Applied Biosystems, Inc., 1989c). "Low-high" HF cleavages are carried out following the detailed description of the original and later publications (Tam et al., 1983; Tam and Merrifield, 1985; Tam and Merrifield, 1987).

chemistry (Figure 2), i.e., Bzl (for Asp, Glu, Ser, Thr, and Tyr), 2-BrZ or 2,6-Cl₂Bzl (for Tyr), cHex (for Asp), 2-ClZ (for Lys), Bom or Tos (for His), Tos or Mts (for Arg), Xan (for Asn and Gln), and Meb (for Cys) (Tam and Merrifield, 1987). HF cleavages are always carried out in the presence of a carbonium ion scavenger, usually 10% anisole. For cleavages of Cys-containing peptides, further addition of 1.8% 4-thiocresol is recommended. Additional scavengers, such as dimethylsulfide, 4-cresol, and 4-thiocresol, are used in conjunction with a two-stage "low-high HF" cleavage method that provides extra control and thereby better product purities (Tam and Merrifield, 1987). Both Trp(CHO) and Met(O) can be deprotected under "low HF" conditions [20 to 25% HF–0 to 5% 4-thiocresol–70 to 80% dimethylsulfide, 0 °C for 1 hour; 4-thiocresol is necessary only for Trp(CHO)] (Tam and Merrifield, 1987). In the presence of the large levels of dimethylsulfide used in "low HF" conditions, Tyr(Bzl) undergoes little C-alkylation (Tam and Merrifield, 1987).

In strong acid, the γ-carboxyl group of Glu can become protonated and lose water. The resulting acylium ion is then trapped either intramolecularly by the N^{α}-amide nitrogen to give a pyrrolidone or (more likely) intermolecularly with the commonly used scavenger anisole (Feinberg and Merrifield, 1975). This serious problem can be controlled by attenuation of the acid strength (i.e., "low HF" conditions) and caution with regard to cleavage temperature (Tam and Merrifield, 1987). Strong acid can also cause an N → O acyl shift in Ser- and Thr-containing peptides, resulting in the thermodynamically less stable O-acyl species (Fujino et al., 1978). This process can be reversed for simple cases by treating the cleaved, deprotected peptide with 5% aqueous NH_4HCO_3, pH 7.5, at 25 °C for several hours or in 2% aqueous NH_4OH at 0 °C for 0.5 hour (Barany and Merrifield, 1979); reversal of the $N \rightarrow O$ acyl shift in multiple Ser- and Thr-containing peptides may be more problematic. HF-liberated Boc groups can modify Met residues (Noble et al., 1976); therefore, the N^{α}-Boc group should be removed prior to HF cleavage. HF deprotection of His(Bom) liberates formaldehyde, resulting in methylation of susceptible side chains and cyclization of N-terminal Cys residues to a thiazolidine (Mitchell et al., 1990; Gesquière et al., 1990; Kumagaye et al., 1991). These side reactions may be inhibited by including in the HF cleavage mixture a formaldehyde scavenger, e.g., resorcinol (0.27 M), or Cys or $CysNH_2$ (30 to 90 equiv), and purifying the peptide immediately after cleavage (Mitchell et al., 1990; Kumagaye et al., 1991). Serious Trp alkylation side reactions have been observed during workup after HF cleavage of peptides containing Cys or Met adjacent to Trp; the problem may be mitigated by adding free Trp (10 equiv) as a scavenger during cleavage (D. Hudson, unpublished results) or during the initial lyophilization (Ponsati et al., 1990a).

Other Strong Acids

The alternative strong acids listed here, and with further examples elsewhere (Barany and Merrifield, 1979), are very likely to promote the same side reactions just described for HF. TFMSA (1 M)-thioanisole (1 M) in TFA cleaves PAM and MBHA linkers (Tam and Merrifield, 1987; Bergot et al., 1987), and removes many side-chain protecting groups used in Boc chemistry, e.g., Mts (for Arg), Bzl (for Asp, Glu, Ser, Thr), cHex (for Asp), Meb (for Cys), 2-ClZ (for Lys), 2-BrZ or 2,6-Cl_2Bzl (for Tyr), and Bom or Tos (for His) (Tam and Merrifield, 1987), without requiring a special apparatus. A "low-high" method can be used with TFMSA, in similar fashion to HF (Tam and Merrifield, 1987). Tetrafluoroboric acid (HBF_4) (1 M)-thioanisole (1 M) in TFA offers a similar range of side-chain deprotection as TFMSA (Kiso et al., 1989; Akaji et al., 1990b).

Trimethylsilyl bromide (TMSBr) and trimethylsilyl trifluoromethanesulfonate (TMSOTf) have also been used for strong acid cleavage and deprotection reactions, which are accelerated by the presence of thioanisole as a "soft" nucleophile (Yajima et al., 1988; Nomizu et al., 1991). TMSBr (1 M)-thioanisole (1 M) removes Mts (for Arg), Bzl (for Asp, Glu, Ser, Thr, and Tyr), 2,6-Cl_2Bzl (for Tyr), and 2-ClZ (for Lys) as well as reducing Met(O) to Met (Yajima et al., 1988). Although not specifically stated, TMSBr-thioanisole probably deprotects His(Tos), Cys(Meb), and Tyr(2-BrZ). TMSOTf (1 M)-thioanisole (1 M) additionally removes Bom from His (Yajima et al., 1988) and efficiently

For TFMSA cleavages, peptide resin (100 mg) is stirred in thioanisole (187 µL) and 1,2-ethanedithiol (EDT) (64 µL) in an ice bath for 10 minutes. TFA (1.21 mL) is added, and after equilibration for 10 minutes, TFMSA (142 µL) is added *slowly*. Cleavage and deprotection proceeds for 1 hour (unless the MBHA linker is being cleaved, in which case cleavage proceeds for 1.5 to 2.5 hours). During cleavage and deprotection, the ice bath is removed, but precautions are taken to ensure that the temperature of the reaction does not increase rapidly. Subsequently, the cleavage mixture is filtered through a fritted glass funnel directly into methyl *t*Bu ether, to rapidly precipitate the peptide and remove the acid and scavengers. The precipitated peptide should be washed with methyl *t*Bu ether and dried under vacuum overnight (Bergot et al., 1987; Fields and Fields, 1991).

cleaves PAM and MBHA linkages (Akaji et al., 1989; Nomizu et al., 1991). Stable Cys side-chain protection should be used during TMSBr-thioanisole cleavages.

Trifluoroacetic Acid (TFA)

The combination of side-chain protecting groups, e.g., *t*Bu (for Asp, Glu, Ser, Thr, and Tyr), Boc (for His and Lys), Bum (for His), Tmob (for Asn, Cys, and Gln), and Trt (for Asn, Cys, Gln, and His), and anchoring linkages, e.g., HMP/PAB or PAL, commonly used in Fmoc chemistry (see Figure 3), are simultaneously deprotected and cleaved by TFA. Such cleavage of *t*Bu and Boc groups results in *tert*-butyl cations and *tert*-butyl trifluoroacetate formation (Jaeger et al., 1978a; Jaeger et al., 1978b; Löw et al., 1978a; Löw et al., 1978b; Lundt et al., 1978; Masui et al., 1980). These species are responsible for *tert*-butylation of the indole ring of Trp, the thioether group of Met, and, to a very low degree (0.5 to 1.0%), the 3'-position of Tyr. Modifications can be minimized during TFA cleavage by utilizing effective *tert*-butyl scavengers. An early comprehensive study showed the advantages of 1,2-ethanedithiol (EDT) (Lundt et al., 1978); this thiol has the additional virtue of protecting Trp from oxidation that occurs due to acid-catalyzed ozonolysis (Scoffone et al., 1966). To avoid acid-catalyzed oxidation of Met to its sulfoxide, a thioether scavenger, such as dimethylsulfide, ethylmethylsulfide, or thioanisole, should be added (Guttmann and Boissonnas, 1959; King et al., 1990). TFA deprotection of Cys(Trt) is reversible in the absence of a scavenger, and it can occur readily following TFA cleavage if solutions of crude peptide in TFA are concentrated by rotary evaporation or

Cleavage and side-chain deprotection of peptide resins assembled by Fmoc chemistry is carried out in TFA in the presence of carefully chosen scavengers. The text discussion should be consulted with respect to peptide sequence and potential side reactions. It is recommended that small-scale cleavages (<10 mg peptide resin) be performed and analyzed before proceeding to large-scale cleavages. Standard cleavages of HMP/PAB, Dod, and PAL linkers and simultaneous side-chain deprotection proceed by stirring peptide resin (50 to 200 mg) in 2 mL of the appropriate, *freshly* prepared cleavage cocktail for 1.5 to 2.5 hours at 25 °C. The resin is filtered over a medium fritted glass filter and rinsed with 1 mL of TFA. The combined filtrate and TFA rinse are either (a) precipitated with methyl *t*Bu ether (~50 mL) or (b) dissolved in ~40 mL H_2O and extracted six times with ~40 mL methyl *t*Bu ether. Following (a), the mixture is centrifuged at 3000 rpm for 2 minutes and decanted. The peptide pellet is dispersed with a rubber policeman, washed thoroughly with methyl *t*Bu ether, and dried overnight in a lyophilizer (King et al., 1990; Albericio et al., 1990a). Following (b), the H_2O layer is loaded directly to a semipreparative HPLC column and the peptide is purified.

lyophilization (Photaki et al., 1970; D. Hudson, unpublished results). EDT, triethylsilane, or triisopropylsilane are recommended as efficient scavengers to prevent Trt reattachment to Cys (Pearson et al., 1989); this recommendation extends to prevent reattachment of Tmob as well (Munson et al., 1992). TFA deprotection of His(Bum) liberates formaldehyde, in a similar fashion to HF deprotection of His(Bom) (Gesquière et al., 1992). Cyclization of *N*-terminal Cys residues to a thiazolidine is only partially (60%) inhibited by even complex TFA/scavenger mixtures, such as reagent K (see following discussion).

The indole ring of Trp can be alkylated irreversibly by Mtr and Pmc groups from Arg (Sieber 1987b; Harrison et al., 1989; Riniker and Hartmann, 1990; King et al., 1990), Tmob groups from Asn, Gln, or Cys (Gausepohl et al., 1989b; Sieber and Riniker, 1990), and even by some TFA-labile ester and amide linkers (Atherton et al., 1988a; Riniker and Kamber, 1989; Albericio et al., 1990a; Gesellchen et al., 1990). Cleavage of the Pmc group may also result in *O*-sulfation of Ser, Thr, and Tyr (Riniker and Hartmann, 1990; Jaeger et al., 1992). Two efficient cleavage "cocktails" for Mtr/Pmc/Tmob quenching and preservation of Trp, Tyr, Ser, Thr, and Met integrity are TFA-phenol-thioanisole-EDT-H_2O (82.5:5:5:2.5:5) (reagent K) and TFA-thioanisole-EDT-anisole (90:5:3:2) (reagent R) (Albericio et al., 1990a). Recent studies on Trp preservation during amino acid analysis (Bozzini et al., 1991) has led to the development of reagent K', where EDT is replaced by 1-dodecanethiol (Fields and Fields, unpublished results). H_2O is an essential component of reagents K and K′, but phenol is necessary only with multiple Trp-containing peptides (King et al., 1990). Thioanisole, a soft nucleophile, accelerates TFA deprotection of both Arg(Mtr) and Arg(Pmc). Triethylsilane (4 equiv) in MeOH-TFA (1:9) has been reported to efficiently cleave and scavenge Pmc groups (Chan and Bycroft, 1992). Given a choice for Arg protection, Pmc is preferred because it is more labile and it gives less Trp alkylation during unscavenged TFA cleavage; the recommendation for Pmc is particularly appropriate for sequences containing multiple Arg residues (Green et al., 1988; Harrison et al., 1989; King et al., 1990). The Trt group, instead of Tmob, is suggested for Asn/Gln side-chain protection in Trp-containing peptides, because Trt cations are easier to scavenge (Sieber and Riniker, 1990).

Nonacidolytic Cleavage Methods

Benzyl ester anchoring linkages can be cleaved usefully under nonacidic conditions. An interesting alternative to standard HF cleavages for Boc chemistry is catalytic transfer hydrogenolysis (CTH), which removes Bzl side-chain protecting groups (from Asp, Glu, Ser, Thr, and Tyr) and cleaves benzyl ester anchors to provide a C-terminal carboxyl group (Anwer and Spatola, 1980; Anwer and Spatola, 1983). Peptide resin (1 g) is treated initially with palladium (II) acetate (1 g) in DMF (13 mL) for 2 hours; ammonium formate (1.3 g) is then added, and the reaction proceeds for an additional 2 hours (Anwer and Spatola, 1983). CTH can reduce Trp to octahydrotryptophan (Méry and Calas, 1988). Benzyl ester linkages also may be cleaved by 2-dimethylaminoethanol (DMAE)-DMF (1:1) for 70 hours, with subsequent treatment of the peptide-DMAE ester by DMF-H_2O (1:5) for 2 hours, yielding the peptide acid (Barton et al., 1973). For some applications when peptide amides are required, benzyl ester-type linkages are treated with NH_3 in anhydrous MeOH, 2-propanol, 2,2,2-trifluoroethanol (TFE), or MeOH-DMF for 2 to 4 days at 25 °C (Atherton et al., 1981c; Stewart and Young, 1984; Story and Aldrich, 1992), although these conditions will also convert Asp(OBzl) and Glu(OBzl) residues to Asn and Gln, respectively. Additionally, ethanolamine in DMF or MeOH cleaves benzyl ester linkages (8 to 40 hours, 45 to 60 °C) to provide an ethanolamidated peptide C-terminus (Prasad et al., 1982; Fields et al., 1988; Fields et al., 1989; Prosser et al., 1991). Nucleophilic cleavages of benzyl esters can be accompanied by side reactions, including racemization of the C-terminal residue (Barany and Merrifield, 1979). On the other hand, base cleavages of the NPE and HMFA linkers appear to be quite safe and general. Peptide acids are obtained upon treatment with either piperidine (15 to 20%)-DMF or DBU (0.1 M)-1,4-dioxane, after 5 minutes (for HMFA) to 2 hours (for NPE) at 25 °C (Liu et al., 1990; Albericio et al., 1991b).

Following synthesis, peptide resins should be well dried, and then stored in a desiccator at 4 °C with the N^{α}-terminus protected. As a general practice, peptides should never be stored in the solid state after being lyophilized from moderate or strong acid; deamidation, among other side reactions, may proceed rapidly (see Auxiliary Issues). If stored in solution following purification, peptides should be used for biological or chemical studies as soon as possible. Met-containing peptides oxidize rapidly upon storage in solution, especially when repeated freeze and thawing occurs (Stewart and Young, 1984), while Asn-containing peptides can hydrolyze spontaneously in solution (see Auxiliary Issues).

Palladium-catalyzed peptide resin cleavage is used for the HYCRAM^{TM} linker (Kunz and Dombo, 1988; Guibé et al., 1989). The fully protected peptide resin (0.1 mmol of peptide) is shaken for 6 to 18 hours under N_2 or argon atmosphere in 8 mL of DMSO-THF-0.5 M aqueous HCl (2:2:1) in the presence of 50 to 190 equiv of either NMM (for Boc-peptides) or dimedone (for Fmoc-peptides) as well as 0.015 equiv of the tetrakis(triphenylphosphino)palladium(0) catalyst. The reaction mixture is filtered and washed with DMF, DMF-0.5 M aqueous HCl (1:1), and DMF. The filtrate and washings are combined and evaporated to minimal volume; the peptide is then precipitated with methyl tBu ether (Orpegen, 1990; Lloyd-Williams et al., 1991b). The crude peptide acid (0.7 mmol scale) can be converted to an amide by dissolving in dry DMF (50 mL) at 25 °C, adding NMM (80 μL), cooling to −20 °C, adding isobutylchloroformate (90 μL) and, after 8 minutes, 25% aqueous NH$_4$OH (0.3 mL), and stirring for 2 to 12 hours. The solution is evaporated and the peptide amide dried over P_2O_5 in vacuo (Orpegen, 1990).

Photolytic cleavage at 350 nm under inert (N_2, Ar) atmosphere is used for the ONb, 2-bromopropionyl, and Nonb linkers. The most efficient photolysis of the ONb and Nonb linkers is achieved when peptide resins are swollen with 20 to 25% 2,2,2-trifluoroethanol in either DCM or toluene (Giralt et al., 1986; Kneib-Cordonier et al., 1990; Hammer et al., 1990). Photolytic cleavage yields after 9 to 16 hours range from 45 to 70 percent for relatively small peptides (5 to 9 residues). The 2-bromopropionyl linker is cleaved in DMF with a yield of 70 percent after 72 hours for a tetrapeptide (Wang, 1976).

POST-TRANSLATIONAL MODIFICATIONS AND UNNATURAL STRUCTURES

Peptides of biological interest often include structural elements beyond the 20 genetically encoded amino acids. This section summarizes the best current methods to duplicate by chemical synthesis the post-translational modifications achieved in nature, including the alignment of half-cystine residues in disulfide bonds. This section also covers a set of unnatural structures that are of considerable interest for peptide drug design, namely side-chain to side-chain lactams, and lastly describes the steps needed to elicit good antibody production from synthetic peptides.

Hydroxylated Residues

Hydroxyproline (Hyp) has been incorporated successfully without side-chain protection in both Boc (Felix et al., 1973; Stewart et al., 1974) and Fmoc (Fields et al., 1987; Netzel-Arnett et al., 1991) SPPS. Alternatively, the usual hydroxyl protecting groups Bzl (Cruz et al., 1989) for Boc and tBu (Becker et al., 1989) for Fmoc can be used. Fmoc SPPS of unprotected Hyp-containing peptides can be carried out without affecting the homogeneity of the product (Fields and Noble, 1990).

Hydroxylysine (Hyl) has been incorporated in SPPS as FmocHyl-(Boc,O-Tbdms). This derivative was prepared by protecting the N^ε-amino group of acetyl-Hyl by Boc-N_3, removing the N^α-acetyl group enzymatically with acylase I, adding the Fmoc group, and, finally, blocking the side-chain hydroxyl group with tert-butyldimethylsilyl chloride (Penke et al., 1989).

γ-Carboxyglutamate

Acid-sensitive γ-carboxyglutamate (Gla) residues have been identified in a number of diverse biomolecules, such as prothrombin and the "sleeper" peptide from the venomous fish-hunting cone snail (*Conus geographus*). Fmoc chemistry has been utilized for the efficient SPPS of the 17-residue sleeper peptide, with the five Gla residues incorporated as Fmoc-Gla(OtBu)$_2$ (Rivier et al., 1987). Cleavage and side-chain deprotection of the peptide resin by TFA-DCM (2:3) for 6 hours resulted in no apparent conversion of Gla to Glu.

Phosphorylation

Incorporation of side-chain phosphorylated Ser and Thr by SPPS is especially challenging, because the phosphate group is decomposed by strong acid and lost with base in a β-elimination process (Perich, 1990). Boc-Ser(PO$_3$Ph$_2$) and Boc-Thr(PO$_3$Ph$_2$) have been used, where HF or hydrogenolysis cleaves the peptide resin, and hydrogenolysis removes the phenyl groups from the phosphate (Perich et al., 1986; Arendt et al., 1989). Alternatively, peptide resins that were built up by Fmoc chemistry to include unprotected Ser or Thr side chains may be treated with a suitable phosphorylating reagent, e.g., N,N-diisopropyl-bis(4-chlorobenzyl)phosphoramidite or dibenzylphosphochloridate. The desired phosphorylated peptide is then obtained in solution following simultaneous deprotection and cleavage with TFA in the presence of scavengers (Otvös et al., 1989a; de Bont et al., 1990).

Side-chain phosphorylated Tyr is less susceptible to strong acid decomposition, and it is not at all base-labile. Thus, SPPS has been used to incorporate directly Fmoc-Tyr(PO_3Me_2) (Kitas et al., 1989), Fmoc-Tyr(PO_3Bzl_2) (Kitas et al., 1991), Fmoc-Tyr(PO_3tBu_2) (Perich and Reynolds, 1991), and Boc-Tyr(PO_3^{2-}) (Zardeneta et al., 1990). Syntheses incorporating Fmoc-Tyr(PO_3Bzl_2) use 2% DBU in DMF for N^α-amino deprotection, because piperidine was found to remove the benzyl protecting groups from phosphate (Kitas et al., 1991). TFMSA or TMSBr can be used for peptide resin cleavage and removal of the methyl phosphate groups without O-dephosphorylation (Kitas et al., 1989; Zardeneta et al., 1990), while TFA is used for removal of the benzyl and *tert*-butyl phosphate groups (Kitas et al., 1991).

Sulfation

Gastrin, cholecystokinin, and related hormones contain sulfated Tyr; thus, incorporation of this residue into synthetic peptides is of great interest. Synthesis of Tyr sulfate-containing peptides is difficult, as a result of the substantial acid lability of the sulfate ester; also, most sulfating agents are more reactive toward the hydroxyls of Ser or Thr with respect to the phenol of Tyr. While there is an elegant history of success in solution chemistry (Beacham et al., 1967; Ondetti et al., 1970; Pluscec et al., 1970; Wünsch et al., 1981), this brief discussion focuses on the best SPPS approaches. Side-chain unprotected Tyr can be incorporated by Boc or Fmoc chemistry, and sulfation is carried out while the otherwise fully protected peptide remains anchored to the support, achieved by use of pyridinium acetyl sulfate (Fournier et al., 1989). Base- or acid-promoted deprotection/cleavage follows under conditions that are carefully optimized to avoid or minimize desulfation. Alternatively, sulfated Tyr can be incorporated directly by use of Fmoc-Tyr($SO_3^-Na^+$)-OPfp, Fmoc-Tyr($SO_3^-Na^+$), or Fmoc-Tyr($SO_3^-Ba_{1/2}^{2+}$) in situ with BOP/HOBt (Penke and Rivier, 1987; Penke and Nyerges, 1989; Penke and Nyerges, 1991; Bontems et al., 1992). A brief and carefully optimized acidolytic cleavage/deprotection is then used to minimize desulfation.

Glycosylation

Methodology for site-specific incorporation of carbohydrates during chemical synthesis of peptides has developed rapidly. The mild conditions of Fmoc chemistry are more suited for glycopeptide syntheses than Boc chemistry, because repetitive acid treatments can be detrimental to sugar linkages (Kunz, 1987). Fmoc-Ser, -Thr, -Hyp, and -Asn have all been incorporated successfully with glycosylated side chains. Side-chain glycosylation is performed with glycosyl bromides or glycose-$BF_3 \cdot Et_2O$

for Ser, Thr, and Hyp, and glycosylamines for Asp (to produce a glycosylated Asn). The side-chain glycosyl is usually hydroxyl protected by either the Bzl or acetyl group (Paulsen et al., 1988; Torres et al., 1989; Paulsen et al., 1990; Jansson et al., 1990; de la Torre et al., 1990; Bardají et al., 1991; Biondi et al., 1991), although some SPPS have been successful with no protection of glycosyl hydroxyl groups (Otvös et al., 1989b; Otvös et al., 1990; Filira et al., 1990). Glycosylated residues are incorporated as preformed Pfp esters or in situ with DCC/HOBt (Paulsen et al., 1988; Paulsen et al., 1990; Bardají et al., 1991; Meldal and Jensen, 1990; Filira et al., 1990; Jansson et al., 1990; Otvös et al., 1990; Biondi et al., 1991). These sugars are relatively stable to Fmoc deprotection by piperidine or morpholine (Paulsen et al., 1988; Paulsen et al., 1990; Meldal and Jensen, 1990; Jansson et al., 1990; Filira et al., 1990; Otvös et al., 1990; Bardají et al., 1991; Biondi et al., 1991), brief treatments with TFA for side-chain deprotection and peptide resin cleavage (Paulsen et al., 1988; Filira et al., 1990; Paulsen et al., 1990; Meldal and Jensen, 1990; Jansson et al., 1990; Otvös et al., 1990; Bardají et al., 1991; Biondi et al., 1991), and palladium treatment for peptide resin cleavage from HYCRAM™ (Kunz and Dombo, 1988). Deacetylation and debenzylation are performed with hydrazine-MeOH (4:1) prior to glycopeptide resin cleavage (Kunz, 1987; Bardají et al., 1991).

Disulfide Bond Formation

In the majority of cases, intramolecular disulfides or simple intermolecular homodimers have been formed from purified linear precursors by nonspecific oxidations in dilute solutions. An even number of Cys residues are brought to the free thiol form by removal of the same S-protecting group, following which disulfide formation is mediated by molecular O_2 (from air), potassium ferricyanide [$K_3Fe(CN)_6$], DMSO, or others from a lengthy catalogue of reagents (Hiskey, 1981; Stewart and Young, 1984; Gariépy et al., 1987; McCurdy, 1989; Tam et al., 1991b). Accomplishing the same end, but proceeding by a different mechanism, the polythiol can be treated with a mixture of reduced and oxidized glutathione, which catalyzes the net oxidation by thiol-disulfide exchange reactions (Ahmed et al., 1975; Lin et al., 1988; Pennington et al., 1991). These procedures, which require scrupulous attention to experimental details, have often given the desired disulfide-containing peptide products in acceptable yields. However, even under the best conditions, significant levels of dimeric, oligomeric, or polymeric materials are observed. The nonmonomeric material has usually proved to be difficult to "recycle" by alternating reduction and reoxidation steps.

A more sophisticated approach, which also requires dilute solutions, involves selective pairwise co-oxidations of two designated free or protected sulfhydryl groups. Such reactions are best carried out in intramolecular fashion, because if the paired groups are on separate chains, there is the problem that homodimers form along with the desired heterodimer. If the thiols have already been deblocked, oxidation follows using procedures mentioned earlier. The prototype oxidative deprotections involve I_2 treatments on Cys(Trt) or Cys(Acm) residues (Kamber et al., 1980). These reactions are carried out in neat or mixtures of the solvents TFE, MeOH, 1,1,1,3,3,3-hexafluoroisopropanol, HOAc, DCM, and chloroform and often proceed in modest to high yield; however, side reactions have been observed at Trp residues, resulting in Trp-2'-thioethers (Sieber et al., 1980) and β-3-oxindolylalanine (Casaretto and Nyfeler, 1991). Pairwise oxidative removal of appropriate Acm or Tacm Cys protecting groups with $Tl(Tfa)_3$ or methyltrichlorosilane (in the presence of diphenylsulfoxide) also furnishes the disulfide directly (Fujii et al., 1987; Kiso et al., 1990; Akaji et al., 1991). However, Trp and Met must be side-chain protected during such treatments. As a final example in this category, Cys(Fm) residues form disulfides directly upon treatment with piperidine (Ruiz-Gayo et al., 1988; Ponsati et al., 1990b).

Most general, but also most demanding in terms of the range of selectively removable sulfhydryl protecting groups required, are unsymmetrical directed methods of disulfide bridge formation (Barany and Merrifield, 1979; Hiskey, 1981). For example, Cys(Acm) or Cys(Trt) residues in peptides can be reacted with methoxy- or ethoxycarbonylsulfenyl chloride to form Cys(Scm) or Cys(Sce) residues, respectively, which are attacked by the free thiol of a deprotected Cys residue to form a disulfide bond (Kullmann and Gutte, 1978; Mott et al., 1986; Ten Kortenaar and van Nispen, 1988). Disulfide bonds may also be formed by a free thiol attack of Cys(NBoc-NHBoc) residues, which are prepared by treatment of Cys with azodicarboxylic acid di-tBu ester (Romani et al., 1987). Cys(Npys) residues form disulfides upon reaction with deprotected Cys residues (Bernatowicz et al., 1986). Directed methods are particularly suited for linking two separate chains.

As already alluded to, directed methods require at least two classes of selectively removable Cys protecting groups; the same holds true for experiments aimed at controlled formation of multiple disulfide bridges by sequential pairwise deprotection/co-oxidations. An overriding concern with all such chemical approaches is to develop conditions that avoid scrambling (disulfide exchange). Oxidation by I_2 in TFE allows for selective disulfide bond formation between Cys(Trt) residues in the presence of Cys(Acm) residues; in DMF, I_2 oxidation is preferred between Cys(Acm) residues in the presence of Cys(Trt) residues (Kamber et al., 1980). Direct I_2 oxidation of Cys(Acm) or Cys(Trt) residues is particularly advantageous in that existing disulfides are not exchanged (Barany and Merrifield, 1979; Kamber et al., 1980; Hiskey, 1981; Gray et al., 1984; Atherton et al., 1985b; Ponsati et al., 1990b). Since the Acm group is essentially stable to HF, an Acm/Meb combination of protecting groups facilitates selective disulfide formation in Boc chemistry (Gray et al., 1984; Tam et al., 1991a).

The alternative of carrying out deprotection/oxidation of the Cys residues while the peptide chain remains anchored to a polymeric support is of obvious interest and has received some recent attention. Such an approach takes advantage of *pseudo-dilution,* which is a kinetic phenomenon expected to favor facile intramolecular processes and thereby minimize dimeric and oligomeric by-products (Barany and Merrifield, 1979). Disulfide bond formation on peptide resins has been demonstrated by $K_3Fe(CN)_6$, air, dithiobis(2-nitrobenzoic acid), or diiodoethane oxidation of free sulfhydryls, direct deprotection/oxidation of Cys(Acm) residues by $Tl(Tfa)_3$ or I_2, direct conversion of Cys(Fm) residues by piperidine, and nucleophilic attack by a free sulfhydryl on either Cys(Npys) or Cys(Scm) (Gray et al., 1984; Mott et al., 1986; Buchta et al., 1986; Eritja et al., 1987; Ploux et al., 1987; Ten Kortenaar and van Nispen, 1988; García-Echeverría et al., 1989; Albericio et al., 1991a, and references to earlier work cited in these papers). The most generally applicable and efficient of these methods is direct conversion of Cys(Acm) or Cys(Trt) residues by I_2 (10 equiv in DMF), Cys(Acm) residues by $Tl(Tfa)_3$ (1.5 equiv in DMF) (Albericio et al., 1991a), and Cys(Fm) residues by (a) piperidine-DMF (1:1) for 3 hours at 25 °C (Ponsati et al., 1990b; Albericio et al., 1991a) or (b) piperidine-DMF-2-mercaptoethanol (10:10:0.7) treatment for 1 hour at 25 °C, followed by air oxidation in pH 8.0 DMF for 1 hour at 25 °C (Albericio et al., 1991a). The best solid-phase yields were at least as good and, in some cases, better than the results from corresponding solution oxidations.

Side-Chain Lactams

Intrachain lactams are formed between the side chains of Lys or Orn and Asp or Glu to conformationally restrain synthetic peptides, with the goal of increasing biological potency and/or specificity. The residues used to form intrachain lactams must be selectively side-chain deprotected, while all side-chain protecting groups of other residues remain intact. Selective deprotection is best achieved by using orthogonal side-chain protection, such as Fmoc and Fm for Lys/Orn and Asp/Glu, respectively, in combination with a Boc/Bzl strategy (Felix et al., 1988a; Felix et al., 1988b; Hruby et al., 1990). A more complicated but equally efficient approach is to use side-chain protection based on graduated acid-lability (Schiller et al., 1985; Sugg et al., 1988; Hruby et al., 1990). Cyclization is carried out most efficiently with BOP (3 to 6 equiv, 2 hours, 20 °C) in the presence of DIEA (6 to 7.5 equiv) while the peptide is still attached to the resin (Felix et al., 1988a; Plaué, 1990), taking advantage of the pseudo-dilution phenomenon discussed in the previous section.

Peptide Antigens

For the production of antipeptide antibodies, the peptide must be attached to a carrier. The simplest, although not necessarily most effective, way to accomplish this goal is to make direct use of peptide resins in which the side chains have been freed but where the anchoring linkage is stable to the appropriate deprotection conditions. Polyamide-type and polyethylene glycol-polystyrene resins have been applied according to this approach (Chanh et al., 1986; Kennedy et al., 1987; Goddard et al., 1988; Fischer et al., 1989; Bayer, 1991). Peptides may also be synthesized on a scaffold that, following cleavage and deprotection, is used directly for immunization. This scaffold, consisting of branched Lys residues, is referred to as a "multiple antigen peptide" system (MAP) (Tam, 1988; Tam and Lu, 1990). MAPs may be prepared by either Boc (Tam, 1988) or Fmoc (Pessi et al., 1990; Drijfhout and Bloemhoff, 1991; Biancalana et al., 1991) chemistry.

The more traditional (and still most common) methods for preparing peptide antigens start with free synthetic peptides that have been cleaved from the support and deprotected. In one variation, the peptide is conjugated to a protein carrier, e.g., bovine serum albumin, by means of a water soluble carbodiimide (see Deen et al., 1990, for a recent example). Alternatively, the peptide can be extended at the C- or N-terminus with a Cys residue, which is subsequently used to form a disulfide bridge with a free sulfhydryl on the carrier or is attached to the carrier using a heterobifunctional cross-linking reagent. Cystine formation is best achieved by the direct attack of a carrier protein thiol onto a peptide Cys(Npys) residue. Thiol groups are introduced on the carrier protein either by reduction to form free Cys (Albericio et al., 1989b; Ponsati et al., 1989) or by functionalization of Lys using S-acetylmercaptosuccinic anhydride (Drijfhout et al., 1988). Peptide-carrier conjugation is quantitated by monitoring the liberation of the Npys group at 329 nm (Drijfhout et al., 1988). For peptides synthesized by Fmoc chemistry, Boc-Cys(Npys) is incorporated as the N-terminal residue, thus avoiding additional piperidine treatments that would remove the Npys group (Albericio et al., 1989b). A heterobifunctional reagent, such as m-maleimidobenzoyl-N-hydroxysuccinimide ester (MBS) (Lerner et al., 1981), can also be used to link sulfhydryl-containing synthetic peptides to lysine-containing carrier proteins such as keyhole limpet hemocyanin and bovine serum albumin. Maleimide activated carrier protein is available commercially (Pierce Chemical Co.).

AUXILIARY ISSUES

This final section of the chapter covers a variety of practical considerations that researchers in SPPS should be aware of. Included are some potential side reactions that do not fit neatly with categories covered earlier, and ways to mitigate the extent of their occurrence. A logical culmination of expertise in SPPS is the successful preparation of long sequences, and we outline current achievements and possible ways to improve on them in the future.

Diketopiperazine Formation

The free N^α-amino group of an anchored dipeptide is poised for a base-catalyzed intramolecular attack on the C-terminal carbonyl (Gisin and Merrifield, 1972; Barany and Merrifield, 1979; Pedroso et al., 1986). Base deprotection (Fmoc) or neutralization (Boc) can thus release a cyclic diketopiperazine while a hydroxymethyl-handle leaving group remains on the resin. With residues that can form *cis* peptide bonds, e.g., Gly, Pro, N-methylamino acids, or D-amino acids, in either the first or second position of the (C → N) synthesis, diketopiperazine formation can be substantial (Albericio and Barany, 1985; Pedroso et al., 1986; Gairi et al., 1990). For most other sequences, the problem can be adequately controlled. In Boc SPPS, the level of diketopiperazine formation can be suppressed either by removing the Boc group with HCl and coupling the NMM salt of the third Boc amino acid without neutralization (Suzuki et al., 1975) or else by deprotecting the Boc group with TFA and coupling the third Boc amino acid in situ using BOP, DIEA, and HOBt without neutralization (Gairi et al., 1990). For susceptible sequences being addressed by Fmoc chemistry, the use of piperidine-DMF (1:1) deprotection for 5 minutes (Pedroso et al., 1986) or deprotection for 2 minutes with a 0.1 M solution in DMF of tetrabutylammonium fluoride ("quenched" by MeOH) (Ueki and Amemiya, 1987) has been recommended to minimize cyclization. Alternatively, the second and third amino acids may be coupled as a preformed N^α-protected-dipeptide, avoiding the diketopiperazine-inducing deprotection/neutralization at the second amino acid.

For continuous-flow Fmoc SPPS, diketopiperazine formation is suppressed by deprotecting for 1.5 minutes with piperidine-DMF (1:4) at an increased flow rate (15 mL/min), washing for 3 minutes with DMF at the same flow rate, and coupling the third Fmoc amino acid in situ with BOP, NMM, and HOBt in DMF (MilliGen/Biosearch, 1990). For batchwise SPPS, rapid (a maximum of 5 minutes) treatments by piperidine-DMF (1:1) should be used, followed by DMF washes and then in situ acylations mediated by BOP or HBTU (Pedroso et al., 1986).

Capping

Some workers choose to "cap" unreacted chains, thereby substituting a family of terminated peptides for a family of deletion peptides. In either case, these by-products must ultimately be separated from the desired product. Intentional termination of chains may be carried out when there is an indication of unreacted sites. In the simplest case, capping is carried out with reactive acetylating agents, such as acetic anhydride (Ac$_2$O) or N-acetylimidazole in the presence of tertiary base (Bayer et al., 1970; Stewart and Young, 1984). An alternative capping reagent is 2-sulfobenzoic acid cyclic anhydride (OSBA). Application of OSBA/tribenzylamine results in a negatively charged amino terminus; the desired product with the positively charged amino terminus may then be isolated after purification by ion-exchange chromatography (Drijfhout and Bloemhoff, 1988). In a reciprocal strategy, chains capped by acetylation are separated by modifying the N-terminus of the desired peptide with Fmoc-derivatives. These derivatives include the 9-(2-sulfo)fluorenylmethyloxycarbonyl (Sulfmoc) (Merrifield and Bach, 1978) or 9-(hydroxymethyl)fluorene-4-carboxylate group, where the carboxylate at the 4-position is in turn derivatized with Lys, Glu, or 2-aminodecanoic acid (Ball et al., 1990; Ball et al., 1991). After ion-exchange or reverse-phase purification, the modified Fmoc group is removed by base.

Deletions

The standard explanation for deletion peptides relates to incomplete couplings, which can usually be diagnosed by qualitative or quantitative monitoring tests (see Monitoring). In contrast, a chemically plausible side reaction that suggests a different reason for deletion peptides has been elucidated (Kent, 1983). Resin-bound aldehyde groups can form a Schiff's base with deprotected amino groups of the peptide chain. Those amino groups that are involved in a Schiff's base are prevented from acylation by the next incoming protected amino acid. The blockage is not permanent, i.e., terminated peptides are not formed. Rather, ready amine exchange of the Schiff's base renders a *different* set of amino groups temporarily inaccessible at a subsequent coupling, and thus deletion peptides result. The side reaction is *not* minimized by capping steps; in fact, amines blocked as Schiff's bases contribute to negative ninhydrin tests. The formation of aldehyde sites on polystyrene resins can be minimized by using a strong acid Friedel-Crafts catalyst during the original functionalization step. In any case, it is crucial to quantitate aldehyde concentrations of prepared or purchased resins (Kent, 1983).

Acid-Sensitive Side Chains and Bonds

Trp is quite sensitive to acid conditions, undergoing reactions with carbonium ions and molecular oxygen (see Cleavage). Synthesis of Trp-containing peptides is thus best approached by means of Fmoc chemistry, where acidolysis is kept at a minimum. Gramicidins A, B, and C (where either 3 or 4 of the 15 residues are Trp) have been synthesized efficiently by Fmoc chemistry; acid was avoided entirely throughout the synthesis and final nucleophilic cleavage was achieved by ethanolamine (Fields et al., 1988; Fields et al., 1989). Efficient Fmoc SPPS of indolicidin (which contains 5 Trp out of 13 residues) used an optimized TFA-scavenger mixture (reagent K) to prevent modification of Trp during acidolytic cleavage and side-chain deprotection (King et al., 1990; Selsted et al., 1992). Fmoc chemistry also has been suggested for incorporation of ^{2}H-labeled amino acids, because the repetitive acidolyses of Boc chemistry can exchange out the ^{2}H label (Fields and Noble, 1990; Prosser et al., 1991). Finally, Fmoc chemistry may be the better choice for the synthesis of peptides containing the acid-labile Asp-Pro bond. SPPS of baboon β-chorionic gonadotropin 109-145, which contains 2 Asp-Pro bonds, was reported to be successful by Fmoc chemistry only, because Boc chemistry resulted in acid-promoted cleavage of the Asp-Pro bonds (Wu et al., 1988).

Imide Formation

Asn residues can cyclize to form a succinimide, which can yield both α-and β-Asp peptides. Imide formation is largely sequence-dependent, with Asn-Gly showing the greatest tendency to rearrange (Stephenson and Clarke, 1989). Succinimide formation can be significant for peptides stored in solution (Stephenson and Clarke, 1989; Patel and Borchardt, 1990a; Patel and Borchardt, 1990b). In addition, peptides containing Asn (and even Gln) stored in the solid state with residual acid can undergo deamidation (Ten Kortenaar et al., 1990). Therefore, Asn- and Gln-containing peptides should never be stored in a solid form with residual acid present, and samples stored in solution should be monitored carefully for deamidation and decomposition.

Solvent Preferences

Effective solvation of the peptide resin is perhaps the most crucial condition for efficient chain assembly. Under proper solvent conditions, there is no decrease in synthetic efficiency up to 60 amino acid residues in Boc SPPS (Sarin et al., 1984). The ability of the peptide resin to swell increases with increasing peptide length due to a net decrease in free energy from solvation of the linear peptide chains (Sarin et al., 1980). Therefore, there is no theoretical upper limit to efficient amino acid couplings, provided that proper solvation conditions exist (Pickup et al., 1990). In practice, obtaining these conditions is not always straightforward. "Difficult couplings" during SPPS have been attributed to poor solvation of the growing chain by DCM. Infrared and NMR spectroscopies have shown that intermolecular β-sheet aggregates are responsible for lowering coupling efficiencies (Live and Kent, 1983; Mutter et al., 1985; Ludwick et al., 1986). A scale of β-sheet structure-stabilizing potential has been developed for Boc amino acid derivatives (Narita and Kojima, 1989). Enhanced coupling efficiencies are seen upon the addition of polar solvents, such as DMF, TFE, and NMP (Yamashiro et al., 1976; Live and Kent, 1983; Geiser et al., 1988; Narita et al., 1989; Fields et al., 1990; Fields and Fields, 1991). It has been suggested that chaotropic salts may be added to organic solvents in order to disrupt β-sheet aggregates (Stewart and Klis, 1990; Thaler et al., 1991).

Aggregation also occurs in regions of apolar side-chain protecting groups, sometimes resulting in a collapsed gel structure (Atherton et al., 1980; Atherton and Sheppard, 1985). In cases where aggregation occurs due to apolar side-chain protecting groups, increased solvent polarity may not be sufficient to disrupt the aggregate. A relatively unstudied problem of Fmoc chemistry is that the lack of polar side-chain protecting groups could, during the course of an extended peptide synthesis, inhibit proper solvation of the peptide resin (Atherton et al., 1980; Fields and Fields, 1991). To alleviate this problem, the use of solvent mixtures containing both a polar and nonpolar component, such as THF-NMP (7:13) or TFE-DCM (1:4), is recommended (Fields and Fields, 1991). The partial substitution or complete replacement of *t*Bu-based side-chain protecting groups for carboxyl, hydroxyl, and amino side chains by more polar groups would also aid peptide resin solvation (Atherton et al., 1980; Fields and Fields, 1991).

Long Syntheses (>50 Residues)

Many impressive long-chain syntheses (>50 residues), including ribonuclease A (124 residues) (Gutte and Merrifield, 1971), human parathyroid hormone (84 residues) (Fairwell et al., 1983), interleukin-3 (140 residues) (Clark-Lewis et al., 1986), HIV-1 aspartyl protease (99 residues) (Schneider and Kent, 1988; Nutt et al., 1988), HIV-1 vpr protein (95 residues) (Gras-Masse et al., 1990), and insulin-like growth factor (70 residues) (Bagley et al., 1990), have been carried out using Boc methodology. There have also been recent successful long-chain syntheses by Fmoc chemistry, including HIV-1 Tat protein (86 residues) (Cook et al., 1989; Chun et al., 1990), preprocecropin A (64 residues) (Pipkorn and Bernath, 1990), ubiquitin (76 residues) (Ramage et al., 1989; Ogunjobi and Ramage, 1990), yeast actin-binding protein 539-588 (50 residues) (King et al., 1990), pancreastatin (52 residues) (Funakoshi et al., 1988), and human β-chorionic gonadotropin 1-74 (Wu et al., 1989). Both chemistries appear susceptible to the same difficult couplings (Meister and Kent, 1983; J. Young et al., 1990; van Woerkom and van Nispen, 1991), and side-by-side syntheses for moderate-length chains (~30 residues) are comparable (Atherton et al., 1983; Wade et al., 1986). However, there are two additional considerations when using Fmoc, rather than Boc, chemistry for long-chain syntheses. First, the efficient solvation of hydrophobic side-chain protecting groups used in conjunction with Fmoc chemistry, which was discussed previously, can become more critical for extended syntheses (Fields and Fields, 1991). Second, deprotection of the Fmoc group can proceed slowly in certain sequences (Atherton and Sheppard, 1985; Larsen et al., 1991). By monitoring deprotection as the synthesis proceeds, one can extend base deprotection times and/or alter solvation conditions as necessary (Ogunjobi and Ramage, 1990).

Segment Condensation

The advantages of segment condensation procedures for the synthesis of large peptides have been well described (Barany and Merrifield, 1979; Kaiser et al., 1989; Kneib-Cordonier et al., 1990), but to date there are relatively few examples for polymer-supported procedures. A significant aspect of the problem involves ready access to pure partially protected peptide segments, which are needed as building blocks. The application of solid-phase synthesis to prepare the requisite intermediates depends on several levels of selectively cleavable protecting groups and anchoring linkages. Combination of the Boc/Bzl strategy with the 4-nitrobenzophenone oxime resin (DeGrado and Kaiser, 1982; Kaiser et al., 1989; Landsbury et al., 1989; Sasaki and Kaiser, 1990), base-labile linkers (Liu et al., 1990; Albericio et al., 1991b), palladium-labile linkers (Kunz and Dombo, 1988; Guibé et al., 1989; Lloyd-Williams et al., 1991b) or photolabile linkers (Rich and Gurwara, 1975; Albericio et al., 1987b; Lloyd-Williams et al., 1991a), and of the Fmoc/tBu strategy with dilute acid-labile linkers (Mergler et al., 1988b; Barlos et al., 1989; Atherton et al., 1990; Albericio and Barany, 1991; Barlos et al., 1991b) or photolabile linkers (Kneib-Cordonier et al., 1990) has proved successful for the generation of N^{α}-amino and side-chain protected segments with free C^{α}-carboxyl groups. Methods for subsequent solubilization and purification of the protected segments are nontrivial (Atherton et al., 1990; Lloyd-Williams et al., 1991a) and beyond the scope of this review.

In recent years, solid-phase assembly of protected segments has proved successful for a 44-residue model of apolipoprotein A-1 (Nakagawa et al., 1985), human cardiodilatin 99-126 (Nokihara et al., 1989), human gastrin-I (17 residues) (Kneib-Cordonier et al., 1990), *Androctonus australis Hector* toxin II (64 residues) (Grandas et al., 1989b), λ-Cro DNA binding protein (66 residues) (Atherton et al., 1990), and prothymosin α (109 residues) (Barlos et al., 1991b). One-percent 1,3-divinylbenzene cross-linked polystyrene and polyamide resins have been shown to be suitable supports for solid-phase segment condensations

(Albericio et al., 1989c). Individual rates for coupling segments are much lower than for activated amino acid species by stepwise synthesis, and there is always a risk of racemization at the C-terminus of each segment. Careful attention to synthetic design and execution may minimize these problems.

SUMMARY

The solid-phase method has made the synthesis of peptides widely accessible. With the increased sophistication of commercial automated instrumentation, the appeal is ever broadening. The majority of solid-phase peptide syntheses of less than 50 residues can be performed with high efficiencies by either Boc or Fmoc chemistry. The Bzl-based side-chain protecting group strategy is used routinely for Boc chemistry and the tBu-based side-chain protection strategy is usually used for Fmoc chemistry. Manufacturers generally suggest specific chemistry packages with their instruments, which represent some variation of either of these two strategies.

It is important not to lose sight of the fact that each synthetic procedure has limitations and that even in the hands of highly experienced workers, certain sequences defy facile preparation. Common residue-specific side reactions that may lead to failed syntheses include (a) dehydration of Asn and Gln to the respective nitriles (see Protection Schemes, In Situ Reagents, and Active Esters), (b) racemization of His (see Protection Schemes), (c) aspartimide formation from Asp (see Protection Schemes), (d) alkylation of Trp and Glu (see Cleavage), and (e) acidolysis of Asp-Pro bonds (see Auxiliary Issues). Additional sources of synthetic problems are sterically hindered couplings and diketopiperazine-forming sequences (see Auxiliary Issues). In all of the aforementioned examples, deleterious side reactions or other difficulties can be minimized somewhat by careful examination of a peptide sequence prior to synthesis. Appropriate precautions as outlined in this chapter (alternative side-chain protecting groups, use of additional reagents during coupling or cleavage, etc.) can be taken.

The maturation of high-performance liquid chromatography (HPLC) has been a major boon to modern peptide synthesis, because the resolving power of this technique facilitates removal of many of the systematic low-level by-products that accrue during chain assembly and upon cleavage. Nowadays, the homogeneity of synthetic materials should be checked by at least two chromatographic or electrophoretic techniques, e.g., reverse-phase and ion-exchange HPLC and capillary zone electrophoresis. Also, determination of a molecular ion by fast atom bombardment mass spectrometry (FABMS) or a related mild ionization method is almost *de rigueur* for proof of structure. Synthetic peptides must be checked routinely for the proper amino acid composition, and, in some cases, sequencing data are helpful. Spectroscopic measurements, particularly through the use of one-and two-dimensional nuclear magnetic resonance (NMR), at the least provide insights on structure and purity, and they can often give conformational information as well.

Improvements in the chemistry of SPPS continue apace. This chapter has touched on most of the key issues and discussed the recent status for each of them. There is every reason to be optimistic that peptide synthesis will continued to play an important role in the elucidation of biological processes.

ACKNOWLEDGMENTS

This article is dedicated to Professors Miklos Bodanzsky and R.B. Merrifield in recognition of their continued high standards in the development of peptide chemistry. We wish to thank Drs. Fernando Albericio, Cynthia Fields, Heinrich Gausepohl, Robert Hammer, Derek Hudson, Stephen Kent, Ken Otteson, and Robert Van Abel for helpful discussions and unpublished results, and Abimed, Applied Biosystems, Inc., Milli-Gen/Biosearch, and Orpegen for technical notes. Work in the authors' laboratories is supported by NIH (GM 28934, 42722, and 43522) and The Minnesota Medical Foundation (MMF) (Grant CRF-152-91).

ABBREVIATIONS

Abbreviations used for amino acids and the designations of peptides follow the rules of the IUPAC-IUB Commission of Biochemical Nomenclature in J. Biol. Chem. 247: 977-983 (1972). The following additional abbreviations are used: AA, amino acid; Ac$_4$βGal, 2,3,4,6-tetra-O-acetyl-β-D-galactopyranosyl; Acm, acetamidomethyl; Ada, adamantyl; Al, allyl; Alloc, allyloxycarbonyl; Boc, *tert*-butyloxycarbonyl; Boc-ON, 2-*tert*-butyloxycarbonyloximino-2-phenylacetonitrile; Bom, benzyloxymethyl; BOP, benzotriazolyl *N*-oxytris(dimethylamino)phosphonium hexafluorophosphate; 2-BrZ, 2-bromobenzyloxycarbonyl; Bum, *tert*-butoxymethyl; Bzl, benzyl; cHex, cyclohexyl; Cs, cesium salt; 2,6-Cl2Bzl, 2,6-dichlorobenzyl; DBU, 1,8-diazabicyclo[5.4.0]undec-7-ene; DCBC, 2,6-dichlorobenzoyl chloride; DCC, *N,N*'-dicyclohexylcarbodiimide; DCE, 1,2-dichloroethane; DCM, dichloromethane (methylene chloride); DIEA, *N,N*-diisopropylethylamine; DIPCDI, *N,N*'-diisopropylcarbodiimide; DMA, *N,N*-dimethylacetamide; DMAP, 4-dimethylaminopyridine; DMF, *N,N*-dimethylformamide; Dnp, 2,4-dinitrophenyl; Dod, 4-(4'-methoxybenzhydryl)phenoxyacetic acid; EDT, 1,2-ethanedithiol; Et$_3$N, triethylamine; Fm, 9-fluorenylmethyl; Fmoc, 9-fluorenylmethyloxycarbonyl; Fmoc-OSu, fluorenylmethyl succinimidyl carbonate; HAL, 5-(4-hydroxymethyl-3,5-dimethoxyphenoxy)valeric acid; HBTU, 2-(1H-benzotriazol-1-yl)-1,1,3,3-tetramethyl uronium hexafluorophosphate; HF, hydrogen fluoride; HMFA, 9-(hydroxymethyl)-2-fluorene-acetic acid; HMP, 4-hydroxymethylphenoxy; HOAc, acetic acid; HOBt, 1-hydroxybenzotriazole; HOPip, *N*-hydroxypiperidine; HPLC, high performance liquid chromatography; Hpp, 1-(4-nitrophenyl)-2-pyrazolin-5-one; HYCRAM^{TM}, hydroxycrotonylaminomethyl; MBHA, 4-methylbenzhydrylamine (resin); Meb, 4-methylbenzyl; MeIm, 1-methylimidazole; MeOH, methanol; MMA, *N*-methylmercaptoacetamide; Mob, 4-methoxybenzyl; MSNT, 2,4,6-mesitylene-sulfonyl-3-nitro-1,2,4-triazolide; Mtr, 4-methoxy-2,3,6-trimethylbenzenesulfonyl; Mts, mesitylene-2-sulfonyl; NCA, *N*-carboxyanhydride; NMM, *N*-methylmorpholine; NMP, *N*-methylpyrrolidone; Nonb, 3-nitro-4-aminomethylbenzoic acid; NPE, 4-(2-hydroxyethyl)-3-nitrobenzoic acid; Npp, 3-methyl-1-(4-nitrophenyl)-2-pyrazolin-5-one; N.R., not reported; Nvoc, 6-nitroveratryloxycarbonyl (4,5-dimethoxy-2-nitroben-zyloxycarbonyl); ODhbt, 1-oxo-2-hydroxydihydrobenzotriazine; ONb, 2-nitrobenzyl ester; ONo, 2-nitrophenyl; ONp, 4-nitrophenyl; OPfp, pentafluorophenyl; Orn, ornithine; OSu, *N*-hydroxysuccinimidyl; OTDO, 2,5-diphenyl-2,3-dihydro-3-oxo-4-hydroxythiophene dioxide; PAB, 4-alkoxybenzyl; PAL, 5-(4-aminomethyl-3,5-dimethoxyphenoxy)valeric acid; PAM, 4-hydroxymethylphenylacetic acid; PEG, polyethylene glycol; Pmc, 2,2,5,7,8-pentamethylchroman-6-sulfonyl; Pnp, 3-phenyl-1-(4-nitrophenyl)-2-pyrazolin-5-one; PSA, preformed symmetrical anhydride; PyBOP, benzotriazole-yl *N*-oxy-tris-pyrrolidinophosphonium hexafluorophosphate; SASRIN^{TM}, 2-methoxy- 4-alkoxybenzyl alcohol; Scm, *S*-carboxymethylsulfenyl; SPPS, solid-phase peptide synthesis; S*t*Bu, *tert*-butylsulfenyl; Tacm, trimethylacetamidomethyl; *t*Bu, *tert*-butyl; TFA, trifluoroacetic acid; TFE, 2,2,2-trifluoroethanol; TFMSA, trifluoromethanesulfonic acid; Tl(Tfa)3, thallium (III) trifluoroacetate; Tmob, 2,4,6-trimethoxybenzyl; TMSBr, trimethylsilyl bromide; TMSOTf, trimethylsilyl trifluoromethanesulfonate; Tos, 4-toluenesulfonyl; Trt, triphenylmethyl; XAL, 5-(9-aminoxanthen-2-oxy)valeric acid; Xan, 9-xanthenyl; Z, benzyloxycarbonyl. Amino acid symbols denote the L-configuration where applicable, unless indicated otherwise.

REFERENCES

Adamson, J.G., Blaskovich, M.A., Groenevelt, H., and Lajoie, G.A. 1991. Simple and convenient synthesis of *tert*-butyl ethers of Fmoc-serine, Fmoc-threonine, and Fmoc-tyrosine. J. Org. Chem. 56:3447-3449.

Ahmed, A.K., Schaffer, S.W., and Wetlaufer, D.B. 1975. Nonenzymic reactivation of reduced bovine pancreatic ribonuclease by air oxidation and by glutathione oxidoreduction buffers. J. Biol. Chem. 250:8477-8482.

Akaji, K., Fujii, N., Tokunaga, F., Miyata, T., Iwanaga, S., and Yajima, H. 1989. Studies on peptides CLXVIII: Syntheses of three peptides isolated from horseshoe crab hemocytes, tachyplesin I, tachyplesin II, and polyphemusin I. Chem. Pharm. Bull. 37:2661-2664.

Akaji, K., Tanaka, H., Itoh, H., Imai, J., Fujiwara, Y., Kimura, T., and Kiso, Y. 1990a. Fluoren-9-ylmethyloxycarbonyl (Fmoc) amino acid chloride as an efficient reagent for anchoring Fmoc amino acid to 4-alkoxybenzyl alcohol resin. Chem. Pharm. Bull. 38:3471-3472.

Akaji, K., Yoshida, M., Tatsumi, T., Kimura, T., Fujiwara, Y., and Kiso, Y. 1990b. Tetrafluoroboric acid as a useful deprotecting reagent in Fmoc-based solid-phase peptide synthesis. J. Chem. Soc. Chem. Commun. 288-290.

Akaji, K., Tatsumi, T., Yoshida, M., Kimura, T., Fujiwara, Y., and Kiso, Y. 1991. Synthesis of cystine-peptide by a new disulphide bond-forming reaction using the silyl chloride-sulphoxide system. J. Chem. Soc. Chem. Commun. 167-168.

Albericio, F., and Barany, G. 1984. Application of N,N-dimethylformamide dineopentyl acetal for efficient anchoring N^α-9-fluorenylmethyloxycarbonylamino acids as p-alkoxybenzyl esters in solid-phase peptide synthesis. Int. J. Pept. Protein Res. 23:342-349.

Albericio, F., and Barany, G. 1985. Improved approach for anchoring N^α-9-fluorenylmethyloxycarbonylamino acids as p-alkoxybenzyl esters in solid-phase peptide synthesis. Int. J. Pept. Protein Res. 26:92-97.

Albericio, F. and Barany, G. 1987a. Mild, orthogonal solid-phase peptide synthesis: Use of N^α-dithiasuccinoyl (Dts) amino acids, and N-(iso-propyl-dithio)carbonylproline, together with p-alkoxybenzyl ester anchoring linkages. Int. J. Pept. Protein Res. 30:177-205.

Albericio, F., and Barany, G. 1987b. Acid-labile anchoring linkage for solid-phase synthesis of C-terminal peptide amides under mild conditions. Int. J. Pept. Protein Res. 30:206-216.

Albericio, F., and Barany, G. 1991. Hypersensitive acid-labile (HAL) tris(alkoxy)benzyl ester anchoring for solid-phase synthesis of protected peptide segments. Tetrahedron Lett 32:1015-1018.

Albericio, F., Grandas, A., Porta, A., Pedroso, E., and Giralt, E. 1987a. One-pot synthesis of S-acetamidomethyl-N-fluorenylmethoxycarbonyl-L-cysteine (Fmoc-Cys(Acm)-OH). Synthesis 271-272.

Albericio, F., Nicolás, E., Josa, J., Grandas, A., Pedroso, E., Giralt, E., Granier, C., and van Rietschoten, J. 1987b. Convergent solid phase peptide synthesis V: Synthesis of the 1-4, 32-34, and 53-59 protected segments of the toxin II of $Androctonus$ $australis$ Hector. Tetrahedron 43:5961-5971.

Albericio, F., Ruiz-Gayo, M., Pedroso, E., and Giralt, E. 1989a. Use of polystyrene-1% divinylbenzene and Kel-F-g-styrene for the simultaneous synthesis of peptides. Reactive Polymers 10:259-268.

Albericio, F., Andreu, D., Giralt, E., Navalpotro, C., Pedroso, E., Ponsati, B., and Ruiz-Gayo, M. 1989b. Use of the Npys thiol protection in solid phase peptide synthesis. Int. J. Pept. Protein Res. 34:124-128.

Albericio, F., Pons, M., Pedroso, E., and Giralt, E. 1989c. Comparative study of supports for solid-phase coupling of protected-peptide segments. J. Org. Chem. 54:360-366.

Albericio, F., Kneib-Cordonier, N., Biancalana, S., Gera, L., Masada, R.I., Hudson, D., and Barany, G. 1990a. Preparation and application of the 5-(4-(9-fluorenylmethyloxycarbonyl)aminomethyl-3,5-dimethoxyphenoxy)valeric acid (PAL) handle for the solid-phase synthesis of C-terminal peptide amides under mild conditions. J. Org. Chem. 55:3730-3743.

Albericio, F., Van Abel, R., and Barany, G. 1990b. Solid-phase synthesis of peptides with C-terminal asparagine or glutamine. Int. J. Pept. Protein Res. 35:284-286.

Albericio, F., Nicolás, E., Rizo, J., Ruiz-Gayo, M., Pedroso, E., and Giralt, E. 1990c. Convenient syntheses of fluorenylmethyl-based side chain derivatives of glutamic and aspartic acids, lysine, and cysteine. Synthesis:119-122.

Albericio, F., Hammer, R.P., García-Echeverría, C., Molins, M.A., Chang, J.L., Munson, M.C., Pons, M., Giralt, E., and Barany, G. 1991a. Cyclization of disulfide-containing peptides in solid-phase synthesis. Int. J. Pept. Protein Res. 37:402-413.

Albericio, F., Giralt, E., and Eritja, R. 1991b. NPE-resin, a new approach to the solid-phase synthesis of protected peptides and oligonucleotides II: Synthesis of protected peptides. Tetrahedron Lett. 32:1515-1518.

Al-Obeidi, F., Sanderson, D.G., and Hruby, V.J. 1990. Synthesis of β- and γ-fluorenylmethyl esters of respectively N^{α}-Boc-L-aspartic acid and N^{α}-Boc-L-glutamic acid. Int. J. Pept. Protein Res. 35:215-218.

Ambrosius, D., Casaretto, M., Gerardy-Schahn, R., Saunders, D., Brandenburg, D., and Zahn, H. 1989. Peptide analogues of the anaphylatoxin C3a; synthesis and properties. Biol. Chem. HoppeSeyler 370:217-227.

AminoTech. 1991. Biochemicals and Reagents for Peptide Synthesis. Amino-Tech Catalogue, AminoTech, Nepean, Ontario.

Anwer, M.K., and Spatola, A.F. 1980. An advantageous method for the rapid removal of hydrogenolysable protecting groups under ambient conditions; synthesis of leucine-enkephalin. Synthesis, 929-932.

Anwer, M.K., and Spatola, A.F. 1983. Quantitative removal of a pentadecapeptide ACTH fragment analogue from a Merrifield resin using ammonium formate catalytic transfer hydrogenation: Synthesis of $[Asp^{25},Ala^{26},Gly^{27},Gln^{30}]$-ACTH-(25-39)-OH. J. Am. Chem. Soc. 48:3503-3507.

Applied Biosystems, Inc. 1989a. Removal of 2,4-dinitrophenyl (Dnp) protection from peptides synthesized with Boc-His(Dnp). Peptide Synthesizer User Bulletin 28, Applied Biosystems, Inc., Foster City, Calif.

Applied Biosystems, Inc. 1989b. Use of the ninhydrin reaction to monitor Fmoc solid-phase peptide synthesis. Peptide Synthesizer User Bulletin 29, Applied Biosystems, Inc., Foster City, Calif.

Applied Biosystems, Inc. 1989c. Model 431A Peptide Synthesizer User's Manual, Applied Biosystems, Inc., Foster City, Calif.

Arad, O., and Houghten, R.A. 1990. An evaluation of the advantages and effectiveness of picric acid monitoring during solid phase peptide synthesis. Peptide Res. 3:42-50.

Arendt, A., and Hargrave, P.A. Optimization of peptide synthesis on polyethylene rods. In Twelfth American Peptide Symposium Program and Abstracts, Massachusetts Institute of Technology, Cambridge, Mass., 1991, p. 269.

Arendt, A., Palczewski, K., Moore, W.T., Caprioli, R.M., McDowell, J.H., and Hargrave, P.A. 1989. Synthesis of phosphopeptides containing O-phosphoserine or O-phosphothreonine. Int. J. Pept. Protein Res. 33:468-476.

Arshady, R., Atherton, E., Clive, D.L.J., and Sheppard, R.C. 1981. Peptide synthesis part 1: Preparation and use of polar supports based on poly(dimethylacrylamide). J. Chem. Soc. Perkin Trans. I:529-537.

Atherton, E., and Sheppard, R.C. Detection of problem sequences in solid phase synthesis. In Peptides: Structure and Function, C.M. Deber, V.J. Hruby, and K.D. Kopple, eds., Pierce Chemical Co., Rockford, Illinois, 1985: pp. 415-418.

Atherton, E., and Sheppard, R.C. Solid Phase Peptide Synthesis: A Practical Approach, IRL Press, Oxford, 1989.

Atherton, E., Fox, H., Harkiss, D., Logan, C.J., Sheppard, R.C., and Williams, B.J. 1978a. A mild procedure for solid phase peptide synthesis: Use of fluorenylmethoxycarbonylamino-acids. J. Chem. Soc. Chem. Commun. 537-539.

Atherton, E., Fox, H., Harkiss, D., and Sheppard, R.C. 1978b. Application of polyamide resins to polypeptide synthesis: An improved synthesis of β-endorphin using fluorenylmethoxycarbonylamino-acids. J. Chem. Soc. Chem. Commun.:539-540.

Atherton, E., Bury, C., Sheppard, R.C., and Williams, B.J. 1979. Stability of fluorenylmethoxycarbonylamino groups in peptide synthesis: Cleavage by hydrogenolysis and by dipolar aprotic solvents. Tetrahedron Lett. 3041-3042.

Atherton E., Woolley, V., and Sheppard, R.C. 1980. Internal association in solid phase peptide synthesis: Synthesis of cytochrome C residues 66-104 on polyamide supports. J. Chem. Soc. Chem. Commun. 970-971.

Atherton, E., Benoiton, N.L., Brown, E., Sheppard, R.C., and Williams, B.J. 1981a. Racemisation of activated, urethane-protected amino-acids by p-dimethylaminopyridine: Significance in solid-phase peptide synthesis. J. Chem. Soc. Chem. Commun. 336-337.

Atherton, E., Brown, E., Sheppard, R.C., and Rosevear, A. 1981b. A physically supported gel polymer for low pressure, continuous flow solid phase reactions: Application to solid phase peptide synthesis. J. Chem. Soc. Chem. Commun. 1151-1152.

Atherton, E., Logan, C.J., and Sheppard, R.C. 1981c. Peptide synthesis, part 2: Procedures for solid-phase synthesis using Na-fluorenylmethoxycarbonylamino-acids on polyamide supports: Synthesis of substance P and of acyl carrier protein 65-74 decapeptide. J. Chem. Soc. Perkin Trans. I:538-546.

Atherton, E., Caviezel, M., Fox, H., Harkiss, D., Over, H., and Sheppard, R.C. 1983. Peptide synthesis, part 3: Comparative solid-phase syntheses of human β-endorphin on polyamide supports using t-butoxycarbonyl and fluorenylmethoxycarbonyl protecting groups. J. Chem. Soc. Perkin Trans. I:65-73

Atherton, E., Cammish, L.E., Goddard, P., Richards, J.D., and Sheppard, R.C. The Fmoc-polyamide solid phase method: New procedures for histidine and arginine. In Peptides 1984, U. Ragnarsson, ed., Almqvist and Wiksell Int., Stockholm, 1984, pp. 153-156.

Atherton, E., Pinori, M., and Sheppard, R.C. 1985a. Peptide synthesis, part 6: Protection of the sulphydryl group of cysteine in solid-phase synthesis using N^{α}-fluorenylmethoxycarbonylamino acids: Linear oxytocin derivatives. J. Chem. Soc. Perkin Trans. I:2057-2064.

Atherton, E., Sheppard, R.C., and Ward, P. 1985b. Peptide synthesis, part 7: Solid-phase synthesis of conotoxin G1. J. Chem. Soc. Perkin Trans. I:2065-2073.

Atherton, E., Cameron, L.R., and Sheppard, R.C. 1988a. Peptide synthesis, part 10: Use of pentafluorophenyl esters of fluorenylmethoxycarbonylamino acids in solid phase peptide synthesis. Tetrahedron 44:843-857.

Atherton, E., Holder, J.L., Meldal, M., Sheppard, R.C., and Valerio, R.M. 1988b. Peptide synthesis, part 12: 3,4-Dihydro-4-oxo-1,2,3-benzotriazin-3-yl esters of fluorenylmethoxycarbonyl amino acids as self-indicating reagents for solid phase synthesis. J. Chem. Soc. Perkin Trans. I:2887-2894.

Atherton, E., Cameron, L.R., Cammish, L.E., Dryland, A., Goddard, P., Priestley, G.P., Richards, J.D., Sheppard, R.C., Wade, J.D., and Williams, B.J. Solid phase fragment condensation—the problems. In Innovation and Perspectives in Solid Phase Synthesis, R. Epton, ed., Solid Phase Conference Coordination, Ltd., Birmingham, U.K., 1990, pp. 11-25.

Atherton, E., Hardy, P.M., Harris, D.E., and Matthews, B.H. Racemization of C-terminal cysteine during peptide assembly. In Peptides 1990, E. Giralt, and D. Andreu, eds., ESCOM, Leiden, The Netherlands, 1991, pp. 243-244.

Azuse, I., Tamura, M., Kinomura, K., Okai, H., Kouge, K., Hamatsu, F., and Koizumi, T. 1989. Peptide synthesis in aqueous solution IV. Bull. Chem. Soc. Jpn. 62:3103-3108.

Bagley, C.J., Otteson, K.M., May, B.L., McCurdy, S.N., Pierce, L., Ballard, F.J., and Wallace, J.C. 1990. Synthesis of insulin-like growth factor I using N-methyl pyrrolidinone as the coupling solvent and trifluromethane sulphonic acid cleavage from the resin. Int. J. Pept. Protein Res. 36:356-361.

Baker, P.A., Coffey, A.F., and Epton, R. Accelerated continuous flow ultra-high load solid (gel) phase oligopeptide assembly. In Innovation and Perspectives in Solid Phase Synthesis, R. Epton, ed., Solid Phase Conference Coordination, Ltd., Birmingham, U.K., 1990, pp. 435-440.

Ball, H.L., Grecian, C., Kent, S.B.H., and Mascagni, P. Affinity methods for purifying large synthetic peptides. In Peptides: Chemistry, Structure and Biology, J.E. Rivier and G.R. Marshall, eds., ESCOM, Leiden, The Netherlands, 1990, pp. 435-436.

Ball, H.L., Kent, S.B.H., and Mascagni, P. Selective purification of large synthetic peptides using removable chromatographic probes. In Peptides 1990, E. Giralt and D. Andreu, eds., ESCOM, Leiden, The Netherlands, 1991, pp. 323-325.

Barany, G., and Albericio, F. 1985. A three-dimensional orthogonal protection scheme for solid-phase peptide synthesis under mild conditions. J. Am. Chem. Soc. 107:4936-4942.

Barany, G., and Albericio, F. Mild orthogonal solid-phase peptide synthesis. In Peptides 1990, E. Giralt and D. Andreu, eds., ESCOM, Leiden, The Netherlands, 1991a, pp. 139-142.

Barany, G., and Albericio, F. Peptide synthesis for biotechnology in the 1990's. In Biotechnology International 1990/1991, S. Bond, ed., Century Press Ltd., London, 1991b, pp. 155-163.

Barany, G., and Merrifield, R.B. 1973. An ATP-binding peptide. Cold Spring Harbor Symp. Quant. Biol. 37:121-125.

Barany, G., and Merrifield, R.B. 1977. A new amino protecting group removable by reduction: Chemistry of the dithiasuccinoyl (Dts) function. J. Am. Chem. Soc. 99:7363-7365.

Barany, G., and Merrifield, R.B. Solid-phase peptide synthesis. In The Peptides, Vol. 2, E. Gross and J. Meienhofer, eds., Academic Press, New York, 1979, pp. 1-284.

Barany, G., Kneib-Cordonier, N., and Mullen, D.G. 1987. Solid-phase peptide synthesis: A silver anniversary report. Int. J. Pept. Protein Res. 30:705-739.

Barany, G., Kneib-Cordonier, N., and Mullen, D.G. Polypeptide synthesis, solid-phase method. In Encyclopedia of Polymer Science and Engineering, Vol. 12, 2nd Ed., J.I. Kroschwitz, ed., John Wiley and Sons, New York, 1988, pp. 811-858.

Barany, G., Albericio, F., Biancalana, S., Bontems, S.L., Chang, J.L., Eritja, R., Ferrer, M., Fields, C.G., Fields, G.B., Lyttle, M.H., Solé, N.A., Tian, Z., Van Abel, R.J., Wright, P.B., Zalipsky, S., and Hudson, D. Biopolymer syntheses on novel polyethylene glycol-polystyrene (PEG-PS) graft supports. In Peptides: Chemistry and Biology, J.A. Smith and J.E. Rivier, eds., ESCOM, Leiden, The Netherlands, 1992, pp. 603-604.

Bardají, E., Torres, J.L., Clapés, P., Albericio, F., Barany, G., Rodríguez, R.E., Sacristán, M.P., and Valencia, G. 1991. Synthesis and biological activity of O-glycosylated morphiceptin analogues. J. Chem. Soc. Perkin Trans. 1:1755-1759.

Barlos, K., Papaioannou, D., and Theodoropoulos, D. 1982. Efficient "one-pot" synthesis of N-trityl amino acids. J. Org. Chem. 47:1324-1326.

Barlos, K., Gatos, D., Kallitsis, J., Papaphotiu, G., Sotiriu, P., Wenqing, Y., and Schäfer, W. 1989. Darstellung geschützter Peptid-Fragmente unter Einsatz substituierter Triphenylmethyl-harze. Tetrahedron Lett. 30:3943-3946.

Barlos, K., Chatzi, O., Gatos, D., and Stavropoulos, G. 1991a. 2-Chlorotrityl chloride resin: Studies on anchoring of Fmoc amino acids and peptide cleavage. Int. J. Pept. Protein Res. 37:513-520.

Barlos, K., Gatos, D., and Schäfer, W. 1991b. Synthesis of prothymosin α (ProTα)—A protein consisting of 109 amino acid residues. Angew. Chem. Int. Ed. Engl. 30:590-593..

Barton, M.A., Lemieux, R.U., and Savoie, J.Y. 1973. Solid-phase synthesis of selectively protected peptides for use as building units in the solid-phase synthesis of large molecules. J. Am. Chem. Soc. 95:4501-4506.

Bayer, E. 1991. Towards the chemical synthesis of proteins. Angew. Chem. Int. Ed. Engl. 30:113-129.

Bayer, E., and Rapp, W. New polymer supports for solid-liquid-phase peptide synthesis. In Chemistry of Peptides and Proteins, Vol. 3, W. Voelter, E. Bayer, Y.A. Ovchinnikov, and V.T. Ivanov, eds., Walter de Gruyter and Co., Berlin, 1986, pp. 3-8.

Bayer, E., Eckstein, H., Hägele, K., König, W.A., Brüning, W., Hagenmaier, H., and Parr, W. 1970. Failure sequences in the solid phase synthesis of polypeptides. J. Am. Chem. Soc. 92:1735-1738.

Bayer, E., Albert, K., Willisch, H., Rapp, W., and Hemmasi, B. 1990. ^{13}C NMR relaxation times of a tripeptide methyl ester and its polymer-bound analogues. Macromolecules 23:1937-1940.

Beacham, J., Bentley, P.H., Kenner, G.W., MacLeod, J.K., Mendive, J.J., and Sheppard, R.C. 1967. Peptides part XXV: The structure and synthesis of human gastrin. J. Chem. Soc. (C) 2520-2529.

Beck-Sickinger, A.G., Dürr, H., and Jung, G. 1991. Semiautomated T-bag peptide synthesis using 9-fluorenylmethoxycarbonyl strategy and benzotriazol-1-yl-tetramethyluronium tetrafluoroborate activation. Peptide Res. 4:88-94.

Becker, S., Atherton, E., and Gordon, R.D. 1989. Synthesis and characterization of μ-conotoxin IIIa. Eur. J. Biochem. 185:79-84.

Belshaw, P.J., Mzengeza, S., and Lajoie, G.A. 1990. Chlorotrimethylsilane mediated formation of ω-allyl esters of aspartic and glutamic acids. Synth. Commun. 20:3157-3160.

Benoiton, N.L., and Chen, F.M.F. Symmetrical anhydride rearrangement leads to three different dipeptide products. In Peptides 1986, D. Theodoropoulos, ed., Walter de Gruyter and Co., Berlin, 1987, pp. 127-130.

Berg, R.H., Almdal, K., Pedersen, W.B., Holm, A., Tam, J.P., and Merrifield, R.B. 1989. Long-chain polystyrene-grafted polyethylene film matrix:A new support for solid-phase peptide synthesis. J. Am. Chem. Soc. 111:8024-8026.

Bergot, B.J., Noble, R.L., and Geiser, T. TFMSA/TFA cleavage and deprotection in SPPS. In Peptides 1986, D. Theodoropoulos, ed., Walter de Gruyter, and Co., Berlin, 1987, pp. 97-101.

Bernatowicz, M.S., Matsueda, R., and Matsueda, G.R. 1986. Preparation of Boc-[S-(3-nitro-2-pyridinesulfenyl)]-cysteine and its use for unsymmetrical disulfide bond formation. Int. J. Pept. Protein Res. 28:107-112.

Bernatowicz, M.S., Daniels, S.B., Coull, J.M., Kearney, T., Neves, R.S., Coassin, P.J., and Köster, H. Recent developments in solid phase peptide synthesis using the 9-fluorenylmethyloxycarbonyl (Fmoc) protecting group strategy. In Current Research in Protein Chemistry: Techniques, Structure, and Function, J.J. Villafranca, ed., Academic Press, San Diego, 1990, pp. 63-77.

Bertho, J.-N., Loffet, A., Pinel, C., Reuther, F., and Sennyey, G. 1991. Amino acid fluorides: Their preparation and use in peptide synthesis. Tetrahedron Lett. 32:1303-1306.

Biancalana, S., Hayes, T., Toll, L., and Hudson, D. Immunological and biological activity of octameric peptides prepared by Fmoc methodology. In Twelfth American Peptide Symposium Program and Abstracts, Massachusetts Institute of Technology, Cambridge, Mass., 1991, pp. P-234.

Biondi, L., Filira, F., Gobbo, M., Scolaro, B., Rocchi, R., and Cavaggion, F. 1991. Synthesis of glycosylated tuftsins and tuftsin-containing IgG fragment undecapeptide. Int. J. Pept. Protein Res. 37:112-121.

Birr, C. Aspects of the Merrifield Peptide Synthesis, Springer-Verlag, Heidelberg, 1978.

Birr, C. 1990a. The transition to solid-phase production of pharmaceutical peptides. Biochem. Soc. Trans. 18:1313-1316.

Birr, C. Scale-up, and production potential of current SPPS strategies. In Innovation and Perspectives in Solid Phase Synthesis, R. Epton, ed., Solid Phase Conference Coordination, Ltd., Birmingham, U.K., 1990b, pp. 155-181.

Birr, C., Lochinger, W., Stahnke, G., and Lang, P. 1972. Der α,α-dimethyl-3.5-dimethoxybenzyloxycarbonyl (Ddz)-rest, eine photo- und säurelabile Stickstoff-Schutzgruppe für die Peptidchemie. Justus Liebigs Ann. Chem. 763:162-172.

Blankemeyer-Menge, B., Nimtz, M., and Frank, R. 1990. An efficient method for anchoring Fmoc amino acids to hydroxyl-functionalised solid supports. Tetrahedron Lett. 31:1701-1704.

Bodanszky, M., and Bodanszky, A. The Practice of Peptide Synthesis, Springer-Verlag, Berlin, 1984.

Bodanszky, M., and Kwei, J.Z. 1978. Side reactions in peptide synthesis VII: Sequence dependence in the formation of aminosuccinyl derivatives from β-benzyl-aspartyl peptides. Int. J. Pept. Protein Res. 12:69-74.

Bodanszky, M., Funk, K.W., and Fink, M.L. 1973. o-Nitrophenyl esters of tert-butyloxycarbonylamino acids and their application in the stepwise synthesis of peptide chains by a new technique. J. Org. Chem. 38:3565-3570.

Bodanszky, M., Fink, M.L., Klausner, Y.S., Natarajan, S., Tatemoto, K., Yiotakis, A.E., and Bodanszky, A. 1977. Side reactions in peptide synthesis 4: Extensive O-acylation by active esters in histidine containing peptides. J. Org. Chem. 42:149-152.

Bodanszky, M., Tolle, J.C., Deshmane, S.S., and Bodanszky, A. 1978. Side reactions in peptide synthesis VI: A reexamination of the benzyl group in the protection of the side chains of tyrosine and aspartic acid. Int. J. Pept. Protein Res. 12:57-68.

Bodanszky, A., Bodanszky, M., Chandramouli, N., Kwei, J.Z., Martinez, J., and Tolle, J.C. 1980. Active esters of 9-fluorenylmethyloxycarbonyl amino acids and their application in the stepwise lengthening of a peptide chain. J. Org. Chem. 45:72-76.

Bolin, D.R., Wang, C.-T., and Felix, A.M. 1989. Preparation of N-t-butyloxycarbonyl-O$^{\omega}$-9-fluorenylmethyl esters of asparatic and glutamic acids. Org. Prep. Proc. Int. 21:67-74.

Bontems, R.J., Hegyes, P., Bontems, S.L., Albericio, F., and Barany, G. Synthesis and applications of XAL, a new acid-labile handle for solid-phase synthesis of peptide amides. In Peptides: Chemistry and Biology, J.A. Smith and J.E. Rivier, eds., ESCOM, Leiden, The Netherlands, 1992, pp. 601-602.

Bozzini, M., Bello, R., Cagle, N., Yamane, D., and Dupont, D. Tryptophan recovery from autohydrolyzed samples using dodecanethiol. Applied Biosystems Research News, February 1991, Applied Biosystems, Inc., Foster City, Calif..

Brady, S.F., Paleveda, W.J., and Nutt, R.F. Studies of acetamidomethyl as cysteine protection: Application in synthesis of ANF analogs. In Peptides: Chemistry and Biology, G.R. Marshall, ed., ESCOM, Leiden, The Netherlands, 1988, pp. 192-194.

Breipohl, G., Knolle, J., and Stüber, W. 1989. Synthesis and application of acid labile anchor groups for the synthesis of peptide amides by Fmoc-solid-phase peptide synthesis. Int. J. Pept. Protein Res. 34:262-267.

Breipohl, G., Knolle, J., and Stüber, W. 1990. Facile SPS of peptides having C-terminal Asn and Gln. Int. J. Pept. Protein Res. 35:281-283.

Brown, T., Jones, J.H., and Richards, J.D. 1982. Further studies on the protection of histidine side chains in peptide synthesis: The use of the π-benzyloxymethyl group. J. Chem. Soc. Perkin Trans. I:1553-1561.

Buchta, R., Bondi, E., and Fridkin, M. 1986. Peptides related to the calcium binding domains II and III of calmodulin: Synthesis and calmodulin-like features. Int. J. Pept. Protein Res. 28:289-297.

Büttner, K., Zahn, H., and Fischer, W.H. Rapid solid phase peptide synthesis on a controlled pore glass support. In Peptides: Chemistry and Biology, G.R. Marshall, ed., ESCOM, Leiden, The Netherlands, 1988, pp. 210-211.

Butwell, F.G.W., Haws, E.J., and Epton, R. 1988. Advances in ultra-high load polymer supported peptide synthesis with phenolic supports 1: A selectively-labile C-terminal spacer group for use with a base-mediated N-terminal deprotection strategy and Fmoc amino acids. Makromol. Chem. Macromol. Symp. 19:69-77.

Cameron, L.R., Holder, J.L., Meldal, M., and Sheppard, R.C. 1988. Peptide synthesis, part 13: Feedback control in solid phase synthesis: Use of fluorenylmethoxycarbonyl amino acid 3,4-dihydro-4-oxo-1,2,3-benzotriazin-3-yl esters in a fully automated system. J. Chem. Soc. Perkin Trans. I:2895-2901.

Carpino, L.A., and Han, G.Y. 1972. The 9-fluorenylmethoxycarbonyl amino-protecting group. J. Org. Chem. 37:3404-3409.

Carpino, L.A., Cohen, B.J., Stephens, Jr., K.E., Sadat-Aalaee, S.Y., Tien, J.-H., and Langridge, D.C. 1986. [(9-Fluorenylmethyl)oxy]carbonyl (Fmoc) amino acid chlorides: Synthesis, characterization, and application to the rapid synthesis of short peptide. J. Org. Chem. 51:3732-3734.

X Carpino, L.A., Sadat-Aalaee, D., Chao, H.G., and DeSelms, R.H. 1990. [(9-Fluorenylmethyl)oxy]carbonyl (Fmoc) amino acid fluorides: Convenient new peptide coupling reagents applicable to the Fmoc/tert-butyl strategy for solution and solid-phase syntheses. J. Am. Chem. Soc. 112:9651-9652.

X Carpino, L.A., Mansour, E.-S.M.E., and Sadat-Aalaee, D. 1991a. tert-Butyloxycarbonyl and benzyloxycarbonyl amino acid fluorides: New, stable rapid-acting acylating agents for peptide synthesis. J. Org. Chem. 56:2611-2614.

X Carpino, L.A., Chao, H.G., Beyermann, M., and Bienert, M. 1991b. [(9-Fluorenylmethyl)oxy]carbonyl amino acid chlorides in solid-phase peptide synthesis. J. Org. Chem. 56:2635-2642.

Casaretto, R., and Nyfeler, R. Isolation, structure and activity of a side product from the synthesis of human endothelin. In Peptides 1990, E. Giralt and D. Andreu, eds., ESCOM, Leiden, The Netherlands, 1991, pp. 181-182.

Chan, W.C., and Bycroft, B.W. Deprotection of Arg(Pmc) containing peptides using TFA-trialkylsilane-methanol-EMS; application to the synthesis of propeptides of nisin. In Twelfth American Peptide Symposium Program and Abstracts, Massachusetts Institute of Technology, Cambridge, Mass., 1992, pp. 613-614.

Chang, C.D., Felix, A.M., Jimenez, M.H., and Meienhofer, J. 1980a. Solid-phase peptide synthesis of somatostatin using mild base cleavage of N^α-fluorenylmethyloxycarbonylamino acids. Int. J. Pept. Protein Res. 15:485-494.

Chang, C.D., Waki, M., Ahmad, M., Meienhofer, J., Lundell, E.O., and Haug, J.D. 1980b. Preparation and properties of N^α-9-fluorenylmethyloxycarbonylamino acids bearing tert.-butyl side chain protection. Int. J. Pept. Protein Res. 15:59-66.

Chanh, T.C., Dreesman, G.R., Kanda, P., Linette, G.P., Sparrow, J.T., Ho, D.D., and Kennedy, R.C. 1986. Induction of anti-HIV neutralizing antibodies by synthetic peptides. EMBO J. 5:3065-3071.

Chen, S.-T., Weng, C.-S., and Wang, K.-T. 1987. The synthesis of p-methoxybenzyloxycarbonyl amino acids. J. Chin. Chem. Soc. 34:117-123.

Chen, S.-T., Wu, S.-H., and Wang, K.-T. 1989. A new synthesis of O-benzyl-L-threonine. Synth. Commun. 19:3589-3593.

Chillemi, F., and Merrifield, R.B. 1969. Use of N^{im}-dinitrophenylhistidine in the solid-phase synthesis of the tricosapeptides 124-146 of human hemoglobin β chain. Biochemistry 8:4344-4346.

Chun, R., Glabe, C.G., and Fan, H. 1990. Chemical synthesis of biologically active *tat trans*-activating protein of human immunodeficiency virus type 1. J. Virol. 64:3074-3077.

Clark-Lewis, I., and Kent, S. Chemical synthesis, purification, and characterization of peptides and proteins. In Receptor Biochemistry and Methodology, Vol. 14: The Use of HPLC in Protein Purification and Characterization, A.R. Kerlavage, J.C. Venter and L.C. Harrison, eds., Alan R. Liss, New York, 1989, pp. 43-75.

Clark-Lewis, I., Aebersold, R., Ziltener, H., Schrader, J.W., Hood, L.E., and Kent, S.B. 1986. Automated chemical synthesis of a protein growth factor for hemopoietic cells, interleukin-3. Science 231:134-139.

Colombo, R., Atherton, E., Sheppard, R.C., and Woolley, V. 1983. 4-Chloromethylphenoxyacetyl polystyrene and polyamide supports for solid-phase peptide synthesis. Int. J. Pept. Protein Res. 21:118-126.

Colombo, R., Colombo, F., and Jones, J.H. 1984. Acid-labile histidine side-chain protection: The π-benzyloxymethyl group. J. Chem. Soc. Chem. Commun. 292-293.

Cook, R.M., Hudson, D., Tsou, D., Teplow, D.B., Wong, H., Zou, A.Q., and Wickstrom, E. Fmoc-mediated solid phase assembly of HIV Tat protein. In Peptides 1988, G. Jung and E. Bayer, eds., Walter de Gruyter, and Co., Berlin, 1989, pp. 187-189.

Coste, J., Le-Nguyen, D., and Castro, B. 1990. PyBOP: A new peptide coupling reagent devoid of toxic by-product. Tetrahedron Lett. 31:205-208.

Cruz, L.J., Kupryszewski, G., LeCheminant, G.W., Gray, W.R., Olivera, B.M., and Rivier, J. 1989. μ-Conotoxin GIIIA, a peptide ligand for muscle sodium channels: Chemical synthesis, radiolabeling, and receptor characterization. Biochemistry 28:3437-3442.

de Bont, H.B.A., van Boom, J.H., and Liskamp, R.M.J. 1990. Automatic synthesis of phophopeptides by phosphorylation on the solid phase. Tetrahedron Lett. 31:2497-2500.

de la Torre, B.G., Torres, J.L., Bardají, E., Clapés, P., Xaus, N., Jorba, X., Calvet, S., Albericio, F., and Valencia, G. 1990. Improved method for the synthesis of *o*-glycosylated Fmoc amino acids to be used in solid-phase glycopeptide synthesis. J. Chem. Soc. Chem. Commun. 965-967.

Deen, C., Claassen, E., Gerritse, K., Zegers, N.D., and Boersma, W.J.A. 1990. A novel carbodiimide coupling method for synthetic peptides: Enhanced anti-peptide antibody responses. J. Immunol. Methods 129:119-125.

DeGrado, W.F., and Kaiser, E.T. 1980. Polymer-bound oxime esters as supports for solid-phase peptide synthesis: Preparation of protected peptide fragments. J. Org. Chem. 45:1295-1300.

DeGrado, W.F., and Kaiser, E.T. 1982. Solid-phase synthesis of protected peptides on a polymer-bound oxime: Preparation of segments comprising the sequence of a cytotoxic 26-peptide analogue. J. Org. Chem. 47:3258-3261.

DiMarchi, R.D., Tam, J.P., Kent, S.B.H., and Merrifield, R.B. 1982. Weak acid-catalyzed pyrrolidone carboxylic acid formation from glutamine during solid phase peptide synthesis. Int. J. Pept. Protein Res. 19:88-93.

Dorman, L.C., Nelson, D.A., and Chow, R.C.L. Solid phase synthesis of glutamine-containing peptides. In Progress in Peptide Research, Vol. 2, S. Lande, ed., Gordon and Breach, New York, 1972, pp. 65-68.

Dourtoglou, V., Gross, B., Lambropoulou, V., and Zioudrou, C. 1984. O-benzotriazolyl-N,N,N',N'-tetramethyluronium hexafluorophosphate as coupling reagent for the synthesis of peptides of biological interest. Synthesis:572-574.

Drijfhout, J.W., and Bloemhoff, W. Capping with O-sulfobenzoic acid cyclic anhydride (OSBA) in solid-phase peptide synthesis enables facile product purification. In Peptide Chemistry 1987, T. Shiba and S. Sakakibara, eds., Protein Research Foundation, Osaka, 1988, pp. 191-194.

Drijfhout, J.W., and Bloemhoff, W. 1991. A new synthetic functionalized antigen carrier. Int. J. Pept. Protein Res. 37:27-32.

Drijfhout, J.W., Perdijk, E.W., Weijer, W.J., and Bloemhoff, W. 1988. Controlled peptide-protein conjugation by means of 3-nitro-2-pyridinesulfenyl protection-activation. Int. J. Pept. Protein Res. 32:161-166.

Eichler, J., Beyermann, M., and Bienert, M. 1989. Application of cellulose paper as support material in simultaneous solid phase peptide synthesis. Collect. Czech. Chem. Commun. 54:1746-1752.

Epton, R., Goddard, P., and Ivin, K.J. 1980. Gel phase ^{13}C n.m.r. spectroscopy as an analytical method in solid (gel) phase peptide synthesis. Polymer 21:1367-1371.

Epton, R., Wellings, D.A., and Williams, A. 1987. Perspectives in ultra-high load solid (gel) phase peptide synthesis. Reactive Polymers 6:143-157.

Erickson, B.W., and Merrifield, R.B. 1973a. Acid stability of several benzylic protecting groups used in solid-phase peptide synthesis: Rearrangement of O-benzyltyrosine to 3-benzyltyrosine. J. Am. Chem. Soc. 95:3750-3756.

Erickson, B.W., and Merrifield, R.B. 1973b. Use of chlorinated benzyloxycarbonyl protecting groups to eliminate N^ϵ-branching at lysine during solid-phase peptide synthesis. J. Am. Chem. Soc. 95:3757-3763.

Erickson, B.W., and Merrifield, R.B. Solid-phase peptide synthesis. In The Proteins, Vol. II, 3rd Ed., H. Neurath and R.L. Hill, eds., Academic Press, New York, 1976, pp. 255-527.

Eritja, R., Ziehler-Martin, J.P., Walker, P.A., Lee, T.D., Legesse, K., Albericio, F., and Kaplan, B.E. 1987. On the use of S-t-butylsulphenyl group for protection of cysteine in solid-phase peptide synthesis using Fmoc amino acids. Tetrahedron 43:2675-2680.

Eritja, R., Robles, J., Fernandez-Forner, D., Albericio, F., Giralt, E., and Pedroso, E. 1991. NPE-resin, a new approach to the solid-phase synthesis of protected peptides and oligonucleotides I: Synthesis of the supports and their application to oligonucleotide synthesis. Tetrahedron Lett. 32:1511-1514.

Fairwell, T., Hospattankar, A.V., Ronan, R., Brewer, Jr., H.B., Chang, J.K., Shimizu, M., Zitzner, L., and Arnaud, C.D. 1983. Total solid-phase synthesis, purification, and characterization of human parathyroid hormone-(1-84). Biochemistry 22:2691-2697.

Feinberg, R.S., and Merrifield, R.B. 1975. Modification of peptides containing glutamic acid by hydrogen-fluoride-anisole mixtures: γ-Acylation of anisole or the glutamyl nitrogen. J. Am. Chem. Soc. 97:3485-3496.

Felix, A.M., Jiminez, M.H., Vergona, R., and Cohen, M.R. 1973. Synthesis and biological studies of novel bradykinin analogues. Int. J. Pept. Protein Res. 5:201-206.

Felix, A.M., Wang, C.-T., Heimer, E.P., and Fournier, A. 1988a. Applications of BOP reagent in solid phase synthesis II: Solid phase side-chain cyclization using BOP reagent. Int. J. Pept. Protein Res. 31:231-238.

Felix, A.M., Heimer, E.P., Wang, C.-T., Lambros, T.J., Fournier, A., Mowles, T.F., Maines, S., Campbell, R.M., Wegrzynski, B.B., Toome, V., Fry, D., and Madison, V.S. 1988b. Synthesis, biological activity and conformational analysis of cyclic GRF analogs. Int. J. Pept. Protein Res. 32:441-454.

Fields, C.G., and Fields, G.B. New approaches to prevention of side reactions in Fmoc solid phase peptide synthesis. In Peptides: Chemistry, Structure, and Biology, J.E. Rivier and G.R. Marshall, eds., ESCOM, Leiden, The Netherlands, 1990, pp. 928-930.

Fields, C.G., Fields, G.B., Noble, R.L., and Cross, T.A. 1989. Solid phase peptide synthesis of [^{15}N]-gramicidins A, B, and C and high performance liquid chromatographic purification. Int. J. Pept. Protein Res. 33:298-303.

Fields, C.G., Lloyd, D.H., Macdonald, R.L., Otteson, K.M., and Noble, R.L. 1991. HBTU activation for automated Fmoc solid-phase peptide synthesis. Peptide Res. 4:95-101.

Fields, G.B., and Fields, C.G. 1991. Solvation effects in solid-phase peptide synthesis. J. Am. Chem. Soc. 113:4202-4207.

Fields, G.B., and Noble, R.L. 1990. Solid phase peptide synthesis utilizing 9-fluorenylmethoxycarbonyl amino acids. Int. J. Pept. Protein Res. 35:161-214.

Fields, G.B., Van Wart, H.E., and Birkedal-Hansen, H. 1987. Sequence specificity of human skin fibroblast collagenase: Evidence for the role of collagen structure in determining the collagenase cleavage site. J. Biol. Chem. 262:6221-6226.

Fields, G.B., Fields, C.G., Petefish, J., Van Wart, H.E., and Cross, T.A. 1988. Solid phase synthesis and solid state NMR spectroscopy of [^{15}N-Ala3]-Val-gramicidin A. Proc. Natl. Acad. Sci. USA 85:1384-1388.

Fields, G.B., Otteson, K.M, Fields, C.G., and Noble, R.L. The versatility of solid phase peptide synthesis. In Innovation and Perspectives in Solid Phase Synthesis, R. Epton, ed., Solid Phase Conference Coordination, Ltd., Birmingham, U.K., 1990, pp. 241-260.

Filira, F., Biondi, L., Cavaggion, F., Scolaro, B., and Rocchi, R. 1990. Synthesis of O-glycosylated tuftsins by utilizing threonine derivatives containing an unprotected monosaccharide moiety. Int. J. Pept. Protein Res. 36:86-96.

Findeis, M.A., and Kaiser, E.T. 1989. Nitrobenzophenone oxime based resins for the solid-phase synthesis of protected peptide segments. J. Org. Chem. 54:3478-3482.

Finn, F.M., and Hofmann, K. The synthesis of peptides by solution methods with emphasis on peptide hormones. In The Proteins, Vol. II, 3rd Ed., H. Neurath and R.L. Hill, eds., Academic Press, New York, 1976, pp. 105-253.

Fischer, P.M., Comis, A., and Howden, M.E.H. 1989. Direct immunization with synthetic peptidyl-polyamide resin: Comparison with antibody production from free peptide and conjugates with carrier proteins. J. Immunol. Methods 118:119-123.

Flegel, M., and Sheppard, R.C. 1990. A sensitive, general method for quantitative monitoring of continuous flow solid phase peptide synthesis. J. Chem. Soc. Chem. Commun. 536-538.

Fletcher, A.R., Jones, J.H., Ramage, W.I., and Stachulski, A.V. 1979. The use of the N(π)-phenacyl group for the protection of the histidine side chain in peptide synthesis. J. Chem. Soc. Perkin Trans. I:2261-2267.

Flörsheimer, A., and Riniker, B. Solid-phase synthesis of peptides with the highly acid-sensitive HMPB linker. In Peptides 1990, E. Giralt and D. Andreu, eds., ESCOM, Leiden, The Netherlands, 1991, pp. 131-133.

Fodor, S.P.A., Read, J.L., Pirrung, M.C., Stryer, L., Lu, A.T., and Solas, D. 1991. Light-directed, spatially addressable parallel chemical synthesis. Science 251:767-773.

Forest, M., and Fournier, A. 1990. BOP reagent for the coupling of pGlu and Boc-His(Tos) in solid phase peptide synthesis. Int. J. Pept. Protein Res. 35:89-94.

Fournier, A., Wang, C.-T., and Felix, A.M. 1988. Applications of BOP reagent in solid phase peptide synthesis: Advantages of BOP reagent for difficult couplings exemplified by a synthesis of [Ala^{15}]-GRF(1-29)-NH$_2$. Int. J. Pept. Protein Res. 31:86-97.

Fournier, A., Danho, W., and Felix, A.M. 1989. Applications of BOP reagent in solid phase peptide synthesis III: Solid phase peptide synthesis with unprotected aliphatic and aromatic hydroxyamino acids using BOP reagent. Int. J. Pept. Protein Res. 33:133-139.

Fox, J., Newton, R., Heegard, P., and Schafer-Nielsen, C. A novel method of monitoring the coupling reaction in solid phase synthesis. In Innovation and Perspectives in Solid Phase Synthesis, R. Epton, ed., Solid Phase Conference Coordination, Ltd., Birmingham, U.K., 1990, pp. 141-153.

Frank, R., and Döring, R. 1988. Simultaneous multiple peptide synthesis under continuous flow conditions on cellulose paper discs as segmental solid supports. Tetrahedron 44:6031-6040.

Frank, R., and Gausepohl, H. Continuous flow peptide synthesis. In Modern Methods in Protein Chemistry, Vol. 3, H. Tschesche, ed., Walter de Gruyter and Co., Berlin, 1988, pp. 41-60.

Fujii, N., Otaka, A., Funakoshi, S., Bessho, K., Watanabe, T., Akaji, K., and Yajima, H. 1987. Studies on peptides CLI: Synthesis of cystine-peptides by oxidation of S-protected cysteine-peptides with thallium (III) trifluoroacetate. Chem. Pharm. Bull. 35:2339-2347.

Fujino, M., Wakimasu, M., Shinagawa, S., Kitada, C., and Yajima, H. 1978. Synthesis of the nonacosapeptide corresponding to mammalian glucagon. Chem. Pharm. Bull. 26:539-548.

Fujino, M., Wakimasu, M., and Kitada, C. 1981. Further studies on the use of multi-substituted benzenesulfonyl groups for protection of the guanidino function of arginine. Chem. Pharm. Bull. 29:2825-2831.

Fuller, W.D., Cohen, M.P., Shabankareh, M., Blair, R.K., Goodman, M., and Naider, F.R. 1990. Urethane-protected amino acid N-carboxy anhydrides and their use in peptide synthesis. J. Am. Chem. Soc. 112:7414-7416.

Funakoshi, S., Tamamura, H., Fujii, N., Yoshizawa, K., Yajima, H., Miyasaki, K., Funakoshi, A., Ohta, M., Inagaki, Y., and Carpino, L.A. 1988. Combination of a new amide-precursor reagent and trimethylsilyl bromide deprotection for the Fmoc-based solid phase synthesis of human pancreastatin and one of its fragments. J. Chem. Soc. Chem. Commun. 1588-1590.

Futaki, S., Taike, T., Akita, T., and Kitagawa, K. 1990a. A new approach for the synthesis of tyrosine sulphate containing peptides: Use of the *p*-(methylsulphinyl)benzyl group as a key protecting group of serine. J. Chem. Soc. Chem. Commun. 523-524.

Futaki, S., Yajami, T., Taike, T., Akita, T., and Kitagawa, K. 1990b. Sulphur trioxide/thiol: A novel system for the reduction of methionine sulphoxide. J. Chem. Soc. Perkin Trans. 1:653-658.

Gaehde, S.T., and Matsueda, G.R. 1981. Synthesis of *N-tert*-butoxycarbonyl-(α-phenyl)aminomethylphenoxyacetic acid for use as a handle in solid-phase synthesis of peptide a-carboxamides. Int. J. Pept. Protein Res. 18:451-458.

Gairi, M., Lloyd-Williams, P., Albericio, F., and Giralt, E. 1990. Use of BOP reagent for the suppression of diketopiperazine formation in Boc/Bzl solid-phase peptide synthesis. Tetrahedron Lett. 31:7363-7366.

García-Echeverría, C., Albericio, F., Pons, M., Barany, G., and Giralt, E. 1989. Convenient synthesis of a cyclic peptide disulfide: A type II β-turn structural model. Tetrahedron Lett. 30:2441-2444.

Gausepohl, H., Kraft, M., and Frank, R. In situ activation of Fmoc amino acids by BOP in solid phase peptide synthesis. In Peptides 1988, G. Jung and E. Bayer, eds., Walter de Gruyter and Co., Berlin, 1989a, pp. 241-243.

Gausepohl, H., Kraft, M., and Frank, R.W. 1989b. Asparagine coupling in Fmoc solid phase peptide synthesis. Int. J. Pept. Protein Res. 34:287-294.

Gausepohl, H., Kraft, M., Boulin, Ch., and Frank, R.W. Automated multiple peptide synthesis with BOP activation. In Peptides: Chemistry, Structure and Biology, J.E. Rivier and G.R. Marshall, eds., ESCOM, Leiden, The Netherlands, 1990, pp. 1003-1004.

Gausepohl, H., Kraft, M., Boulin, C., and Frank, R.W. A multiple reaction system for automated simultaneous peptide synthesis. In Peptides 1990, E. Giralt and D. Andreu, eds., ESCOM, Leiden, The Netherlands, 1991a, pp. 206-207.

Gausepohl, H., Pieles, U., and Frank, R.W. Schiffs base analog formation during in situ activation by HBTU and TBTU. In Peptides: Chemistry and Biology, J.A. Smith and J.E. Rivier, eds., ESCOM, Leiden, The Netherlands, 1991b, pp. 523-524.

Geiser, T., Beilan, H., Bergot, B.J., and Otteson, K.M. Automation of solid-phase peptide synthesis. In Macromolecular Sequencing and Synthesis: Selected Methods and Applications, D.H. Schlesinger, ed., Alan R. Liss, New York, 1988, pp. 199-218.

Gesellchen, P.D., Rothenberger, R.B., Dorman, D.E., Paschal, J.W., Elzey, T.K., and Campbell, C.S. A new side reaction in solid-phase peptide synthesis: Solid support-dependent alkylation of tryptophan. In Peptides: Chemistry, Structure and Biology, J.E. Rivier and G.R. Marshall, eds., ESCOM, Leiden, The Netherlands, 1990, pp. 957-959.

Gesquière, J.-C., Diesis, E., and Tartar, A. 1990. Conversion of *N*-terminal cysteine to thiazolidine carboxylic acid during hydrogen fluoride deprotection of peptides containing *N*-p-Bom protected histidine. J. Chem. Soc. Chem. Commun. 1402-1403.

Gesquière, J.-C., Najib, J., Diesis, E., Barbry, D., and Tartar, A. Investigations of side reactions associated with the use of Bom and Bum groups for his-tidine protection. In Peptides: Chemistry and Biology, J.A. Smith and J.E. Rivier, eds., ESCOM, Leiden, The Netherlands, 1992, pp. 641-642.

Geysen, H.M., Meloen, R.H., and Barteling, S.J. 1984. Use of peptide synthesis to probe viral antigens for epitopes to a resolution of a single amino acid. Proc. Natl. Acad. Sci. USA 81:3998-4002.

Giralt, E., Albericio, F., Pedroso, E., Granier, C., and Van Rietschoten, J. 1982. Convergent solid phase peptide synthesis II: Synthesis of the 1-6 apamin protected segment on a Nbb-resin: synthesis of apamin. Tetrahedron 38:1193-1208.

Giralt, E., Rizo, J., and Pedroso, E. 1984. Application of gel-phase ^{13}C-NMR to monitor solid phase peptide synthesis. Tetrahedron 40:4141-4152.

Giralt, E., Eritja, R., Pedroso, E., Granier, C., and van Rietschoten, J. 1986. Convergent solid phase peptide synthesis III: Synthesis of the 44-52 protected segment of the toxin II of Androctonus australis Hector. Tetrahedron 42:691-698.

Gisin, B.F. 1973. The preparation of Merrifield-resins through total esterifica-tion with cesium salts. Helv. Chim. Acta 56:1476-1482.

Gisin, B.F., and Merrifield, R.B. 1972. Carboxyl-catalyzed intramolecular aminolysis: A side reaction in solid-phase peptide synthesis. J. Am. Chem. Soc. 94:3102-3106.

Goddard, P., McMurray, J.S., Sheppard, R.C., and Emson, P. 1988. A solubilisable polymer support suitable for solid phase peptide synthesis and for injection into experimental animals. J. Chem. Soc. Chem. Commun. 1025-1027.

Grandas, A., Jorba, X., Giralt, E., and Pedroso, E. 1989a. Anchoring of Fmoc amino acids to hydroxymethyl resins. Int. J. Pept. Protein Res. 33:386-390.

Grandas, A., Albericio, F., Josa, J., Giralt, E., Pedroso, E., Sabatier, J.M., and van Rietschoten, J. 1989b. Convergent solid phase peptide synthesis VII: Good yields in the coupling of protected segments on a solid support. Tetrahedron 45:4637-4648.

Gras-Masse, H., Ameisen, J.C., Boutillon, C., Gesquière, J.C., Vian, S., Neyrinck, J.L., Drobecq, H., Capron, A., and Tartar, A. 1990. A synthetic protein corresponding to the entire vpr gene product from the human im-munodeficiency virus HIV-1 is recognized by antibodies from HIV-infected patients. Int. J. Pept. Protein Res. 36:219-226.

Gray, W.R., Luque, A., Galyean, R., Atherton, E., Sheppard, R.C., Stone, B.L., Reyes, A., Alford, J., McIntosh, M., Olivera, B.M., Cruz, L.J., and Rivier, J. 1984. Contoxin GI: Disulfide bridges, synthesis, and preparation of iodinated derivatives. Biochemistry 23:2796-2802.

Green, J., Ogunjobi, O.M., Ramage, R., Stewart, A.S.J., McCurdy, S., and Noble, R. 1988. Application of the N^G-(2,2,5,7,8-pentamethylchroman-6-sulphonyl) derivative of Fmoc-arginine to peptide synthesis. Tetrahedron Lett. 29:4341-4344.

Green, M., and Berman, J. 1990. Preparation of pentafluorophenyl esters of Fmoc protected amino acids with pentafluorophenyl trifluoroacetate. Tetrahedron Lett. 31:5851-5852.

Greene, T.W. Protective Groups in Organic Synthesis. John Wiley and Sons, New York, 1991.

Groginsky, C. 1990. Independent simultaneous multiple peptide synthesis. Am. Biotech. Lab. 8(13):40-43.

Gross, H., and Bilk, L. 1968. Zur Reaktion von N-hydroxysuccinimid mit Dicyclohexylcarbodiimid. Tetrahedron 24:6935-6939.

Guibé, F., Dangles, O., Balavoine, G., and Loffet, A. 1989. Use of an allylic anchor group and of its palladium catalyzed hydrostannolytic cleavage in the solid phase synthesis of protected peptide fragments. Tetrahedron Lett. 30:2641-2644.

Gutte, B., and Merrifield, R.B. 1971. The synthesis of ribonuclease A. J. Biol. Chem. 246:1922-1941.

Guttman, St., and Boissonnas, R.A. 1959. Synthése de l'α-mélanotropine (α-MSH) de porc. Helv. Chim. Acta 42: 1257-1264.

Hahn, K.W., Klis, W.A., and Stewart, J.M. 1990. Design and synthesis of a peptide having chymotrypsin-like esterase activity. Science 248:1544-1547.

Hammer, R.P., Albericio, F., Gera, L., and Barany, G. 1990. Practical approach to solid-phase synthesis of C-terminal peptide amides under mild conditions based on photolysable anchoring linkage. Int. J. Pept. Protein Res. 36:31-45.

Harrison, J.L., Petrie, G.M., Noble, R.L., Beilan, H.S., McCurdy, S.N., and Culwell, A.R. Fmoc chemistry: Synthesis, kinetics, cleavage, and deprotection of arginine-containing peptides. In Techniques in Protein Chemistry, T.E. Hugli, ed., Academic Press, San Diego, 1989, pp. 506-516.

Heimer, E.P., Chang, C.-D., Lambros, T., and Meienhofer, J. 1981. Stable isolated symmetrical anhydrides of N^α-fluorenylmethyloxycarbonylamino acids in solid-phase peptide synthesis. Int. J. Pept. Protein Res. 18:237-241.

Hellermann, H., Lucas, H.-W., Maul, J., Pillai, V.N.R., and Mutter, M. 1983. Poly(ethylene glycol)s grafted onto crosslinked polystyrenes, 2: Multidetachably anchored polymer systems for the synthesis of solubilized peptides. Makromol. Chem. 184:2603-2617.

Hiskey, R.G. Sulfhydryl group protection in peptide synthesis. In The Peptides, Vol. 3, E. Gross, and J. Meienhofer, eds., Academic Press, New York, 1981, pp. 137-167.

Hodges, R.S., and Merrifield, R.B. 1975. Monitoring of solid phase peptide synthesis by an automated spectrophotometric picrate method. Anal. Biochem. 65:241-272.

Hoeprich, P.D., Jr., Langton, B.C., Zhang, J.-w., and Tam, J.P. 1989. Identification of immunodominant regions of transforming growth factor α: Implications of structure and function. J. Biol. Chem. 264:19086-19091.

Horn, M., and Novak, C. 1987. A monitoring and control chemistry for solid-phase peptide synthesis. Am. Biotech. Lab. 5(Sept./Oct.):12-21.

Houghten, R.A., and Li, C.H. 1979. Reduction of sulfoxides in peptides and proteins. Anal. Biochem. 98:36-46.

Houghten, R.A., DeGraw, S.T., Bray, M.K., Hoffmann, S.R., and Frizzell, N.D. 1986. Simultaneous multiple peptide synthesis: The rapid preparation of large numbers of discrete peptides for biological, immunological, and methodological studies. BioTechniques 4:522-528.

Hruby, V.J., Al-Obeidi, F., Sanderson, D.G., and Smith, D.D. Synthesis of cyclic peptides by solid phase methods. In Innovation and Perspectives in Solid Phase Synthesis, R. Epton, ed., Solid Phase Conference Coordination, Ltd., Birmingham, U.K., 1990, pp. 197-203.

Hudson, D. 1988a. Methodological implications of simultaneous solid-phase peptide synthesis 1: Comparison of different coupling procedures. J. Org. Chem. 53:617-624.

Hudson, D. 1988b. 2,4,6-Trimethoxybenzyl (Tmob) protection for asparagine and glutamine in Fmoc solid-phase peptide synthesis. Biosearch Technical Bulletin 9000-01, MilliGen/Biosearch Division of Millipore, Bedford, Mass.

Hudson, D. New logically developed active esters for solid-phase peptide synthesis. In Peptides: Chemistry, Structure and Biology, J.E. Rivier and G.R. Marshall, eds., ESCOM, Leiden, The Netherlands, 1990a, pp. 914-915.

+ Hudson, D. 1990b. Methodological implications of simultaneous solid-phase peptide synthesis: A comparison of active esters. Peptide Res. 3:51-55.

Hudson, D., Kain, D., and Ng, D. High yielding fully automatic synthesis of ceropin A amide and analogues. In Peptide Chemistry 1985, Y. Kiso, ed., Protein Research Foundation, Osaka, 1986, pp. 413-418.

Ikeda, S., Yokota, T., Watanabe, K., Kan, M., Takahashi, Y., Ichikawa, T., Takahashi, K., and Matsueda, R. Solid phase peptide synthesis by the use of 3-nitro-2-pyridinesulfenyl (Npys)-amino acids. In Peptide Chemistry 1985, Y. Kiso, ed., Protein Research Foundation, Osaka, 1986, pp. 115-120.

Ishiguro, T., and Eguchi, C. 1989. Unexpected chain-terminating side reaction caused by histidine and acetic anhydride in solid-phase peptide synthesis. Chem. Pharm. Bull. 37:506-508.

Jaeger, E., Thamm, P., Knof, S., Wünsch, E., Löw, M., and Kisfaludy, L. 1978a. Nebenreaktionen bei Peptidsynthesen III: Synthese und Charakterisierung von N^{in}-tert-butylierten Tryptophan-Derivaten. Hoppe-Seyler's Z. Physiol. Chem. 359:1617-1628.

Jaeger, E., Thamm, P., Knof, S., and Wünsch, E. 1978b. Nebenreaktionen bei Peptidsynthesen IV: Charakterisierung von C- und C,N-tert-butylierten Tryptophan-Derivaten. HoppeSeyler Z. Physiol. Chem. 359:1629-1636.

Jaeger, E., Jung, G., Remmer, H.A., and Rücknagel, P. Formation of hydroxy amino acid-O-sulfates during removal of the Pmc-group from arginine residues in solid phase synthesis. In Peptides: Chemistry and Biology, J.A. Smith and J.E. Rivier, eds., ESCOM, Leiden, The Netherlands, 1992, pp. 629-630.

Jansson, A.M., Meldal, M., and Bock, K. 1990. The active ester N-Fmoc-3-O-[Ac4-α-D-Manp-(1→2)-Ac3-128Ma-D-Manp-1-]-threonine-O-Pfp as a building block in solid-phase synthesis of an O-linked dimannosyl glycopeptide. Tetrahedron Lett. 31:6991-6994.

Johnson, C.R., Biancalana, S., Hammer, R.P., Wright, P.B., and Hudson, D. New active esters and coupling reagents based on pyrazolinones. In Peptides: Chemistry and Biology, J.A. Smith and J.E. Rivier, eds., ESCOM, Leiden, The Netherlands, 1992, pp. 585-586.

Jones, J.H., Ramage, W.I., and Witty, M.J. 1980. Mechanism of racemization of histidine derivatives in peptide synthesis. Int. J. Pept. Protein Res. 15:301-303.

Kaiser, E., Colescott, R.L., Bossinger, C.D., and Cook, P.I. 1970. Color test for detection of free terminal amino groups in the solid-phase synthesis of peptides. Anal. Biochem. 34:595-598.

Kaiser, E., Bossinger, C.D., Colescott, R.L., and Olsen, D.B. 1980. Color test for terminal prolyl residues in the solid-phase synthesis of peptides. Anal. Chim. Acta 118:149-151.

Kaiser, E.T., Mihara, H., Laforet, G.A., Kelly, J.W., Walters, L., Findeis, M.A., and Sasaki, T. 1989. Peptide and protein synthesis by segment synthesis-condensation. Science 243:187-192.

Kamber, B., Hartmann, A., Eisler, K., Riniker, B., Rink, H., Sieber, P., and Rittel, W. 1980. The synthesis of cystine peptides by iodine oxidation of S-trityl-cysteine and S-acetamidomethyl-cysteine peptides. Helv. Chim. Acta 63:899-915.

Kawasaki, K., Miyano, M., Murakami, T., and Kakimi, M. 1989. Amino acids and peptides XI: Simple preparation of N^α-protected histidine. Chem. Pharm. Bull. 37:3112-3113.

Kearney, T., and Giles, J. 1989. Fmoc peptide synthesis with a continuous flow synthesizer. Am. Biotech. Lab. 7(9):34-44.

Kemp, D.S., Fotouhi, N., Boyd, J.G., Carey, R.I., Ashton, C., and Hoare, J. 1988. Practical preparation and deblocking conditions for N-α-[2-(p-biphenylyl)-2-propyloxycarbonyl]-amino acid (N-α-Bpoc-Xxx-OH) derivatives. Int. J. Pept. Protein Res. 31:359-372.

Kennedy, R.C., Dreesman, G.R., Chanh, T.C., Boswell, R.N., Allan, J.S., Lee, T.-H., Essex, M., Sparrow, J.T., Ho, D.D., and Kanda, P. 1987. Use of a resin-bound synthetic peptide for identifying a neutralizing antigenic determinant associated with the human immunodeficiency virus envelope. J. Biol. Chem. 262:5769-5774.

Kenner, G.W., and Seely, J.H. 1972. Phenyl esters for C-terminal protection in peptide synthesis. J. Am. Chem. Soc. 94:3259-3260.

Kenner, G.W., Galpin, I.J., and Ramage, R. Synthetic studies directed towards the synthesis of a lysozyme analog. In Peptides: Structure and Biological Function, E. Gross and J. Meienhofer, eds., Pierce Chemical Co., Rockford, Illinois, 1979, pp. 431-438.

Kent, J.J., Alewood, P., and Kent, S.B.H. Peptide synthesis using Fmoc amino acids stored in DMF solution for prolonged periods. In Twelfth American Peptide Symposium Program and Abstracts, Massachusetts Institute of Technology, Cambridge, Mass., 1991, pp. P-386.

Kent, S.B.H. Chronic formation of acylation-resistant deletion peptides in stepwise solid phase peptide synthesis: Chemical mechanism, occurrence, and prevention. In Peptides: Structure and Function, V.J. Hruby and D.H. Rich, eds., Pierce Chemical Co., Rockford, Illinois, 1983, pp. 99-102.

Kent, S.B.H. 1988. Chemical synthesis of peptides, and proteins. Ann. Rev. Biochem. 57:957-989.

Kent, S.B.H., and Merrifield, R.B. 1978. Preparation and properties of tert-butyloxycarbonylaminoacyl-4-(oxymethyl)phenylacetamidomethyl-(Kel F-g-styrene) resin, an insoluble, noncrosslinked support for solid phase peptide synthesis. Isr. J. Chem. 17:243-247.

Kent, S.B.H., and Parker, K.F. The chemical synthesis of therapeutic peptides and proteins. In Banbury Report 29:Therapeutic Peptides and Proteins: Assessing the New Technologies, D.R. Marshak and D.T. Liu, eds., Cold Spring Harbor, New York, 1988, pp. 3-16.

Kent, S.B.H., Riemen, M., LeDoux, M., and Merrifield, R.B. A study of the Edman degradation in the assessment of the purity of synthetic peptides. In Methods in Protein Sequence Analysis, M. Elzinga, ed., Humana Press, Clifton, New Jersey, 1982, pp. 205-213.

Kent, S.B.H., Hood, L.E., Beilan, H., Meister, S., and Geiser, T. High yield chemical synthesis of biologically active peptides on an automated peptide synthesizer of novel design. In Peptides 1984, U. Ragnarsson, ed., Almqvist and Wiksell Int., Stockholm, 1984, pp. 185-188.

⤙ King, D.S., Fields, C.G., and Fields, G.B. 1990. A cleavage method for minimizing side reactions following Fmoc solid phase peptide synthesis. Int. J. Pept. Protein Res. 36:255-266.

Kirstgen, R., and Steglich, W. Fmoc amino acid-TDO esters as reagents for peptide coupling and anchoring in solid phase synthesis. In Peptides 1988, G. Jung and E. Bayer, eds., Walter de Gruyter and Co., Berlin, 1989, pp. 148-150.

Kirstgen, R., Sheppard, R.C., and Steglich, W. 1987. Use of esters of 2,5-diphenyl-2,3-dihydro-3-oxo-4-hydroxythiophene dioxide in solid phase peptide synthesis: A new procedure for attachment of the first amino acid. J. Chem. Soc. Chem. Commun. 1870-1871.

Kirstgen, R., Olbrich, A., Rehwinkel, H., and Steglich, W. 1988. Ester von N-(9-fluorenylmethyloxycarbonyl)aminosäuren mit 4-hydroxy-3-oxo-2,5-diphenyl-2,3-dihydrothiophen-1,1-dioxid (Fmoc-aminosäure-TDO-ester) und ihre Verwendung zur Festphasenpeptidsynthese. Liebigs Ann. Chem. 437-440.

Kisfaludy, L., Löw, M., Nyéki, O., Szirtes, T., and Schön, I. 1973. Die Verwendung von Pentafluorophenylestern bei Peptid-Synthesen. Justus Liebigs Ann. Chem., 1421-1429.

Kisfaludy, L., and Schön, I. 1983. Preparation and applications of pentafluorophenyl esters of 9-fluorenylmethyloxycarbonyl amino acids for peptide synthesis. Synthesis 325-327.

Kiso, Y., Yoshida, M., Tatsumi, T., Kimura, T., Fujiwara, Y., and Akaji, K. 1989. Tetrafluoroboric acid, a useful deprotecting reagent in peptide synthesis. Chem. Pharm. Bull. 37:3432-3434.

Kiso, Y., Yoshida, M., Fujiwara, Y., Kimura, T., Shimokura, M., and Akaji, K. 1990. Trimethylacetamidomethyl (Tacm) group, a new protecting group for the thiol function of cysteine. Chem. Pharm. Bull. 38:673-675.

Kitas, E.A., Perich, J.W., Wade, J.D., Johns, R.B., and Tregear, G.W. 1989. Fmoc-polyamide solid phase synthesis of an O-phosphotyrosine-containing tridecapeptide. Tetrahedron Lett. 30:6229-6232.

✦ Kitas, E.A., Wade, J.D., Johns, R.B., Perich, J.W., and Tregear, G.W. 1991. Preparation and use of N^α-fluorenylmethoxycarbonyl-O-dibenzylphosphono-L-tyrosine in continuous flow solid phase peptide synthesis. J. Chem. Soc. Chem. Commun. 338-339.

✗ Kneib-Cordonier, N., Albericio, F., and Barany, G. 1990. Orthogonal solid-phase synthesis of human gastrin-I under mild conditions. Int. J. Pept. Protein Res. 35:527-538.

⤝ Knorr, R., Trzeciak, A., Bannwarth, W., and Gillessen, D. 1,1,3,3-Tetramethyluronium compounds as coupling reagents in peptide and protein chemistry. In Peptides 1990, E. Giralt and D. Andreu, eds., ESCOM, Leiden, The Netherlands, 1991, pp. 62-64.

Kochersperger, M.L., Blacher, R., Kelly, P., Pierce, L., and Hawke, D.H. 1989. Sequencing of peptides on solid phase supports. Am. Biotech. Lab. 7(3):26-37.

König, W., and Geiger, R. 1970a. Eine neue Methode zur Synthese von Peptiden: Aktivierung der Carboxylgruppe mit Dicyclohexylcarbodiimid unter Zusatz von 1-hydroxy-benzotriazolen. Chem. Ber. 103:788-798.

König, W., and Geiger, R. 1970b. Racemisierung bei Peptidsynthesen. Chem. Ber. 103:2024-2033.

König, W., and Geiger, R. 1970c. Eine neue Methode zur Synthese von Peptiden: Aktivierung der Carboxylgruppe mit Dicyclohexylcarbodiimid und 3-hydroxy-4-oxo-3,4-dihydro-1.2.3-benzotriazin. Chem. Ber. 103:2034-2040.

König, W., and Geiger, R. 1973. N-hydroxyverbindungen als Katalysatoren für die aminolyse aktivierter Ester. Chem. Ber. 106:3626-3635.

Krchnák, V., Vágner, J., Safár, P., and Lebl, M. 1988. Noninvasive continuous monitoring of solid-phase peptide synthesis by acid-base indicator. Coll. Czech. Chem. Commun. 53:2542-2548.

Kullmann, W., and Gutte, B. 1978. Synthesis of an open-chain asymmetrical cystine peptide corresponding to the sequence A^{18-21}-B^{19-26} of bovine insulin by solid phase fragment condensation. Int. J. Pept. Protein Res. 12:17-26.

Kumagaye, K.Y., Inui, T., Nakajima, K., Kimura, T., and Sakakibara, S. 1991. Suppression of a side reaction associated with N^{im}-benzyloxymethyl group during synthesis of peptides containing cysteinyl residue at the N-terminus. Peptide Res. 4:84-87.

Kunz, H. 1987. Synthesis of glycopeptides: Partial structures of biological recognition components. Angew. Chem. Int. Ed. Engl. 26:294-308.

Kunz, H. Allylic anchoring groups in the solid phase synthesis of peptides and glycopeptides. In Innovation and Perspectives in Solid Phase Synthesis, R. Epton, ed., Solid Phase Conference Coordination, Ltd., Birmingham, U.K., 1990, pp. 371-378.

Kunz, H., and Dombo, B. 1988. Solid phase synthesis of peptides and glycopeptides on polymeric supports with allylic anchor groups. Angew. Chem. Int. Ed. Engl. 27:711-713.

Kusunoki, M., Nakagawa, S., Seo, K., Hamana, T., and Fukuda, T. 1990. A side reaction in solid phase synthesis: Insertion of glycine residues into peptide chains via $N^{im} \rightarrow N^{\alpha}$ transfer. Int. J. Pept. Protein Res. 36:381-386.

Lajoie, G., Crivici, A., and Adamson, J.G. 1990. A simple and convenient synthesis of ω-tert-butyl esters of Fmoc-aspartic and Fmoc-glutamic acids. Synthesis 571-572.

Lansbury, P.T., Jr., Hendrix, J.C., and Coffman, A.I. 1989. A practical method for the preparation of protected peptide fragments using the Kaiser oxime resin. Tetrahedron Lett. 30: 4915-4918.

Larsen, B.D., Larsen, C., and Holm, A. Incomplete Fmoc-deprotection in solid phase synthesis. In Peptides 1990, E. Giralt and D. Andreu, eds., ESCOM, Leiden, The Netherlands, 1991, pp. 183-185.

Lapatsanis, L., Milias, G., Froussios, K., and Kolovos, M. 1983. Synthesis of N-2,2,2-(trichloroethoxycarbonyl)-L-amino acids and N-(9-fluorenyl-methoxycarbonyl)-L-amino acids involving succinimidoxy anion as a leaving group in amino acid protection. Synthesis 671-673.

Lebl, M., and Eichler, J. 1989. Simulation of continuous solid phase synthesis: Synthesis of methionine enkephalin and its analogs. Peptide Res. 2:297-300.

Lerner, R. A., Green, N., Alexander, H., Liu, F-T., Sutcliffe, J. G., and Shinnick,

T. M. 1981. Chemically synthesized peptides predicted from the nucleotide sequence of the hepatitus B virus genome elicit antibodies reactive with the native envelope protein of Dane particles. Proc. Natl. Acad. Sci. USA 78:3403-3407.

Li, C.H., Lemaire, S., Yamashiro, D., and Doneen, B.A. 1976. The synthesis and opiate activity of β-endorphin. Biochem. Biophys. Res. Commun. 71:19-25.

Lin, Y.-Z., Caporaso, G., Chang, P.-Y., Ke, X.-H., and Tam, J.P. 1988. Synthesis of a biological active tumor growth factor from the predicted DNA sequence of Shope fibroma virus. Biochemistry 27:5640-5645.

Liu, Y.-Z., Ding, S.-H., Chu, J.-Y., and Felix, A.M. 1990. A novel Fmoc-based anchorage for the synthesis of protected peptides on solid phase. Int. J. Pept. Protein Res. 35:95-98.

Live, D.H., and Kent, S.B.H. Fundamental aspects of the chemical applications of cross-linked polymers. In Elastomers and Rubber Elasticity, J.E. Mark and J. Lal, eds., American Chemical Society, Washington, D.C., 1982, pp. 501-515.

Live, D.H., and Kent, S.B.H. Correlation of coupling rates with physicochemical properties of resin-bound peptides in solid phase synthesis. In Peptides: Structure and Function, V.J. Hruby and D.H. Rich, eds., Pierce Chemical Co., Rockford, Illinois, 1983, pp. 65-68.

Lloyd-Williams, P., Albericio, F., and Giralt, E. 1991a. Convergent solid-phase peptide synthesis VIII: Synthesis, using a photolabile resin, and purification of a methionine-containing protected peptide. Int. J. Pept. Protein Res. 37:58-60.

Lloyd-Williams, P., Jou, G., Albericio, F., and Giralt, E. 1991b. Solid-phase synthesis of peptides using allylic anchoring groups: An investigation of their palladium-catalysed cleavage. Tetrahedron Lett. 32:4207-4210.

Loffet, A., Galeotti, N., Jouin, P., and Castro, B. 1989. Tert-butyl esters of N-protected amino acids with tert-butyl fluorocarbonate (Boc-F). Tetrahedron Lett. 30:6859-6860.

Löw, M., Kisfaludy, L., Jaeger, E., Thamm, P., Knof, S., and Wünsch, E. 1978a. Direkte tert-Butylierung des Tryptophans: Herstellung von 2,5,7-tri-tert-butyltryptophan. Hoppe-Seyler's Z. Physiol. Chem. 359:1637-1642.

Löw, M., Kisfaludy, L., and Sohár, P. 1978b. tert-Butylierung des Tryptophan-indolringes während der Abspaltung der tert-Butyloxycarbonyl-Gruppe bei Peptidsynthesen. Hoppe-Seyler's Z. Physiol. Chem. 359:1643-1651.

Lu, G.-s., Mojsov, S., Tam, J.P., and Merrifield, R.B. 1981. Improved synthesis of 4-alkoxybenzyl alcohol resin. J. Org. Chem. 46:3433-3436.

Ludwick, A.G., Jelinski, L.W., Live, D., Kintanar, A., and Dumais, J.J. 1986. Association of peptide chains during Merrifield solid-phase peptide synthesis: A deuterium NMR study. J. Am. Chem. Soc. 108:6493-6496.

Lundt, B.F., Johansen, N.L., Volund, A., and Markussen, J. 1978. Removal of t-butyl and t-butoxycarbonyl protecting groups with trifluoroacetic acid. Int. J. Pept. Protein Res. 12:258-268.

Lyttle, M.H., and Hudson, D. Allyl based side-chain protection for SPPS. In Peptides: Chemistry and Biology, J.A. Smith and J.E. Rivier, eds., ESCOM, Leiden, The Netherlands, 1992, pp. 583-584.

Masui, Y., Chino, N., and Sakakibara, S. 1980. The modification of tryptophyl residues during the acidolytic cleavage of Boc-groups I: Studies with Boc-tryptophan. Bull. Chem. Soc. Jpn. 53:464-468.

Matsueda, G.R., and Haber, E. 1980. The use of an internal reference amino acid for the evaluation of reactions in solid-phase peptide synthesis. Anal. Biochem. 104:215-227.

Matsueda, G.R., and Stewart, J.M. 1981. A p-methylbenzhydrylamine resin for improved solid-phase synthesis of peptide amides. Peptides 2:45-50.

Matsueda, G.R., Haber, E., and Margolies, M.N. 1981. Quantitative solid-phase Edman degradation for evaluation of extended solid-phase peptide synthesis. Biochemistry 20:2571-2580.

Matsueda, R., and Walter, R. 1980. 3-nitro-2-pyridinesulfenyl (Npys) group: A novel selective protecting group which can be activated for peptide bond formation. Int. J. Pept. Protein Res. 16:392-401.

McCurdy, S.N. 1989. The investigation of Fmoc-cysteine derivatives in solid phase peptide synthesis. Peptide Res. 2:147-151.

McFerran, N.V., Walker, B., McGurk, C.D., and Scott, F.C. 1991. Conductance measurements in solid phase peptide synthesis I: Monitoring coupling and deprotection in Fmoc chemistry. Int. J. Pept. Protein Res. 37:382-387.

Meienhofer, J. Protected amino acids in peptide synthesis. In Chemistry and Biochemistry of the Amino Acids, G.C. Barrett, ed., Chapman and Hall, London, 1985, pp. 297-337.

Meienhofer, J., Waki, M., Heimer, E.P., Lambros, T.J., Makofske, R.C., and Chang, C.-D. 1979. Solid phase synthesis without repetitive acidolysis. Int. J. Pept. Protein Res. 13:35-42.

Meister, S.M., and Kent, S.B.H. Sequence-dependent coupling problems in stepwise solid phase peptide synthesis: Occurence, mechanism, and correction. In Peptides: Structure and Function, V.J. Hruby and D.H. Rich, eds., Pierce Chemical Co., Rockford, Illinois, 1983, pp. 103-106.

Meldal, M., and Jensen, K.J. 1990. Pentafluorophenyl esters for the temporary protection of the α-carboxy group in solid phase glycopeptide synthesis. J. Chem. Soc. Chem. Commun. 483-485.

Mergler, M., Tanner, R., Gosteli, J., and Grogg, P. 1988a. Peptide synthesis by a combination of solid-phase and solution methods I: A new very acid-labile anchor group for the solid phase synthesis of fully protected fragments. Tetrahedron Lett. 29:4005-4008.

Mergler, M., Nyfeler, R., Tanner, R., Gosteli, J., and Grogg, P. 1988b. Peptide synthesis by a combination of solid-phase and solution methods II: Synthesis of fully protected peptide fragments on 2-methoxy-4-alkoxy-benzyl alcohol resin. Tetrahedron Lett. 29:4009-4012.

Mergler, M., Nyfeler, R., and Gosteli, J. 1989a. Peptide synthesis by a combination of solid-phase and solution methods III: Resin derivatives allowing minimum-racemization coupling of N^α-protected amino acids. Tetrahedron Lett. 30:6741-6744.

Mergler, M., Nyfeler, R., Gosteli, J., and Tanner, R. 1989b. Peptide synthesis by a combination of solid-phase and solution methods IV: Minimum-racemization coupling of N^α-9-fluorenylmethyloxycarbonyl amino acids to alkoxy benzyl alcohol type resins. Tetrahedron Lett. 30:6745-6748.

Merrifield, R.B. 1986. Solid phase synthesis. Science 232:341-347.

Merrifield, R.B., and Bach, A.E. 1978. 9-(2-Sulfo)fluorenylmethyloxycarbonyl chloride, a new reagent for the purification of synthetic peptides. J. Org. Chem. 43:4808-4816.

Merrifield, R.B., Stewart, J.M., and Jernberg, N. 1966. Instrument for automated synthesis of peptides. Anal. Chem. 38:1905-1914.

Merrifield, R.B., Mitchell, A.R., and Clarke, J.E. 1974. Detection and prevention of urethane acylation during solid-phase peptide synthesis by anhydride methods. J. Org. Chem. 39:660-668.

Merrifield, R.B., Vizioli, L.D., and Boman, H.G. 1982. Synthesis of the antibacterial peptide cecropin A(1-33). Biochemistry 21:5020-5031.

Merrifield, R.B., Singer, J., and Chait, B.T. 1988. Mass spectrometric evaluation of synthetic peptides for deletions and insertions. Anal. Biochem. 174:399-414.

Méry, J., and Calas, B. 1988. Tryptophan reduction and histidine racemization during deprotection by catalytic transfer hydrogenation of an analog of the luteinizing hormone releasing factor. Int. J. Pept. Protein Res. 31:412-419.

MilliGen/Biosearch 1990. Flow rate programming in continuous flow peptide synthesizer solves problematic synthesis. MilliGen/Biosearch Report 7, MilliGen/Biosearch Division of Millipore, Bedford, Mass., pp. 4-5, 11.

Milton, R.C. de L., Becker, E., Milton, S.C.F., Baxter, J.E.J., and Elsworth, J.F. 1987. Improved purities for Fmoc amino acids from Fmoc-ONSu. Int. J. Pept. Protein Res. 30:431-432.

Mitchell, A.R., Kent, S.B.H., Engelhard, M., and Merrifield, R.B. 1978. A new synthetic route to tert-butyloxycarbonylaminoacyl-4-(oxymethyl)phenylacetamidomethyl-resin, an improved support for solid-phase peptide synthesis. J. Org. Chem. 43:2845-2852.

Mitchell, M.A., Runge, T.A., Mathews, W.R., Ichhpurani, A.K., Harn, N.K., Dobrowolski, P.J., and Eckenrode, F.M. 1990. Problems associated with use of the benzyloxymethyl protecting group for histidines: Formaldehyde adducts formed during cleavage by hydrogen fluoride. Int. J. Pept. Protein Res. 36:350-355.

Mojsov, S., Mitchell, A.R., and Merrifield, R.B. 1980. A quantitative evaluation of methods for coupling asparagine. J. Org. Chem. 45:555-560.

Mott, A.W., Slomczynska, U., and Barany, G. Formation of sulfur-sulfur bonds during solid-phase peptide synthesis: Application to the synthesis of oxytocin. In Forum peptides le Cap d'Agde 1984, B. Castro and J. Martinez, eds., Les Impressions Dohr, Nancy, pp. 321-324.

Munson, M.C., García-Echeverría, C., Albericio, F., and Barany, G. 1992. S-2,4,6-trimethoxybenzyl (Tmob): A novel cysteine protecting group for the N^{α}-9-fluorenylmethyloxycarbonyl (Fmoc) strategy of peptide synthesis. J. Org. Chem., in press.

Mutter, M., and Bellof, D. 1984. A new base-labile anchoring group for polymer-supported peptide synthesis. Helv. Chim. Acta 67:2009-2016.

Mutter, M., Altmann, K.-H., Bellof, D., Flörsheimer, A., Herbert, J., Huber, M., Klein, B., Strauch, L., and Vorherr, T. The impact of secondary structure formation in peptide synthesis. In Peptides: Structure and Function, C.M. Deber, V.J. Hruby and K.D. Kopple, eds., Pierce Chemical Co., Rockford, Illinois, 1985, pp. 397-405.

Nakagawa, S.H., Lau, H.S.H., Kézdy, F.J., and Kaiser, E.T. 1985. The use of polymer-bound oximes for the synthesis of large peptides usable in segment condensation: Synthesis of a 44 amino acid amphiphilic peptide model of apolipoprotein A-1. J. Am. Chem. Soc. 107:7087-7092.

Narita, M., and Kojima, Y. 1989. The β-sheet structure-stabilizing potential of twenty kinds of amino acid residues in protected peptides. Bull. Chem. Soc. Jpn. 62:3572-3576.

Narita, M., Umeyama, H., and Yoshida, T. 1989. The easy disruption of the β-sheet structure of resin-bound human proinsulin C-peptide fragments by strong electron-donor solvents. Bull. Chem. Soc. Jpn. 62:3582-3586.

Netzel-Arnett, S., Fields, G.B., Birkedal-Hansen, H., and Van Wart, H.E. 1991. Sequence specificities of human fibroblast and neutrophil collagenases. J. Biol. Chem. 266:6747-6755, 21326.

Nicolás, E., Pedroso, E., and Giralt, E. 1989. Formation of aspartimide peptides in Asp-Gly sequences. Tetrahedron Lett. 30:497-500.

Nielson, C.S., Hansen, P.H., Lihme, A., and Heegaard, P.M.H. 1989. Real time monitoring of acylations during solid phase peptide synthesis:A method based on electrochemical detection. J. Biochem. Biophys. Methods 20:69-75.

Noble, R.L., Yamashiro, D., and Li, C.H. 1976. Synthesis of a nonadecapeptide corresponding to residues 37-55 of ovine prolactin: Detection and isolation of the sulfonium form of methionine-containing peptides. J. Am. Chem. Soc. 98:2324-2328.

Nokihara, K., Hellstern, H., and Höfle, G. Peptide synthesis by fragment assembly on a polymer support. In Peptides 1988, G. Jung and E. Bayer, eds., Walter de Gruyter and Co., Berlin, 1989, pp. 166-168.

Nomizu, M., Inagaki, Y., Yamashita, T., Ohkubo, A., Otaka, A., Fujii, N., Roller, P.P., and Yajima, H. 1991. Two-step hard acid deprotection/cleavage procedure for solid phase peptide synthesis. Int. J. Pept. Protein Res. 37:145-152.

Nutt, R.F., Brady, S.F., Darke, P.L., Ciccarone, T.M., Colton, C.D., Nutt, E.M., Rodkey, J.A., Bennett, C.D., Waxman, L.H., Sigal, I.S., Anderson, P.S., and Veber, D.F. 1988. Chemical synthesis and enzymatic activity of a 99-residue peptide with a sequence proposed for the human immunodeficiency virus protease. Proc. Natl. Acad. Sci. USA 85:7129-7133.

Ogunjobi, O., and Ramage, R. 1990. Ubiquitin: Preparative chemical synthesis, purification and characterization. Biochem. Soc. Trans. 18:1322-1323.

Okada, Y., and Iguchi, S. 1988. Amino acid and peptides, part 19: Synthesis of β-1- and β-2-adamantyl aspartates and their evaluation for peptide synthesis. J. Chem. Soc. Perkin Trans. I:2129-2136.

Ondetti, M.A., Pluscec, J., Sabo, E.F., Sheehan, J.T., and Williams, N. 1970. Synthesis of cholecystokinin-pancreozymin I: The C-terminal dodecapeptide. J. Am. Chem. Soc. 92:195-199.

Orlowska, A., Witkowska, E., and Izdebski, J. 1987. Sequence dependence in the formation of pyroglutamyl peptides in solid phase peptide synthesis. Int. J. Pept. Protein Res. 30:141-144.

Orpegen 1990. HYCRAM^{TM} support: Loading and cleavage. Technical Note, Orpegen, Heidelberg, Germany

Otteson, K.M., Harrison, J.L., Ligutom, A., and Ashcroft, P. 1989. Solid phase peptide synthesis with N-methylpyrrolidone as the solvent for both Fmoc and Boc synthesis. Poster Presentations at the Eleventh American Peptide Symposium, Applied Biosystems, Inc., Foster City, Calif., pp. 34-38.

Otvös, L., Jr., Elekes, I., and Lee, V.M.-Y. 1989a. Solid-phase synthesis of phosphopeptides. Int. J. Pept. Protein Res. 34:129-133.

Otvös, L., Jr., Wroblewski, K., Kollat, E., Perczel, A., Hollosi, M., Fasman, G.D., Ertl, H.C.J., and Thurin, J. 1989b. Coupling strategies in solid-phase synthesis of glycopeptides. Peptide Res. 2:362-366.

✕ Otvös, L., Jr., Urge, L., Hollosi, M., Wroblewski, K., Graczyk, G., Fasmán, G.D., and Thurin, J. 1990. Automated solid-phase synthesis of glycopeptides: Incorporation of unprotected mono- and disaccharide units of N-glycoprotein antennae into T cell epitopic peptides. Tetrahedron Lett. 31:5889-5892.

Pacquet, A. 1982. Introduction of 9-fluorenylmethyloxycarbonyl, trichloroethoxycarbonyl, and benzyloxycarbonyl amine protecting groups into O-unprotected hydroxyamino acids using succinimidyl carbonates. Can. J. Chem. 60:976-980.

Patchornik, A., Amit, B., and Woodward, R.B. 1970. Photosensitive protecting groups. J. Am. Chem. Soc. 92:6333-6335.

✕ Patel, K., and Borchardt, R.T. 1990a. Chemical pathways of peptide degradation II: Kinetics of deamidation of an asparaginyl residue in a model hexapeptide. Pharm. Res. 7:703-711.

✕ Patel, K., and Borchardt, R.T. 1990b. Chemical pathways of peptide degradation III: Effect of primary sequence on the pathways of deamidation of asparaginyl residues in hexapeptides. Pharm. Res. 7:787-793.

⟩ Paulsen, H., Merz, G., and Weichert, U. 1988. Solid-phase synthesis of O-glycopeptide sequences. Angew. Chem. Int. Ed. Engl. 27:1365-1367.

✕ Paulsen, H., Merz, G., Peters, S., and Weichert, U. 1990. Festphasensynthese von O-glycopeptiden. Liebigs Ann. Chem. 1165-1173.

Pearson, D.A., Blanchette, M., Baker, M.L., and Guindon, C.A. 1989. Trialkylsilanes as scavengers for the trifluoroacetic acid deblocking of protecting groups in peptide synthesis. Tetrahedron Lett. 30:2739-2742.

Pedroso, E., Grandas, A., Eritja, R., and Giralt, E. Use of Nbb-resin and 4-hydroxymethylphenoxymethyl-resin in solid phase peptide synthesis. In Peptides 1982, K. Blaha and P. Malon, eds., Walter de Gruyter and Co., Berlin, 1983, pp. 237-240.

Pedroso, E., Grandas, A., de las Heras, X., Eritja, R., and Giralt, E. 1986. Diketopiperazine formation in solid phase peptide synthesis using p-alkoxybenzyl ester resins and Fmoc amino acids. Tetrahedron Lett. 27:743-746.

Penke, B., and Nyerges, L. Preparation and application of a new resin for synthesis of peptide amides via Fmoc-strategy. In Peptides 1988, G. Jung and E. Bayer, eds., Walter de Gruyter and Co., Berlin, 1989, pp. 142-144.

✕ Penke, B., and Nyerges, L. 1991. Solid phase synthesis of porcine cholecystokinin-33 in a new resin via Fmoc strategy. Peptide Res. 4:289-295.

Penke, B., and Rivier, J. 1987. Solid-phase synthesis of peptide amides on a polystyrene support using fluorenylmethoxycarbonyl protecting groups. J. Org. Chem. 52:1197-1200.

Penke, B., and Tóth, G.K. An improved method for the preparation of large amounts of ω-cyclohexylesters of aspartic and glutamic acid. In Peptides 1988, G. Jung and E. Bayer, eds., Walter de Gruyter and Co., Berlin, 1989, pp. 67-69.

Penke, B., Baláspiri, L., Pallai, P., and Kovács, K. 1974. Application of pen-tafluorophenyl esters of Boc amino acids in solid phase peptide synthesis. Acta Phys. Chem. 20:471-476.

Penke, B., Nyerges, L., Klenk, N., Nagy, K., and Asztalos, A. Preparation and application of a new resin for the synthesis of peptide amides with Fmoc strategy. In Peptides: Chemistry, Biology, Interactions with Proteins, B. Penke and A. Torok, eds., Walter de Gruyter and Co., Berlin, 1988, pp. 121-126.

Penke, B., Zsigo, J., and Spiess, J. Synthesis of a protected hydroxylysine derivative for application in peptide synthesis. In The Eleventh American Peptide Symposium Abstracts, The Salk Institute and University of California at San Diego, La Jolla, Calif., 1989, pp. P-335.

Pennington, M.W., Festin, S.M., and Maccecchini, M.L. Comparison of folding procedures on synthetic ω-conotoxin. In Peptides 1990, E. Giralt and D. Andreu, eds., ESCOM, Leiden, The Netherlands, 1991, pp. 164-166.

Perich, J.W. Modern methods of O-phosphoserine- and O-phosphotyrosine-con-taining peptide synthesis. In Peptides and Protein Phosphorylation, B.E. Kemp, ed., CRC Press, Boca Raton, Florida, 1990, pp. 289-314.

Perich, J.W., and Reynolds, E.C. 1991. Fmoc/solid-phase synthesis of Tyr(P)- ✕ containing peptides through t-butyl phosphate protection. Int. J. Pept. Protein Res. 37:572-575.

Perich, J.W., Valerio, R.M., and Johns, R.B. 1986. Solid-phase synthesis of an O-phosphoseryl-containing peptide using phenyl phosphorotriester protec-tion. Tetrahedron Lett. 27:1377-1380.

Pessi, A., Bianchi, E., Bonelli, F., and Chiappinelli, L. 1990. Application of the ✕ continuous-flow polyamide method to the solid-phase synthesis of a multi-ple antigen peptide (MAP) based on the sequence of a malaria epitope. J. Chem. Soc. Chem. Commun. 8-9.

Photaki, I., Taylor-Papadimitriou, J., Sakarellos, C., Mazarakis, P., and Zervas, L. 1970. On cysteine and cystine peptides, part V: S-trityl- and S-diphenyl-methyl-cysteine and -cysteine peptides. J. Chem. Soc. (C):2683-2687.

Pickup, S., Blum, F.D., and Ford, W.T. 1990. Self-diffusion coefficients of Boc amino acid anhydrides under conditions of solid phase peptide synthesis. J. Polym. Sci.: Polym. Chem. Ed. 28:931-934.

Pipkorn, R., and Bernath, E. Solid phase synthesis of preprocecropin A. In In- ✕ novation and Perspectives in Solid Phase Synthesis, R. Epton, ed., Solid Phase Conference Coordination, Ltd., Birmingham, U.K., 1990, pp. 537-541.

Plaué, S. 1990. Synthesis of cyclic peptides on solid support: Application to ✕ analogs of hemagglutinin of influenza virus. Int. J. Pept. Protein Res. 35:510-517.

Ploux, O., Chassaing, G., and Marquet, A. 1987. Cyclization of peptides on a solid support: Application to cyclic analogs of substance P. Int. J. Pept. Protein Res. 29:162-169.

Pluscec, J., Sheehan, J.T., Sabo, E.F., Williams, N., Kocy, O., and Ondetti, M.A. 1970. Synthesis of analogs of the C-terminal octapeptide of cholecys-tokinin-pancreozymin: Structure-activity relationship. J. Med. Chem. 13:349-352.

Ponsati, B., Giralt, E., and Andreu, D. 1989. A synthetic strategy for simultaneous purification-conjugation of antigenic peptides. Anal. Biochem. 181:389-395.

✕ Ponsati, B., Giralt, E., and Andreu, D. Side reactions in post-HF workup of peptides having the unusual Tyr-Trp-Cys sequence: Low-high acidolysis revisited. In Peptides: Chemistry, Structure and Biology, J.E. Rivier and G.R. Marshall, eds., ESCOM, Leiden, The Netherlands, 1990a, pp. 960-962.

✕ Ponsati, B., Giralt, E., and Andreu, D. 1990b. Solid-phase approaches to regiospecific double disulfide formation: Application to a fragment of bovine pituitary peptide. Tetrahedron 46:8255-8266.

Prasad, K.U., Trapane, T.L., Busath, D., Szabo, G., and Urry, D.W. 1982. Synthesis and characterization of 1-^{13}C-D-Leu^{12,14} gramicidin A. Int. J. Pept. Protein Res. 19:162-171.

Prosser, R.S., Davis, J.H., Dahlquist, F.W., and Lindorfer, M.A. 1991. ^{2}H nuclear magnetic resonance of the gramicidin A backbone in a phospholipid bilayer. Biochemistry 30:4687-4696.

Ramage, R., and Green, J. 1987. N$_G$-2,2,5,7,8-pentamethylchroman-6-sulphonyl-L-arginine: A new acid labile derivative for peptide synthesis. Tetrahedron Lett. 28:2287-2290.

Ramage, R., Green J., and Ogunjobi, O.M. 1989. Solid phase peptide synthesis of ubiquitin. Tetrahedron Lett. 30:2149-2152.

Reddy, M.P., and Voelker, P.J. 1988. Novel method for monitoring the coupling efficiency in solid phase peptide synthesis. Int. J. Pept. Protein Res. 31:345-348.

Rich, D.H., and Gurwara, S.K. 1975. Preparation of a new o-nitrobenzyl resin for solid-phase synthesis of tert-butyloxycarbonyl-protected peptide acids. J. Am. Chem. Soc. 97:1575-1579.

Rich, D.H., and Singh, J. The carbodiimide method. In The Peptides, Vol. 1, E. Gross and J. Meienhofer, eds., Academic Press, New York, 1979, pp. 241-314.

Riniker, B., and Hartmann, A. Deprotection of peptides containing Arg(Pmc) and tryptophan or tyrosine:Elucidation of by-products. In Peptides: Chemistry, Structure and Biology, J.E. Rivier and G.R. Marshall, eds., ESCOM, Leiden, The Netherlands, 1990, pp. 950-952.

Riniker, B., and Kamber, B. Byproducts of Trp-peptides synthesized on a p-benzyloxybenzyl alcohol polystyrene resin. In Peptides 1988, G. Jung and E. Bayer, eds., Walter de Gruyter, and Co., Berlin, 1989, pp. 115-117.

Riniker, B., and Sieber, P. Problems and progress in the synthesis of histidine-containing peptides. In Peptides: Chemistry, Biology, Interactions with Proteins, B. Penke and A. Torok, eds., Walter de Gruyter and Co., Berlin, 1988, pp. 65-74.

Rink, H. 1987. Solid-phase synthesis of protected peptide fragments using a tri-alkoxy-diphenyl-methylester resin. Tetrahedron Lett. 28:3787-3790.

✕ Rink, H., and Ernst, B. Glycopeptide solid-phase synthesis with an acetic acid-labile trialkoxy-benzhydryl linker. In Peptides 1990, E. Giralt and D. Andreu, eds., ESCOM, Leiden, The Netherlands, 1991, pp. 418-419.

Rink, H., Sieber, P., and Raschdorf, F. 1984. Conversion of N^G-urethane protected arginine to ornithine in peptide solid phase synthesis. Tetrahedron Lett. 25:621-624.

Rivier, J., Galyean, R., Simon, L., Cruz, L.J., Olivera, B.M., and Gray, W.R.

1987. Total synthesis and further characterization of the γ-carboxyglutamate-containing "sleeper" peptide from *Conus geographus* venom. Biochemistry 26:8508-8512.

Romani, S., Moroder, L., Göhring, W., Scharf, R., Wünsch, E., Barde, Y.A., and Thoenen, H. 1987. Synthesis of the trypsin fragment 10-25/75-88 of mouse nerve growth factor II: The unsymmetrical double chain cystine peptide. Int. J. Pept. Protein Res. 29:107-117.

Rosen, O., Rubinraut, S., and Fridkin, M. 1990. Thiolysis of the 3-nitro-2-   ⅄
pyridinesulfenyl (Npys) protecting group: An approach towards a general deprotection scheme in peptide synthesis. Int. J. Pept. Protein Res. 35:545-549.

Ruiz-Gayo, M., Albericio, F., Pons, M., Royo, M., Pedroso, E., and Giralt, E. 1988. Uteroglobin-like peptide cavities I: Synthesis of antiparallel and parallel dimers of bis-cysteine peptides. Tetrahedron Lett. 29:3845-3848.

Rzeszotarska, B., and Masiukiewicz, E. 1988. Arginine, histidine and tryptophan in peptide synthesis: The guanidino function of arginine. Org. Prep. Proc. Int. 20:427-464.

Sarin, V.K., Kent, S.B.H., and Merrifield, R.B. 1980. Properties of swollen polymer networks: Solvation and swelling of peptide-containing resins in solid-phase peptide synthesis. J. Am. Chem. Soc. 102:5463-5470.

Sarin, V.K., Kent, S.B.H., Tam, J.P., and Merrifield, R.B. 1981. Quantitative monitoring of solid-phase peptide synthesis by the ninhydrin reaction. Anal. Biochem. 117:147-157.

Sarin, V.K., Kent, S.B.H., Mitchell, A.R., and Merrifield, R.B. 1984. A general approach to the quantitation of synthetic efficiency in solid-phase peptide synthesis as a function of chain length. J. Am. Chem. Soc. 106:7845-7850.

Sasaki, T., and Kaiser, E.T. 1990. Synthesis and structural stability of   ✗
helichrome as an artificial hemeproteins. Biopolymers 29:79-88.

Scarr, R.B., and Findeis, M.A. 1990. Improved synthesis and aminoacylation of *p*-nitrobenzophenone oxime polystyrene resin for solid-phase synthesis of protected peptides. Peptide Res. 3:238-241.

Schielen, W.J.G., Adams, H.P.H.M., Nieuwenhuizen, W., and Tesser, G.I. 1991. Use of Mpc-amino acids in solid phase peptide synthesis leads to improved coupling efficiencies. Int. J. Pept. Protein Res. 37:341-346.

Schiller, P.W., Nguyen, T.M.-D., and Miller, J. 1985. Synthesis of side-chain cyclized peptide analogs on solid supports. Int. J. Pept. Protein Res. 25:171-177.

Schneider, J., and Kent, S.B.H. 1988. Enzymatic activity of a synthetic 99 residue protein corresponding to the putative HIV-1 protease. Cell 54:363-368.

Schnolzer, M., Alewood, P.F., and Kent, S.B.H. "In situ" neutralization in Boc chemistry SPPS: High yield assembly of "difficult" sequences. In Peptides: Chemistry and Biology, J.A. Smith and J.E. Rivier, eds., ESCOM, Leiden, The Netherlands, 1992, pp. 623-624.

Schnorrenberg, G., and Gerhardt, H. 1989. Fully automatic simultaneous multiple peptide synthesis in micromolar scale rapid synthesis of series of peptides for screening in biological assays. Tetrahedron 45:7759-7764.

Scoffone, E., Previero, A., Benassi, C.A., and Pajetta, P. Oxidative modification of tryptophan residues in peptides. In Peptides 1963, L. Zervas, ed., Pergamon Press, Oxford, 1966, pp. 183-188.

⤉ Selsted, M.E., Levy, J.N., Van Abel, R.J., Cullor, J.C., Bontems, R.J., and
    Barany, G. Purification, characterization, synthesis and cDNA cloning of in-
    dolicidin: A tryptophan-rich microbicidal tridecapeptide from neutrophils.
    In Peptides: Chemistry and Biology, J.A. Smith and J.E. Rivier, eds.,
    ESCOM, Leiden, The Netherlands, 1992, pp. 905-907.

⤉ Seyer, R., Aumelas, A., Caraty, A., Rivaille, P., and Castro, B. 1990. Repetitive
    BOP coupling (REBOP) in solid phase peptide synthesis: Luliberin syn-
    thesis as model. Int. J. Pept. Protein Res. 35:465-472.

Shao, J., Shekhani, M.S., Krauss, S., Grübler, G., and Voelter, W. 1991. A test
    case for the 1-(1-adamantyl)-1-methylethoxycarbonyl (Adpoc) group: Solid-
    phase synthesis of LH-RH using N^α-Adpoc protection and an acid labile
    handle. Tetrahedron Lett. 32:345-346.

Sheppard, R.C., and Williams, B.J. 1982. Acid-labile resin linkage agents for
    use in solid phase peptide synthesis. Int. J. Pept. Protein Res. 20:451-454.

Sieber, P. 1987a. An improved method for anchoring of 9-fluorenylmethoxycar-
    bonyl-amino acids to 4-alkoxybenzyl alcohol resins. Tetrahedron Lett.
    28:6147-6150.

Sieber, P. 1987b. Modification of tryptophan residues during acidolysis of 4-
    methoxy-2,3,6-trimethylbenzenesulfonyl groups: Effects of scavengers.
    Tetrahedron Lett. 28:1637-1640.

Sieber, P. 1987c. A new acid-labile anchor group for the solid-phase synthesis
    of C-terminal peptide amides by the Fmoc method. Tetrahedron Lett.
    28:2107-2110.

Sieber, P., and Riniker, B. 1987. Protection of histidine in peptide synthesis: A
    reassessment of the trityl group. Tetrahedron Lett. 28:6031-6034.

Sieber, P., and Riniker, B. Side-chain protection of asparagine and glutamine by
    trityl: Application to solid-phase peptide synthesis. In Innovation and
    Perspectives in Solid Phase Synthesis, R. Epton, ed., Solid Phase Con-
    ference Coordination, Ltd., Birmingham, U.K., 1990, pp. 577-583.

⤉ Sieber, P., and Riniker, B. 1991. Protection of carboxamide functions by the trityl
    residue: Application to peptide synthesis. Tetrahedron Lett. 32:739-742.

Sieber, P., Kamber, B., Riniker, B., and Rittel, W. 1980. Iodine oxidation of S-
    trityl- and S-acetamidomethyl-cysteine-peptides containing tryptophan: Con-
    ditions leading to the formation of tryptophan-2-thioethers. Helv. Chim.
    Acta 63:2358-2363.

Sigler, G.F., Fuller, W.D., Chaturvedi, N.C., Goodman, M., and Verlander, M.S.
    1983. Formation of oligopeptides during the synthesis of 9-fluorenyl-
    methyloxycarbonyl amino acid derivatives. Biopolymers 22:2157-2162.

Simmons, J., and Schlesinger, D.H. 1980. High-performance liquid chromatog-
    raphy of side-chain-protected amino acid phenylthiohydantoins. Anal.
    Biochem. 104:254-258.

Small, P.W., and Sherrington, D.C. 1989. Design and application of a new rigid
    support for high efficiency continuous-flow peptide synthesis. J. Chem. Soc.
    Chem. Commun. 1589-1591.

Southard, G.L. Comments. In Peptides 1969, E. Scoffone, ed., North-Holland
    Publishing, Amsterdam, The Netherlands, 1971.

Stanley, M., Tom, J.Y.K., Burdick, D.J., Struble, M., and Burnier, J.P. In
    Twelfth American Peptide Symposium Program and Abstracts, Mas-
    sachusetts Institute of Technology, Cambridge, Mass., 1991, pp. P-328.

Steiman, D.M., Ridge, R.J., and Matsueda, G.R. 1985. Synthesis of side chain-protected amino acid phenylthiohydantoins and their use in quantitative solid-phase Edman degradation. Anal. Biochem. 145:91-95.

Stephenson, R.C., and Clarke, S. 1989. Succinimide formation from aspartyl and asparaginyl peptides as a model for the spontaneous degradation of proteins. J. Biol. Chem. 264:6164-6170.

Stewart, J.M., and Klis, W.A. Polystyrene-based solid phase peptide synthesis: The state of the art. In Innovation and Perspectives in Solid Phase Synthesis, R. Epton, ed., Solid Phase Conference Coordination, Ltd., Birmingham, U.K., 1990, pp. 1-9.

Stewart, J.M., and Young, J.D. Solid Phase Peptide Synthesis, 2nd Ed., Pierce Chemical Co., Rockford, Illinois, 1984.

Stewart, J.M., Knight, M., Paiva, A.C.M., and Paiva, T. Histidine in solid phase peptide synthesis: Thyrotropin releasing hormone and the angiotensins. In Progress in Peptide Research, Vol. 2, S. Lande, ed., Gordon and Breach, New York, 1972, pp. 59-64.

Stewart, J.M., Ryan, J.W., and Brady, A.H. 1974. Hydroxyproline analogs of bradykinin. J. Med. Chem. 17:537-539.

Story, S.C., and Aldrich, J.V. 1992. A resin for the solid phase synthesis of $\times$ protected peptide amides using the Fmoc chemical protocol. Int. J. Pept. Protein Res., 39:87-92.

Stüber, W., Knolle, J., and Breipohl, G. 1989. Synthesis of peptide amides by Fmoc-solid-phase peptide synthesis and acid labile anchor groups. Int. J. Pept. Protein Res. 34:215-221.

Sugg, E.E., Castrucci, A.M. de L., Hadley, M.E., van Binst, G., and Hruby, V.J. 1988. Cyclic lactam analogues of Ac-[Nle^4]α-MSH$_{4-11}$-NH$_2$. Biochemistry 27:8181-8188.

Suzuki, K., Nitta, K., and Endo, N. 1975. Suppression of diketopiperazine formation in solid phase peptide synthesis. Chem. Pharm. Bull. 23:222-224.

Tam, J.P. 1988. Synthetic peptide vaccine design: Synthesis and properties of a high-density multiple antigenic peptide system. Proc. Natl. Acad. Sci. USA 85:5409-5413.

Tam, J.P., and Lu, Y.-A. Synthetic peptide vaccine engineering: Design and synthesis of unambiguous peptide-based vaccines containing multiple peptide antigens for malaria and hepatitis. In Innovation and Perspectives in Solid Phase Synthesis, R. Epton, ed., Solid Phase Conference Coordination, Ltd., Birmingham, U.K., 1990, pp. 351-370.

Tam, J.P., and Merrifield, R.B. 1985. Solid phase synthesis of gastrin I: Comparison of methods utilizing strong acid for deprotection and cleavage. Int. J. Pept. Protein Res. 26:262-273.

Tam, J.P., and Merrifield, R.B. Strong acid deprotection of synthetic peptides: Mechanisms and methods. In The Peptides, Vol. 9, S. Udenfriend and J. Meienhofer, eds., Academic Press, New York, 1987, pp. 185-248.

Tam, J.P., Health, W.F., and Merrifield, R.B. 1983. S$_N$2 deprotection of synthetic peptides with a low concentration of HF in dimethyl sulfide: Evidence and application in peptide synthesis. J. Am. Chem. Soc. 105:6442-6455.

Tam, J.P., Kent, S.B.H., Wong, T.W., and Merrifield, R.B. 1979. Improved synthesis of 4-(Boc-aminoacyloxymethyl)phenylacetic acids for use in solid phase peptide synthesis. Synthesis 955-957.

Tam, J.P., Riemen, M.W., and Merrifield, R.B. 1988. Mechanisms of aspartimide formation: The effects of protecting groups, acid, base, temperature
and time. Peptide Res. 1:6-18.

Tam, J.P., Liu, W., Zhang, J.-W., Galantino, M., and de Castiglione, R. D-
Amino acid and alanine scans of endothelin: An approach to study refolding
intermediates. In Peptides 1990, E. Giralt and D. Andreu, eds., ESCOM,
Leiden, The Netherlands, 1991a, pp. 160-163.

✗ Tam, J.P., Wu, C.-R., Liu, W., and Zhang, J.-W. 1991b. Disulfide bond formation in peptides by dimethyl sulfoxide: Scope and applications. J. Am.
Chem. Soc. 113:6659-6662.

Ten Kortenaar, P.B.W., and van Nispen, J.W. 1988. Formation of open-chain
asymmetrical cystine peptides on a solid support: Synthesis of pGlu-Asn-
Cyt-Pro-Arg-Gly-OH. Coll. Czech. Chem. Commun. 53:2537-2541.

Ten Kortenaar, P.B.W., van Dijk, B.G., Peters, J.M., Raaben, B.J., Adams,
P.J.H.M., and Tessier, G.I. 1986. Rapid and efficient method for the prepara-
tion of Fmoc amino acids starting from 9-fluorenylmethanol. Int. J. Pept.
Protein Res. 27:398-400.

✗ Ten Kortenaar, P.B.W., Hendrix, B.M.M., and van Nispen, J.W. 1990. Acid-
catalyzed hydrolysis of peptide-amides in the solid state. Int. J. Pept. Protein
Res. 36:231-235.

Tesser, M., Albericio, F., Pedroso, E., Grandas, A., Eritja, R., Giralt, E., Granier,
C., and van Rietschoten, J. 1983. Amino-acid condensations in the preparation of N^α-9-fluorenylmethyloxycarbonylamino-acids with 9-fluorenyl-
methylchloroformate. Int. J. Pept. Protein Res. 22:125-128.

Thaler, A., Seebach, D., and Cardinaux, F. 1991. Lithium-salt effects in peptide
synthesis, part II: Improvement of degree of resin swelling and of efficiency
of coupling in solid-phase synthesis. Helv. Chim. Acta 74:628-643.

Torres, J.L., Haro, I., Valencia, G., Reig, F., and Garcia-Anton, J.M. 1989. Synthesis of $O^{1.5}$-(β-D-galactopyranosyl)[DMet^2,Hyp^5] enkephalin amide, a
new highly potent analgesic enkephalin-related glycosyl peptide. Experientia 45:574-576.

Tregear, G.W. Graft copolymers as insoluble supports in peptide synthesis. In
Chemistry and Biology of Peptides, J. Meienhofer, ed., Ann Arbor Sci.
Publ., Ann Arbor, Michigan, 1972, pp. 175-178.

Tregear, G.W., van Rietschoten, J., Sauer, R., Niall, H.D., Keutmann, H.T.,
and Potts, Jr., J.T. 1977. Synthesis, purification, and chemical characterization of the amino-terminal 1-34 fragment of bovine parathyroid
hormone synthesized by the solid-phase procedure. Biochemistry 16:2817-
2823.

Ueki, M., and Amemiya, M. 1987. Removal of 9-fluorenylmethyloxycarbonyl
(Fmoc) group with tetrabutylammonium fluoride. Tetrahedron Lett.
28:6617-6620.

van der Eijk, J.M., Nolte, R.J.M., and Zwikker, J.W. 1980. A simple and mild
method for the removal of the N^{lm}-tosyl protecting group. J. Org. Chem.
45:547-548.

van Nispen, J.W., Polderdijk, J.P., and Greven, H.M. 1985. Suppression of side-
reactions during the attachment of Fmoc amino acids to hydroxymethyl
polymers. Recl. Trav. Chim. Pays-Bas 104:99-100.

✗ van Woerkom, W.J., and van Nispen, J.W. 1991. Difficult couplings in stepwise solid phase peptide synthesis: Predictable or just a guess? Int. J. Pept. Protein Res. 38:103-113.

Veber, D.F., Milkowski, J.D., Varga, S.L., Denkewalter, R.G., and Hirschmann, R. 1972. Acetamidomethyl: A novel thiol protecting group for cysteine. J. Am. Chem. Soc. 94:5456-5461.

Voelter, W., Kalbacher, H., Beni, C., Heinzel, W., and Müller, J. Recently developed amino protecting groups. In Chemistry of Peptides and Proteins, Vol. 2, W. Voelter, E. Bayer, Y.A. Ovchinnikov and E. Wünsch, eds., Walter de Gruyter and Co., Berlin, 1987, pp. 103-114.

Voss, C., and Birr, C. 1981. Synthetic insulin by selective disulfide bridging II: Polymer phase synthesis of the human B chain fragments. Hoppe-Seyler's Z. Physiol. Chem. 362:717-725.

Wade, J.D., Fitzgerald, S.P., McDonald, M.R., McDougall, J.G., and Tregear, G.W. 1986. Solid-phase synthesis of α-human atrial natriuretic factor: Comparison of the Boc-polystyrene and Fmoc-polyamide methods. Biopolymers 25:S21-S37.

Wade, J.D., Bedford, J., Sheppard, R.C., and Tregear, G.W. 1991. DBU as an $\times$ N^{α}-deprotecting reagent for the fluorenylmethoxycarbonyl group in continuous flow solid-phase peptide synthesis. Peptide Res. 4:194-199.

Wallace, C.J.A., Mascagni, P., Chait, B.T., Collawn, J.F., Paterson, Y., Proudfoot, A.E.I., and Kent, S.B.H. 1989. Substitutions engineered by chemical synthesis at three conserved sites in mitochondrial cytochrome c. J. Biol. Chem. 264:15199-15209.

Wang, S.S. 1973. p-Alkoxybenzyl alcohol resin and p-alkoxybenzyloxycarbonylhydrazide resin for solid phase synthesis of protected peptide fragments.. J. Am. Chem. Soc. 95:1328-1333.

Wang, S.S. 1976. Solid phase synthesis of protected peptides via photolytic cleavage of the α-methylphenacyl ester anchoring linkage. J. Org. Chem. 41:3258-3261.

Wang, S.S., and Merrifield, R.B. 1969. Preparation of some new biphenylisopropyloxycarbonyl amino acids and their application to the solid phase synthesis of a tryptophan-containing heptapeptide of bovine parathyroid hormone. Int. J. Pept. Protein Res. 1:235-244.

Wang, S.S., Matsueda, R., and Matsueda, G.R. Automated peptide synthesis under mild conditions. In Peptide Chemistry 1981, T. Shioiri, ed., Protein Research Foundation, Osaka, 1982, pp. 37-40.

Wang, S.S., Chen, S.T., Wang, K.T., and Merrifield, R.B. 1987. 4-methoxybenzyloxycarbonyl amino acids in solid phase peptide synthesis. Int. J. Pept. Protein Res. 30:662-667.

Weygand, F., Steglich, W., and Bjarnason, J. 1968a. Leicht abspaltbare Schutzgruppen für die Säureamidfunktion 3: Derivate des Asparagins und Glutamins mit 2.4-dimethoxy-benzyl-und 2.4.6-trimethoxy-benzyl-geschützten Amidgruppen. Chem. Ber. 101:3642-3648.

Weygand, F., Steglich, W., and Chytil, N. 1968b. Bildung von N-succinimidoxycarbonyl-β-alanin-amiden bei Amidsynthesen mit Dicyclohexylcarbodiimid/N-hydroxysuccinimid. Z. Naturforschg. 23b:1391-1392.

Wolfe, H.R., and Wilk, R.R. 1989. The RaMPS system: Simplified peptide synthesis for life science researchers. Peptide Res. 2:352-356.

Wu, C.-R., Wade, J.D., and Tregear, G.W. 1988. β-Subunit of baboon chorionic gonadotropin: Continuous flow Fmoc-polyamide synthesis of the C-terminal 37-peptide. Int. J. Pept. Protein Res. 31:47-57.

Wu, C.-R., Stevens, V.C., Tregear, G.W., and Wade, J.D. 1989. Continuous-flow solid-phase synthesis of a 74-peptide fragment analogue of human β-chorionic gonadotropin. J. Chem. Soc. Perkin Trans. I:81-87.

Wünsch, E. Synthese von Peptiden. In Houben-Weyl's Methoden der Organischen Chemie, Vol. 15, parts 1 and 2, E. Müller, ed., Thieme, Stuttgart, 1974.

Wünsch, E., and Spangenberg, R. Eine neue S-schutzgruppe für Cystein. In Peptides 1969, E. Scoffone, ed., North-Holland Pub., Amsterdam, 1971, pp. 30-34.

Wünsch, E., Moroder, L., Wilschowitz, L., Göhring, W., Scharf, R., and Gardner, J.D. 1981. Zur Totalsynthese von Cholecystokinin-pankreozymin: Darstellung des verknüpfungsfähigen "Schlüsselfragments" der Sequenz 24-33. Hoppe-Seyler's Z. Physiol. Chem. 362:143-152.

Yajima, H., Takeyama, M., Kanaki, J., and Mitani, K. 1978. The mesitylene-2-sulphonyl group, an acidolytically removable N^G-protecting group for arginine. J. Chem. Soc. Chem. Commun. 482-483.

Yajima, H., Fujii, N., Funakoshi, S., Watanabe, T., Murayama, E., and Otaka, A. 1988. New strategy for the chemical synthesis of proteins. Tetrahedron 44:805-819.

Yamashiro, D. 1987. Preparation and properties of some crystalline symmetrical anhydrides of N^α-tert.-butyloxycarbonyl-amino acids. Int. J. Pept. Protein Res. 30:9-12.

Yamashiro, D., and Li, C.H. 1973. Protection of tyrosine in solid-phase peptide synthesis. J. Org. Chem. 38:591-592.

Yamashiro, D., Blake, J., and Li, C.H. 1976. The use of trifluoroethanol for improved coupling in solid-phase peptide synthesis. Tetrahedron Lett. 1469-1472.

Yoshida, M., Tatsumi, T., Fujiwara, Y., Iinuma, S., Kimura, T., Akaji, K., and Kiso, Y. 1990. Deprotection of the S-trimethylacetamidomethyl (Tacm) group using silver tetrafluoroborate: Application to the synthesis of porcine brain natriuretic peptide-32 (pBNP-32). Chem. Pharm. Bull. 38:1551-1557.

Young, J.D., Huang, A.S., Ariel, N., Bruins, J.B., Ng, D., and Stevens, R.L. 1990. Coupling efficiencies of amino acids in solid phase synthesis of peptides. Peptide Res. 3:194-200.

Young, S.C., White, P.D., Davies, J.W., Owen, D.E.I.A., Salisbury, S.A., and Tremeer, E.J. 1990. Counterion distribution monitoring: A novel method for acylation monitoring in solid-phase synthesis. Biochem. Soc. Trans. 18:1311-1312.

Yu, H.-M., Chen, S.-T., Chiou, S.-H., and Wang, K.-T. 1988. Determination of amino acids on Merrifield resin by microwave hydrolysis. J. Chromatogr. 456:357-362.

Zalipsky, S., Albericio, F., and Barany, G. Preparation and use of an aminoethyl polyethylene glycol-crosslinked polystyrene graft resin support for solid-phase peptide synthesis. In Peptides: Proceedings of the Ninth American Peptide Symposium, C.M. Deber, V.J. Hruby, and K.D. Kopple, eds., Pierce Chemical Co., Rockford, Illinois, 1985, pp. 257-260.

Zalipsky, S., Albericio, F., Somczyńska, U., and Barany, G. 1987. A convenient general method for synthesis of N^α- or N^ω-dithiasuccinoyl (Dts) amino acids and dipeptides: Application of polyethylene glycol as a carrier for funtional purification. Int. J. Pept. Protein Res. 30:740-783.

Zardeneta, G., Chen, D., Weintraub, S.T., and Klebe, R.J. 1990. Synthesis of phosphotyrosyl-containing phosphopeptides by solid-phase peptide synthesis. Anal. Biochem. 190:340-347.

4

Evaluation of the Finished Product

Gregory A. Grant

In 1987, an article appeared in the *International Journal of Peptide and Protein Research* commemorating the twenty-fifth anniversary of the development of solid-phase peptide synthesis (Barany et al., 1987). Although that article dealt with many aspects of peptide synthesis, one statement in particular stands out as exemplifying the rationale for this chapter. It states:

> No synthetic endeavor can be considered complete until the product has been adequately purified and subjected to a battery of analytical tests to verify its structure.

The characterization, or evaluation, of a synthetic peptide is the one step in its production and experimental utilization that will validate the experimental data obtained. Unfortunately, it is also the one step many investigators all too often give too little attention. If the synthetic product, upon which the theory and performance of the experimental investigation is based, is not the intended product, the conclusions will be incorrect.

Without proper characterization, the investigator will have to be either lucky or wrong. Worse yet, he or she will not know which is the case.

Although, today, the synthesis of a given peptide is often considered routine, the product should never be taken for granted. Peptide synthesis chemistry is quite sophisticated, complex, and subject to a variety of problems (see Chapter 3). These problems, which can manifest themselves as unwanted side reactions and decreased reaction efficiency, arise from a variety of factors, such as reagent quality, incompatible chemistries, instrument malfunctions, sequence-specific effects, and even operator error. Although every effort is made to eliminate the causes of these problems and to plan for potential problems in the design and synthesis steps, it is not always successful. Therefore, one must never assume that the final product is the expected one until it has been proved to be so. To do otherwise may seriously jeopardize the outcome of the research.

Used and performed properly, the evaluation stage is where the fruits of the synthesis are scrutinized and a decision is made to use the peptide as intended, submit it for further purification, or resynthesize it, possibly changing design elements or synthesis protocols. The battery of tests, referred to in the quote at the beginning of this chapter, comprise the tools used to assure the investigator that he or she has obtained the intended peptide, and they are the main subject of this chapter.

GENERAL CONSIDERATIONS

Before embarking on a detailed discussion of the techniques and theory of peptide evaluation, it is perhaps useful to discuss a few general considerations for the reader to keep in mind. First, homogeneity and correct covalent structure are the basic goals of the synthetic endeavor and are the two factors that pervade any and all considerations of the state of the synthetic product. Second, a familiarity with the properties of the amino acid residues may have direct bearing on peptide design and handling (see Chapter 2). Third, the question of peptide purity is important with regard to the peptide's use and whether its use requires much characterization. Moreover, that question becomes even more germane from the viewpoint of synthesis facilities which are a major source of synthetic peptides today. Those facilities produce peptides for a large group of diverse investigators. Both time and cost must be considered carefully when deciding how much characterization is necessary or feasible.

Strategy for the Evaluation of Synthetic Peptides

The general strategy for the evaluation of synthetic peptides is presented diagrammatically in Figure 1. This diagram depicts the general course of the evaluation process as it might be carried out and indicates the types of analyses that are available, which might be helpful at any particular

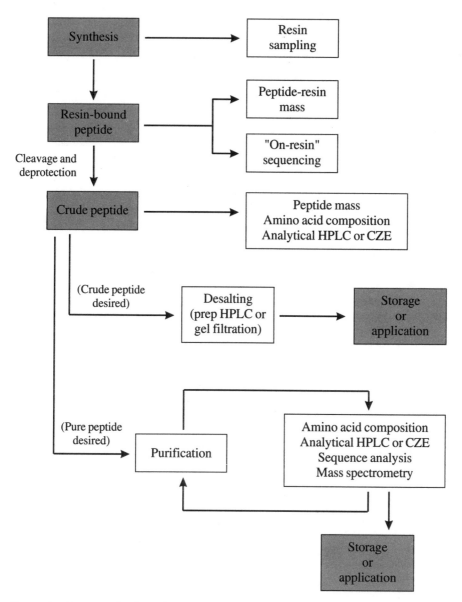

FIGURE 1   A strategy for the evaluation of synthetic peptides.

stage. The individual components of this diagram are discussed in specific sections of this chapter. In addition to detailed structural characterization of the peptide, this strategy includes initial indicators of quality that can often be helpful in the overall evaluation. When an evaluation is

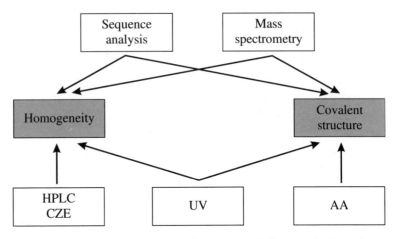

FIGURE 2   The relative contribution of analytical techniques to the determination of homogeneity and covalent structure of synthetic peptides.

approached in a systematic manner, a clear picture of the quality of the synthesis will emerge, which will allow an accurate and clear assessment of the peptide product.

At each stage, starting with the crude product and ending with the homogeneous peptide, the product must be evaluated to ascertain the desired purity and the correct covalent nature of the peptide. The latter stages of the evaluation process comprise a purification–evaluation cycle. This cycle may have to be repeated several times before the desired result is obtained and the synthetic chemist feels confident that the product obtained is the one intended.

The numerous analytical tests generally available for the evaluation of synthetic peptides are the tools that the chemist uses to assess homogeneity and covalent structure. The techniques that are commonly used include analytical high performance (or high-pressure) liquid chromatography (HPLC), capillary zone electrophoresis (CZE), amino acid compositional analysis, UV spectroscopy, sequence analysis, and mass spectrometry. Figure 2 indicates the relative contributions of these tools toward the goals of assessing homogeneity and covalent structure, and Table 1 summarizes some of their advantages and limitations. Each one makes its own specialized contribution to the evaluation process by providing unique information, and, generally, they are all used to one extent or another in a complementary fashion. Taken as a whole, the information provided by these analytic techniques should be sufficient to allow the investigator to correctly judge the state of the synthetic peptide.

Infrared and nuclear magnetic resonance (NMR) spectroscopy can also be useful tools for evaluating synthetic peptides, but they are usually

TABLE 1   Advantages and Limitations of Analytical Techniques

Technique	Advantages	Limitations
Reverse-phase HPLC	Assesses level of heterogeneity Useful for purification	Does not yield structural information
Capillary zone electrophoresis	Assesses level of heterogeneity Provides assessment based on a second physical principle	Not useful for preparative scale Does not yield structural information
Amino acid analysis	Accurately quantitates peptide levels Provides mole ratio of amino acids	Some amino acids not quantitated Accuracy decreases for large peptides Does not distinguish heterogeneous species
Ultraviolet spectroscopy	Assists in monitoring tryptophan integrity Useful for monitoring purification	Does not distinguish heterogeneous species Does not yield structural information other than for tryptophan
Sequence analysis	Determines linear order of amino acids Very good for detecting deletion and addition sequences Detects presence of some adducts	Blocked peptides are invisible Cannot sequence long peptides to the end Not reliable for quantitation May destroy or reverse some adducts
Mass spectroscopy	Provides accurate molecular weight Can indicate heterogeneity	Not quantitative Not readily available on routine basis for most laboratories High mass analysis only recently available and still in developmental stages

necessary only in cases where secondary structures or new chemistries are being investigated. In addition, NMR availability is often limited, particularly for core facilities, and it is expensive and time consuming. Because these two techniques are generally not needed for routine evaluation, they will not be discussed here.

Either analytical HPLC or CZE is usually done as the first level of assessment. These two techniques give a relatively rapid overall picture, or "feel," for the synthetic product's complexity because they reveal the relative number of different species present in the preparation, but they provide no direct information about covalent structure. Amino acid analysis quantitates the amount of the peptide and determines whether the requisite amino acids are present at reasonable ratios. However, it is less reliable for certain unstable amino acids such as cysteine, methionine, and tryptophan, and the compositional accuracy decreases as the peptide gets larger. UV spectroscopy monitors the integrity of the aromatic amino acids, particularly tryptophan, and is very useful in monitoring peptide purification, but the amount of information that can be obtained about the covalent state of the peptide is very limited. Sequence analysis and mass spectrometry come as close as any to addressing both homogeneity and covalent structure in a single step. However, while these techniques give the best assessment of the covalent nature of the peptide, it must be pointed out that they can lead to incorrect conclusions about the sample if they are relied on exclusively. For example, peptides that are blocked at their amino terminus to the Edman chemistry, either as a consequence of the synthesis chemistry employed or because of an unwanted side reaction, will be completely invisible to sequence analysis. Mass spectrometry is not generally found to be a quantitative technique, so conclusions of relative homogeneity based on mass analysis can be misleading and should be avoided.

In the examples that follow throughout this chapter, many instances occur where incorrect conclusions would have been made if they were based only on the limited information provided by a single analytical technique. Indeed, in many cases, information from all of the available analytical techniques was necessary for the correct conclusion to be made.

Homogeneity and Purification

The assessment of peptide homogeneity goes hand in hand with the purification process. First of all, homogeneity is not a characteristic that can be proved directly, because there is no single test that, when met, automatically indicates that a peptide is a clean, single species. Rather, homogeneity is demonstrated by an inability to demonstrate heterogeneity, or the presence of more than one species. This implies, of course, that reasonable attempts have been made to demonstrate heterogeneity under conditions where it should be detected if present and that good judgment, based on experience with the techniques used, has been exercised. It often follows that if a particular technique or procedure indicates the existence of multiple species, it also provides either the means to separate and purify the individual species or the evidence upon which a separation protocol can be based. For example, if more than one peptide peak is found using analytical reverse-phase HPLC, one may have the ability to physically isolate individual peptides by collecting fractions of the column effluent. On the other hand, if two sequences appear on automated Edman degradation, the sequencing procedure cannot be used, in itself, to physically separate the two species. However, the sequence information obtained may suggest a productive separation procedure, e.g., one based on charge differences.

Covalent Structure

In most instances, before the covalent structure of a peptide can be unambiguously determined, the peptide must be relatively homogeneous. Although many techniques used for the evaluation of covalent structure can in themselves indicate the presence of multiple species, those techniques, with the possible exception of mass spectrometry, are not generally able to provide a specific or unambiguous analysis of the covalent nature of each species within a mixture.

The two main items to be evaluated in the assessment of covalent structure of a peptide are the presence of the correct amino acid sequence and the presence, or, preferably, absence, of derivatives or adducts of those amino acids. Probably the most common problem encountered is the occurence of deletion or addition sequences. Deletion-sequence peptides are those containing an internal absence, or deletion, of an amino acid residue brought about by either incomplete coupling to, or incomplete deprotection of, the α-amino group at a particular cycle (Tregear et al., 1977). Addition-sequence peptides are those that contain two residues of the same amino acid adjacent to each other in the sequence when only one was intended. The degree to which deletion and addition sequences occur depends on the conditions of the individual synthesis, but as long as these do not occur too often during a synthesis and do not represent an overwhelming proportion of the total product, the full-length peptide may still be present in appreciable amounts.

Deletion sequences that result from incomplete coupling, but not incomplete deblocking or other causes (see Chapter 3), can be largely eliminated if capping is used. Capping is the chemical derivatization of free α-amino groups, usually by acetylation with acetic anhydride, before deprotection of the residue that was just added takes place. This results in an irreversibly blocked peptide that is no longer able to be elongated. Although capped peptides will represent contaminants that will have to be removed from the final product, they are generally easier to separate from the full-length peptide than are deletion-sequence peptides, which may differ from the full-length peptide by only one or two amino acids. Addition sequences are more problematic and cannot be prevented by a routine procedure such as capping.

Unwanted modifications usually come in two general forms. They are either (1) side-chain protecting groups that were not completely removed or (2) random adducts that were produced as a result of uncontrolled side reactions during the synthesis or cleavage and deprotection steps. Some of the more common side reactions encountered are described in Chapter 3, which deals specifically with the synthesis chemistry. Although identification of the exact nature of the adduct may help to prevent it in subsequent syntheses, it is not always possible, or necessary, to do so. If the modified peptides do not represent a large proportion of the sample, which is usually the case, they often can be removed during the purification procedure. A common observation in peptide synthesis is that the crude product will contain a variety of species in addition to the intended product. An acceptable synthesis is one that, when analyzed by HPLC or CZE, indicates a predominant component, which hopefully is the intended product, and some number of minor components that can be separated away (e.g., see Figure 26). The minor components of the synthesis represent side products that occur as the result of unwanted modifications and inefficiencies in the chemistry. As long as they do not represent an overwhelming proportion of the product, they can be removed without determining their identity. It is only when the unwanted side products become the major species that either the chemical protocol or instrument operation needs to be changed. Today, both the t-Boc and, more recently, the Fmoc chemistries have been refined to such a degree that they are generally reliable for routine use.

Residues that Present Potential Problems

Certain amino acid residues deserve some consideration because they can be the source of problems in the finished peptide. The problems considered here are not meant to include most of those that result from unwanted reactions during the synthesis, cleavage, and deprotecting chemistry, as discussed in Chapter 3, although they can occur during those steps, but rather those that may develop due to the characteristics of certain amino acid residues in the peptide. They may develop during the workup of the peptide or during handling or storage, and they present a particular problem in the evaluation steps or the application of the peptide. In fact, the characteristics of these residues are such that if one has a choice, it might be better to avoid them altogether merely because it simplifies the possibilities. That does not, however, mean that they should not be used at all. The residues in question are cysteine, tryptophan, methionine, and amino-terminal glutamine, and they have certainly been incorporated into thousands of useful peptides without causing serious problems.

The main problem with tryptophan, methionine, and cysteine occurs because of their susceptibility to oxidation. Of these, methionine and cysteine are the least troublesome, because their common oxidation products are usually reversible. When possible, it is perhaps advisable to work below neutral pH and with degassed or deoxygenated solvents when working up peptides that contain these residues.

Methionine

Methionine is easily oxidized to methionine sulfoxide. Methionine sulfoxide can be converted back to methionine by treatment with thiol reagents, such as dithiothreitol or N-methylmercaptoacetamide (Cullwell,

1987; Houghton and Li, 1979), but 2-mercaptoethanol is not as effective for this purpose (Houghton and Li, 1979). Other methods reported to be effective include NH_4-I-dimethylsulfide (Fujii et al., 1987) and $DMF \cdot SO_3$-EDT (Futaki et al., 1990). The latter should be performed only while hydroxyl residues are protected. Peptides that contain methionine sulfoxide tend to elute slightly earlier on reverse-phase HPLC than the corresponding peptide containing methionine because of the more polar nature of the sulfoxide; its presence can often be identified in this way. Many laboratories routinely treat methionine-containing peptides with a thiol reagent prior to HPLC analysis to check for elution shifts due to the presence of the sulfoxide. This procedure is particularly important if more than one predominant peak is originally present in the chromatogram.

Cysteine

Cysteine is prone to the oxidative formation of disulfide bonds which can result in the presence of multiple species. If a single cysteine residue is present in a peptide, a dimer can form. However, if the peptide contains more than one cysteine, disulfide cross-linked oligomeric aggregates can form.

Any disulfides that form can be reduced by treatment with a molar excess (5- to 100-fold) of dithiothreitol. It should be pointed out, however, that some complex disulfide cross-linked aggregates can prove to be very resistant to reduction. Peptides linked by intermolecular disulfides tend to elute later in reverse-phase HPLC than their monomeric counterparts. The relative elution behavior of peptides with intramolecular cross-links is less predictable.

Generally speaking, peptides usually emerge from the deprotection procedure in their fully reduced state. It is only after exposure to air that appreciable amounts of disulfides start to form. The time of air exposure needed can vary from one peptide to another. If one can avoid the presence of two or more cysteine residues in the same peptide, it is generally advisable to do so. Also, when using ammonium bicarbonate as a solvent, disulfide formation can be diminished initially if it is chilled before adding it to the peptide.

Tryptophan

Oxidation products of tryptophan are generally not reversible. Common oxidation products of the indolyl ring of tryptophan are oxyindolyl, formylkynureninyl, and kynureninyl derivatives. The latter two derivatives result in opening of the five-membered ring of tryptophan and shifting

> Tryptophan oxidation can result from the conditions under which the peptide was handled. For example, investigators at different locations observed, by sequence analysis, varying levels of tryptophan oxidation for the same peptide from a common source. One possible cause that has been suggested is exposure of the samples to small rust particles from equipment that was used in processing the sample.

the absorbance maximum toward shorter wavelengths. Significant oxidation of tryptophan residues also can be detected by mass spectrometry and sequencing. In sequence analysis, a common oxidation product of tryptophan elutes just before intact PTH-tryptophan on the HPLC.

Glutamine

When a glutamine residue occurs at the amino-terminus of a peptide, it can cyclize by interaction of the side chain with the α-amino group to form a lactam structure. This structure, often referred to as pyroglutamate, renders the peptide inaccessible to Edman chemistry so that it cannot be sequenced. This occurence is more likely at an acid pH, but it can also happen slowly at a neutral or basic pH when the peptide is stored in aqueous solution. This same phenomenon is sometimes seen with glutamic acid, but it is less common.

Peptide Solubility

The solubility properties of synthetic peptides can have profound effects on their usability and should always be considered before synthesis. In general, it is advisable to incorporate charged amino acids into a synthetic peptide when possible. Peptides composed completely of uncharged amino acids tend to be less soluble in aqueous solutions, and their solubility will tend to decrease as their size increases. The presence of histidine, lysine, and arginine promotes solubility in solutions at a pH where their side chains possess a positive charge. Likewise, the presence of aspartic acid and glutamic acid promotes solubility in solutions at a pH where their side chains will be negatively charged. A general rule of thumb for solubility in aqueous solution is that the number of charges (total) should be at least 20 percent of the total number of residues. The contribution of the charge properties of the α-amino and carboxyl groups should also be included. In this regard, it is useful to remember that modification of these end groups, so that they are no longer able to ionize, can significantly affect the solubility properties, especially of small peptides. Long stretches of hydrophobic residues and the presence of polar uncharged side chains are also factors to be considered in assessing solubility. In a "borderline" peptide, their presence can tip the scale one way or the other. However, remember that exceptions always exist and that these general rules may not always be adequate.

Some of the common aqueous solutions used for the dissolution of peptides are 0.1% trifluoroacetic acid (TFA), 0.1N acetic acid, and 0.1M ammonium bicarbonate. If these solutions are not effective, 25 to 50% TFA, 30% acetic acid, or the inclusion of acetonitrile or ethanol can help. Aqueous formic acid is another excellent solvent that has been used for peptides, but it has the potential drawback that the peptide may become formylated on primary amino groups if formaldehyde, which is a potential contaminant, is present. In some cases, increasing concentrations of dimethyl sulfoxide (DMSO) may be helpful. Dimethylformamide (DMF) and dichloromethane (DCM) are effective nonaqueous organic solubilizers of synthetic peptides.

Crude versus Pure Peptides

Typically, synthetic chemists are called on to produce peptides of two general categories, simply referred to as either *crude* or *pure*. Crude peptides consist of the synthetic product either as it comes directly from the synthesis or after only minimal purification is performed. Pure peptides, as the name implies, are those that have undergone rigorous purification to homogeneity or near homogeneity. Obviously, the evaluation criteria applied to each of these categories is quite different. The desired state of the peptide and the extent of evaluation needed depends on its intended use. If the peptide is intended to be used as an antigen for producing polyclonal antisera, perhaps a lesser degree of purification may be acceptable than for a peptide used for detailed conformational studies. Although ideally it may be desirable to purify all peptides to a high degree of homogeneity, practical limitations often occur in the terms of amount of time and effort required as well as fiscal limitations. Moreover, a core laboratory may get requests for crude peptides from an investigator who intends to do his own purification, in which case the product is released before it is completely purified and evaluated. It must be emphasized, however, that accepting a lesser degree of purification does not mean sacrificing the integrity of the product. The intended covalent structure surely should be present, and some minimal criteria should be assessed to verify the likelihood of that situation. These criteria are discussed in the next section (see Initial Indicators of Quality).

INITIAL INDICATORS OF QUALITY

Several means are available to make an initial assessment of the overall quality of the synthetic product. These procedures can be performed prior to embarking on a full-scale purification and characterization of the peptide. They can often give initial indications of potential problems and save a great deal of subsequent work and expense. Synthesis problems can occur either during the synthesis itself or during the procedure used to cleave the peptide from the resin and deprotect the side chains (see Chapter 3). If a problem does occur, an ability to distinguish the step at which the problem manifests itself can be very useful in correcting it. For example, if a problem occurs during the cleavage or deprotection reaction, it makes no sense to change the synthesis protocol. So, in addition to being early indicators, the following procedures can also help to pinpoint which of the two main steps is suspect.

Those procedures that are performed prior to cleavage of the peptide from the resin include (1) determination of the mass of the resin-bound peptide and its consistency with expected yields, (2) resin sampling during synthesis to assess the coupling efficiency, and (3) "on-resin" sequencing. Those that are done after cleavage from the resin include (1) determination of the mass of the crude product obtained, (2) analytical high-performance liquid chromatography or capillary electrophoresis, and (3) amino acid compositional analysis.

Taken together, these initial indicators represent the primary evaluation of the synthesis and provide the basis upon which all subsequent actions are performed. The potential success of a synthesis can usually be evaluated adequately by the resin-bound peptide mass yield, the cleaved peptide mass yield, analytical HPLC or CZE, and amino acid analysis. These four evaluations are all relatively easy to perform and represent the minimum tests for each synthetic peptide. In addition, resin sampling and on-resin sequencing can also be done, but most laboratories do not routinely perform either of these procedures because of the time and expense involved. These two procedures are often reserved for troubleshooting after a problem is suspected. On the other hand, on-resin sequencing is one of the most powerful methods available for assessing the quality of the synthesis separate from the subsequent contributions of the cleavage and deprotection procedures.

Indicators of Synthesis Quality Prior to Cleavage from the Resin

Peptide-Resin Mass Yield

As the peptide is being synthesized, the mass of the resin-bound polypeptide increases as a function of the initial molar substitution of the resin and the addition of amino acid residues. The theoretical mass of the resin can thus be calculated at any given step during the elongation and compared to the mass of the dried resin. This procedure is usually done at the end of the synthesis and before cleavage from the resin and provides the first indication of the synthesis outcome. For example, if the mass of the peptide-bound resin is extremely low compared to the theoretical mass, it would indicate a significant failure in chain elongation.

Table 2   Resin Substitution

Residue number	Residue ID	Mol. wt. Boc-AA	Total wt. peptide-resin (g)	Substitution mmol/g	Comment
				0.640	Initial sub.
9	Lys	(415)	0.782	0.640	(Cl-Z)
8	Lys	415	0.930	0.537	(Cl-Z)
7	Pro	215	0.979	0.511	
6	Val	217	1.028	0.486	
5	Lys	415	1.177	0.425	(Cl-Z)
4	Gly	175	1.205	0.415	
3	Tyr	494	1.393	0.359	(Br-Z)
2	Gly	175	1.422	0.352	
1	Cys	325	1.525	0.328	(4-MeBzl)

On some automated peptide synthesizers, this calculation is performed automatically so that the operation is as simple as weighing the dried resin-bound peptide. Table 2 shows a resin substitution table similar to that produced automatically on an Applied Biosystems 430A peptide synthesizer. This table depicts a 9-residue peptide that was synthesized, starting with a Lys-PAM resin. Although these calculations were produced by the instrument, it is not difficult to compose a similar table for any peptide either manually or by using a simple computer program. In this case, the Lys-PAM resin used was substituted with t-Boc-L-Lysine (Cl-Z) to the level of 0.640 mmol/g. So, for a 0.5 mmol scale synthesis, 0.782 g of Lys-PAM resin was used. Knowing the molecular weight of the t-Boc-L-Lysine (Cl-Z), it is possible to calculate that the weight of the resin alone (without added amino acid) is 0.583 g. Similarly, the additive molecular weight of the growing resin-bound peptide chain can be calculated at any cycle. The theoretical dry weight of peptide resin at the completion of the synthesis is 1.525 g. Note that this weight is for the t-Boc protected amino-terminus. If the amino terminus is deprotected before drying, as is usually the case, the total weight would be 1.429 g. This value is determined by multiplying the molecular weight of the peptide (in mg/mmol) by the 0.5 mmol scale and adding the weight of the resin [(1691 mg/mmol) (0.5 mmol) + 0.583 g of resin]. (Remember, if you calculate the molecular weight of the peptide based on the t-Boc amino acid weights given in Table 2, you need to subtract 100

It is a good idea to get into the habit of weighing the dried resin-bound peptide before proceeding with the cleavage. This simple step can save the time and expense of performing the cleavage and deblocking step on a resin that is devoid of peptide. It will pinpoint the problem to the synthesis and avoid the uncertainty that the problem could have occurred during cleavage or extraction. It will also give an immediate indication that something is wrong with the synthesizer and will save the investigator from unwittingly performing several more unproductive syntheses before identifying the problem.

for each t-Boc group removed and 17 for the α-acid hydroxyl group removed from each residue upon formation of the peptide bond.)

In practice, a value within 75 percent of the expected calculated mass is not unusual and is considered to be acceptable. This amount allows for loss of resin during transfer from the reaction vessel and weighing as well as for the fact that the chemistry is not 100 percent efficient. If the ratio of actual weight to calculated weight decreases even further, it may be an indication that a significant problem may have occurred during synthesis. In practice, this occurrence can be in a range all the way from no increase in mass over the original resin weight at the start of the synthesis to any value in between. The former reveals that no synthesis took place and indicates catastrophic instrument failure or severe reagent problems. Intermediate ratios can be more problematic. Usually, if the yield is below 50 percent, a significant problem should be suspected, and the system should be checked out thoroughly. If similar results are obtained a second time, it may indicate a sequence-specific chemistry problem that may necessitate a change in the synthesis protocol. It should be pointed out, however, that low ratios, e.g., 40- to 75-percent range, do not always indicate that bona fide product cannot be recovered. If a loss in mass is caused by a single episode of failure and that failure leads to a percentage of the growing polypeptide chains being unable to accept any additional residues, the remaining chains may continue to elongate normally and yield the desired product. The result is a lower ratio of resin weight to expected weight, but a significant population of complete, full-length product still may be recoverable by HPLC separation.

Resin Sampling

Resin sampling is a procedure by which a small amount of resin is removed from the reaction vessel after the completion of each coupling cycle. Its purpose is to provide a measure of the coupling efficiency at each step. In order for a synthesis to proceed smoothly, the coupling efficiency as each new residue is added must remain high. This factor is very important because the effect is cumulative. Even if each cycle has a coupling efficiency of 95 percent, a value considered excellent for most organic reactions, the overall efficiency after 10 couplings is only approximately 60 percent. At 99-percent coupling efficiency for each cycle, the overall efficiency after 10 cycles is 90 percent. If one or more cycles has a significantly reduced coupling efficiency, the outcome of the overall synthesis can be seriously affected even though all the other cycles experienced excellent efficiencies. To effectively attempt to correct the problem, it is important to know which cycle is causing the problem, and resin sampling can help in providing that answer. Some instruments perform resin sampling automatically and deliver the resin to a fraction collector. The sampling is done at the end of the coupling step but before the deblocking step. Once collected, the resin then needs to be analyzed manually for the presence of free α-amino groups. This analysis is done most commonly by the Kaiser test (Kaiser et al., 1970; Kaiser et al., 1980; Sarin et al., 1981), which is a colorimetric ninhydrin analysis that is specific for primary (ninhydrin) and secondary (isatin) amino groups (also see Chapter 3). The drawbacks of resin sampling analysis are that it consumes resin and it is time consuming to perform on a routine basis. Moreover, there are reports (Schroll and Barany, 1989; Fontenot et al., 1991) that in certain sequence contexts, some residues produce a very low color yield with ninhydrin, which can lead to incorrect conclusions concerning coupling efficiency. Some laboratories still employ the Kaiser test routinely, but many do not. It is now often reserved for a synthesis in which a problem was found during the subsequent evaluation and in which one is trying to pinpoint the exact nature of the problem during a resynthesis. An example of the results of a resin sampling procedure is given in Figure 3. It illustrates coupling efficiencies determined from resin sampling during the synthesis of a peptide, where certain residues were coupled more than once at their respective positions. This was necessary because initial couplings were low at those positions and indicates one approach to increasing coupling efficiency as part of the fine tuning of a synthesis.

In addition to the Kaiser test, other resin sampling chemistries have also been described, but they are not as widely used today. These tests include the utilization of picric acid, 2,4,6-trinitrobenzenesulfonic acid (TNBS), fluorescamine, and chloranil. A description of their use can be found in Stewart and Young (1984) and in Chapter 3 of this volume. Most instrument manufacturers suggest a resin sampling protocol for use with their instruments.

On-Resin Sequencing

The growing polypeptide chain can also be analyzed by automated Edman degradation while it is still attached to the resin (Kent et al., 1982). This analysis is often done upon completion of the synthesis, but it can be done at any time during the synthesis to assess the quality of the growing peptide.

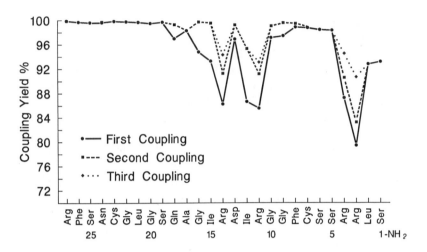

FIGURE 3   Plot of the data of a resin sampling from the synthesis of a 28-residue peptide. Synthesis proceeded from left to right as the data are plotted. (Data kindly provided by Anita Hong of Applied Biosystems, Inc.)

This procedure is often referred to as "preview analysis" because deletion sequences will show up as a preview of residue $i + 1$ in cycle i (see also Chapter 3). For example, such an analysis might be performed several times during the course of synthesis of a very long peptide. It requires very little resin because of the excellent sensitivity (10 to 100 pmol) of modern sequencers. The only requirement is that the resin be removed at a time when the peptide possesses a free α-amino group, such as after removal of the α-amino protecting group but before the beginning of the next coupling cycle. Because the sequencing is performed in the solid phase while the peptide is covalently coupled to an insoluble bead, sample washout from the sequencer is not a problem (see Sequence Analysis). Therefore, the peptide can easily be sequenced to its penultimate residue.

Since peptides sequenced on resin have not yet been subjected to side-chain deprotecting procedures, those amino acids whose side chains contain protecting groups will not elute on an HPLC at the same position as the free PTH amino acids and will not be detected with the standard HPLC program used for routine sequence analysis. Generally, the side-chain protected PTH amino acids elute later than their unprotected counterparts, and most elute well after the last of the normal PTH amino acids. For detection of side-chain protected PTH amino acids, the gradient and run time needs to be extended (Kent et al., 1982; Kochersperger et al., 1989), as illustrated in Figure 4. This figure illustrates the first 11 cycles of an on-resin sequencing of a synthetic peptide, made with t-Boc chemistry, whose synthesis was in progress. The sequence of this peptide is

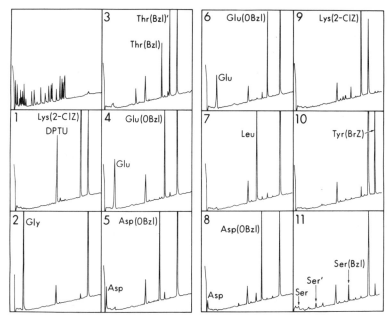

FIGURE 4   Extended program reverse-phase HPLC of the PTH amino acids from the first 11 cycles of an "on-resin" sequencing of a synthetic peptide. The expected sequence of the peptide is shown in Figure 5. The chromatogram in the upper left corner is a standard mixture of PTH amino acids. The late eluting peak seen in all of the chromatograms is an artefact.

shown in Figure 5 and indicates which residues contain protecting groups. With standard programs, most of the side-chain protected amino acids would not show up, but with an extended HPLC program, all the residues can be directly evaluated. Note that with some amino acids, particularly Glu, Asp, and Ser, the side-chain protecting group is unstable to the sequence chemistry, and some regeneration of unprotected amino acids results.

The extended gradients can also be used in routine sequencing of cleaved, deprotected peptides, because it gives a better picture of other components that can be present, particularly side chains that were not successfully deprotected. Some of the more common blocking groups that have been detected still attached to cleaved peptides are tosyl-arginine and formyl-tryptophan.

A similar approach can also be used for peptides synthesized with Fmoc chemistry (Kochersperger et al., 1989), but there are several disadvantages. First, the bond to the resin is not stable under the conditions of the sequencing chemistry, so the advantage of sequencing in the "solid

1				5
Lys(Cl-Z)	—Gly	—Thr(Bzl)	—Glu(OBzl)	—Asp(OBzl)
				10
Glu(OBzl)	—Leu	—Asp(OBzl)	—Lys(Cl-Z)	—Tyr(Br-Z)
				15
Ser(Bzl)	—Glu(OBzl)	—Ala	—Leu	—Lys(Cl-Z)
				20
Asp(OBzl)	—Ala	—Gln	—Glu(OBzl)	—Lys(CL-Z) —Leu

FIGURE 5   The sequence of the resin-bound, side-chain protected peptide, whose automated Edman degradation is shown in Figure 4.

phase" is lost. Second, the side-chain protecting groups are less stable to the sequencing chemistry and are removed more readily. This is not necessarily a disadvantage for on-resin sequencing, but it can be a serious drawback in trying to assess the degree of deprotection in cleaved peptides made with Fmoc chemistry.

Initial Indicators of Synthesis Quality after Cleavage from the Resin

After the peptide is cleaved from the resin and the side chains have been deprotected, it is ready for more extensive and detailed evaluation. However, before spending a great deal of time and expense on a detailed characterization that, in the end, may not produce any meaningful data, there are a few relatively simple analyses that can be done to assess the character of the peptide, which serve as a basis for deciding the feasibility of continuing the evaluation procedure. These procedures consist of (1) determination of the mass yield, (2) analytical chromatography or electrophoresis, and (3) amino acid compositional analysis.

Peptide Mass Yield

The mass of the cleaved, deprotected, dried peptide is another good initial approximation of the success of the synthesis. If the mass is significantly smaller than expected, it may indicate either a chemistry problem or a significant loss of material during the extraction of the peptide subsequent to cleavage. A comparison of the peptide mass after cleavage and the peptide-bound resin mass after synthesis may indicate potential problems subsequent to synthesis. For instance, if the peptide resin mass yield is reasonable, but the recovered mass of peptide cleaved from the resin is disproportionately small, the source of the low yield lies in either the cleavage or the extraction process.

For a 0.5 mmol scale synthesis, considering that the average molecular weight of an amino acid residue is 110, the average expected maximum yield of a synthetic peptide is 55 mg per residue length. Depending on the particular residue, of course, that value may be higher or lower, but for a peptide with average residue distribution, it is a quick and reasonable working rule of thumb. The peptide described in Table 2 has a molecular weight of 978 (after cleavage from the resin and deprotection of the side chains), so its theoretical yield from a 0.5 mmol scale synthesis would be 489 mg. Because of manipulation losses and failure sequences, the maximum value is seldom obtained. A reasonable mass expectation would be in the range of 30 mg of crude product per residue length.

It must also be remembered that a dried peptide is seldom 100-percent peptide. Adsorbed solvents, counterions, salts, and other inert materials may be present in addition to the peptide itself. This occurence may be particularly apparent in the crude product, but it is also a factor with the purified peptide. In practice, the weight of the dried peptide due to nonpeptide components can be as much as 50 percent, but it is more commonly around 20 to 30 percent. This factor is particularly important to remember when calculating the concentration of peptide for an important experiment. An accurate amino acid analysis of an aliquot of the peptide solution will give an accurate determination of the amount of the actual peptide present.

Analytical Electrophoresis and Chromatography

Analytical scale HPLC and CZE are the two most common and effective techniques used for the assessment of heterogeneity of synthetic peptides. In practice, they complement each other, because their respective modes of separation are based on different physical principles. A detailed discussion of the theory and operation of these two techniques, which is beyond the scope of this chapter, can be found in several excellent reviews and books (CZE: Ewing et al., 1989; HPLC: Hancock, 1984; Henschen et al., 1985; Mant and Hodges, 1990; Chicz and Regnier, 1990).

The most common form of depicting the results of either CZE or HPLC is a graph that shows measurement of absorption of light at a particular wavelength plotted against elution time or fraction number. The most common wavelengths used for peptides are in a range from 210 nm to 230 nm, where the polypeptide bond absorbs strongly. If the peptide contains tryptophan or tyrosine, detection at 280 nm also can be used. However, caution must be exercised when following an elution at 280 nm, because peptides that do not contain tryptophan or tyrosine, or where the tryptophan has been extensively oxidized, will not be seen. With both CZE and HPLC, a pure peptide is expected to produce a single symmetrical peak. However, it must be kept in mind that this criterion alone is not sufficient for homogeneity. A single symmetrical peak does not guarantee homogeneity (see Figure 21) and a homogeneous sample does not guarantee the intended product (see Figure 22). Furthermore, reversible oxidation of methionine and cysteine (disulfide bonds) can lead to the observation of multiple peaks. These peaks can be misleading but do not, in themselves, necessarily indicate serious chemistry problems. If they are recognized for what they are, they can usually be dealt with successfully.

Finally, as the size of peptides increases to greater than 50 residues, the expectation that a homogeneous peptide will form a single, sharp, symmetrical peak tends to decrease. With peptides of this size, this finding has been attributed to the onset of the formation of relatively stable, slowly exchanging folded structures that exhibit slightly altered elution characteristics and will result in a broad peak, often with absorption spikes appearing across its profile (Kent, 1988; Regnier, 1987). Moreover, in the case of HPLC, the onset of these structures can be enhanced or induced by interaction with the chromatography supports themselves (Regnier, 1987), particularly the hydrophobic reverse-phase supports. This interaction can result in an elution profile that is broad and irregular. Thus, without independent data, incorrect conclusions regarding heterogeneity can be made.

Capillary Zone Electrophoresis

Capillary zone electrophoresis (CZE), which was first demonstrated in 1981 (Jorgenson and Lukacs, 1981), separates peptides on the basis of their differential migration in solution in an electric field (Ewing et al., 1989). Like ion exchange HPLC, the separation is a function of the charge properties of the peptide, but the physical basis of separation is different. The separation takes place in a liquid-filled capillary tube (<100 microns in diameter) whose ends are placed in buffers exposed to opposite poles of an electric field. Typical running voltages can be from 5 to 30 kV, with times from 1 minute to longer than 1 hour. Migration takes place due to charge attraction and is recorded, usually by UV absorbance, when the individual peptide species pass a small window in the capillary that is placed in line with a detector. Actually, migration is caused by two components. The first component is migration of the charged peptide toward the pole of opposite charge, and the second component is bulk fluid flow within the capillary that is caused by electroendoosmosis. In aqueous solution in fused silica capillaries, electroendoosmotic flow usually occurs toward the cathode, so unless the electrophoretic mobility of an anion in the opposite direction is greater than the osmotic flow, theoretically, both anions and cations can be ana-lyzed in a single run (Young and Merion, 1990). Because the electroen-doosmotic flow occurs at the same rate for all solute molecules (pep-tides), the separation is a function of the electrophoretic mobility of the individual species. A common setup for peptide separation uses a buffer with pH 2.0, so all charged peptides are cationic and thus migrate to the cathode. This setup also minimizes the effect of electroendoosmosis.

CZE is often referred to, somewhat misleadingly, as a technique that has greater sensitivity than analytical HPLC. Although it is true that the actual amount of peptide analyzed is very small when compared to ana-lytical HPLC (nanograms compared to micrograms), it must be remem-bered that the volume of sample injected into the column is also very small (low nanoliters). Thus, the peptide concentration in the original sample needed for analysis is not much different from that needed for HPLC analysis. Unlike HPLC, which can be scaled up for preparative purposes, CZE is mainly useful as an analytical technique at the present time. Although preparative CZE has been reported (Rose and Jorgenson, 1988), commercially available instruments do not yet offer this feature. For the analysis of synthetic peptides, the main advantage of CZE is that it provides a second method, based on a different physical property, for separation of peptide mixtures and thus the assessment of heterogeneity.

The resolving power of CZE is demonstrated in Figures 6 through 8, which illustrate the separation of several synthetic peptides that differ only slightly in their composition. While peptides that differ in the num-ber of charges they carry are resolved by relatively large distances, it can be seen that even peptides with apparent equal charge can be separated. Figure 9 shows the results of a typical CZE run of the crude product from the synthesis of an 11-residue peptide and compares it to a reverse-phase HPLC analysis of the same sample. Although heterogeneity is evident with both analyses, the two profiles are not equivalent, reflecting the different separation principles employed.

Analytical High-Performance Liquid Chromatography

Analytical HPLC separates peptides as a result of their interaction with derivatized insoluble supports. This support is referred to as the *station-ary phase*. In practice, it is completely analogous to conventional "low-pressure" bench-top chromatography. Both conventional chromatogra-phy and "high-pressure" or "high-performance" chromatography employ (1) a cylindrical column that contains the derivatized support upon which separation occurs, (2) a means for flowing a modifying solution (referred to as the mobile phase), by which the separation is developed, through the column, and (3) a means to detect the eluting species as they exit from the column. The major difference between high-pressure and low-pres-sure techniques, and hence the advantage of HPLC, is that HPLC

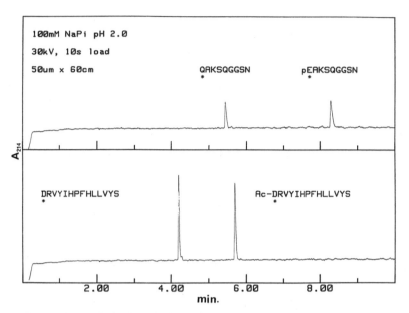

FIGURE 6   Analysis of synthetic peptides by capillary zone electrophoresis, showing the effect of the α-amino group on mobility. The capillary was 50 mm × 60 cm and the sample was run at 30 kV, after a 10-second injection, in 100 mM sodium phosphate buffer, pH 2.0. (Figure kindly provided by Dan Crimmins of the Howard Hughes Medical Institute.)

employs columns and pumping systems that can withstand and deliver very high pressures so that very fine particles (3 to 10 micron mean diameter) can be used as column packing material. The result is superior resolution in relatively short time periods (minutes for HPLC as compared to hours or days for conventional systems).

The two general categories of HPLC supports that are used almost exclusively for peptides are ion exchange and reverse phase. Separation by ion exchange HPLC, like its conventional low-pressure counterpart, is based on direct charge interaction between the peptide and the stationary phase. The column support is derivatized with an ionic species that maintains a particular charge over a pH range in which a peptide (or mixture of peptides) can exhibit an opposite charge, depending on their amino acid compositions. When the sample is loaded onto the column, the charge interaction binds the peptides to the support. The peptides are eluted from the column by changing either the pH or the ionic strength, or both, of the column buffer such that the charge interaction is defeated either through neutralization of the interacting charge groups (pH) or competition with higher concentrations of like-charged molecules (ionic

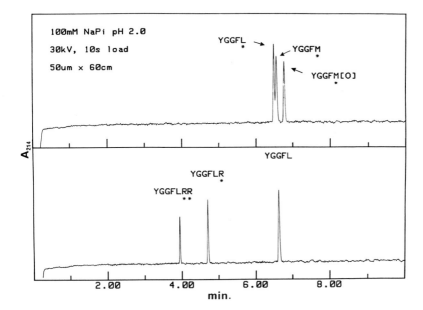

FIGURE 7   Another group of synthetic peptides analyzed by capillary zone electrophoresis, showing the effect of charge and polarity on mobility. Conditions were the same as described in Figure 6. (Figure kindly provided by Dan Crimmins of the Howard Hughes Medical Institute.)

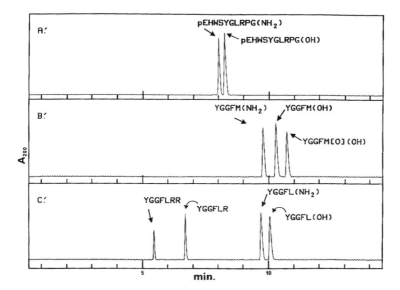

FIGURE 8   A third group of synthetic peptides analyzed by capillary zone electrophoresis, showing the effects of polarity of the α-carboxyl group. Conditions were the same as described in Figure 6. (Figure kindly provided by Dan Crimmins of the Howard Hughes Medical Institute.)

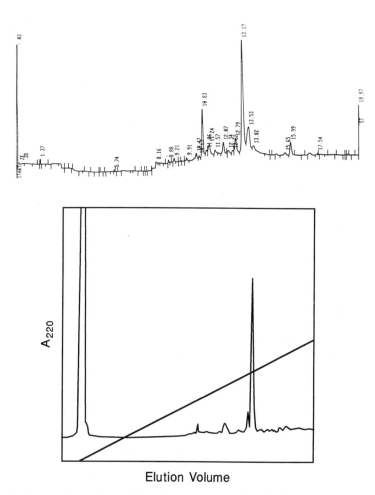

FIGURE 9   Comparison of the separation produced by capillary zone electro-phoresis and reverse-phase HPLC on the same peptide sample. The CZE is the upper panel and the HPLC is the lower panel. The CZE was run at pH 2.0 and the HPLC was on a C18 column in 0.1% trifluoroacetic acid with a 1%/min linear gradient of acetonitrile.

strength). Increasing ionic strength with solutions of a salt either in a stepwise fashion or by a gradually increasing concentration gradient is the most common method. Thus, ion-exchange HPLC, sometimes re-ferred to as "normal-phase" chromatography, binds the peptide at a rela-tively low ionic strength and elution is accomplished by progressing to conditions of higher ionic strength. Peptides are separated because of their differential interaction with the ionic support and hence elute at different levels of ionic strength. One approach to the ion-exchange

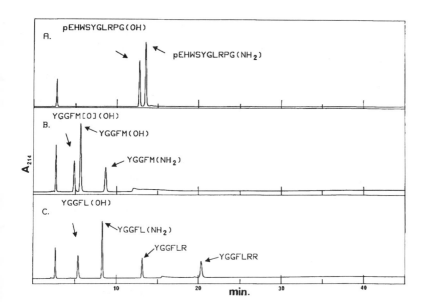

FIGURE 10   Strong cation-exchange chromatography of synthetic peptides on a sulfoethyl aspartimide column. The peptide mixtures are the same as those shown in Figure 8 for CZE separation. The column conditions can be found in Crimmins *et al.*, (1988). (Figure kindly provided by Dan Crimmins of the Howard Hughes Medical Institute.)

separation of synthetic peptides has been the use of strong cation-exchange columns such as sulfoethyl aspartamide columns (Alpert and Andrews, 1988; Crimmins et al., 1988), which separate peptides on the basis of positive charge at acid pH. Sulfoethyl aspartamide SCX columns (obtained from the Nest Group) are particularly useful in this regard, because monomeric peptides elute in a monotonic fashion according to their net positive charge at pH 3.0 (Crimmins et al., 1988). An example of peptide separation using this column is shown in Figure 10. The peptides shown in this figure are the same as those illustrated in Figure 8 for CZE separation. Note that while the patterns are somewhat different and the order of elution is reversed, both give comparable resolution.

Reverse-phase HPLC is so named because the elution conditions are essentially the reverse of normal phase chromatography. Binding occurs through hydrophobic interaction with the column support and elution is accomplished by decreasing the ionic nature, or increasing the hydrophobicity, of the eluant so that it competes for the hydrophobic groups on the column. Column supports are generally hydrocarbon alkane chains ranging from 4 (C-4) to 18 (C-18) carbons in length, or they are aromatic hydrocarbons such as phenyl groups. Although many different buffer

It is generally advisable, when using a buffer system for the first time, to test the compatibility of the components by simple mixing experiments. For example, phosphate is insoluble in high concentrations of acetonitrile, and it is potentially disastrous to use these components together without first determining their solubility limits. Such simple experiments could prevent the loss of an expensive column or a valuable sample.

systems and stationary phases have been reported (Snyder and Kirkland, 1979; Henschen et al., 1985; Hancock, 1984), the most commonly employed system for peptides is a two-buffer system, where the column is equilibrated in 0.1% trifluoroacetic acid (TFA) for sample loading, and elution is produced by an increasing gradient of acetonitrile as a water-miscible organic modifier. A typical recipe has reservoir A containing 0.1% TFA in water and reservoir B containing 80% acetonitrile in water that has been made 0.1% in TFA. The columns are usually eluted with a linear gradient run between 0.5 to 1% B per minute. Most peptides will elute from reverse-phase columns by the time the eluant is 60% acetonitrile. Other buffer recipes and elution protocols can be found in the general references in this section. These techniques include the use of 0.1% heptafluorobutyric acid, 0.1% phosphoric acid, dilute HCl, and 5 to 60% formic acid for low pH (pH 2–4); and 10 to 100mM ammonium bicarbonate, sodium or ammonium acetate, trifluoroacetic acid/triethylamine, sodium or potassium phosphate, and triethylammonium phosphate for pH 4–8. In addition to acetonitrile, methanol, ethanol, propanol, and isopropanol have been employed as water-miscible eluants (Rubinstein, 1979; Mahoney and Hermodson, 1980; Hancock, 1984; Feldhoff, 1991).

Again, the basis for peptide separation is that each peptide interacts differently with the column and will thus elute at a different concentration of organic modifier. The resolution of reverse-phase HPLC can be excellent, but one must be constantly mindful that a single peak does not guarantee homogeneity. This is demonstrated quite convincingly in the chromatograms in Figure 21, where it can be seen that a single symmetrical peak, generated under one set of HPLC conditions, actually contained two peptides that were separated when the conditions of the chromatography were altered. The two peptides present in this sample differed by only a single glutamic acid residue as a result of a partial deletion during synthesis. In this particular example, the pH of the TFA buffer was raised with triethylamine. Under basic conditions, the glutamic acid in the peptide becomes negatively charged which in turn alters its interaction with the stationary phase. This selectivity modulation through buffer changes points out the advantage that can be gained

Most reverse-phase columns available today have a silica-based support to which the bonded phase (i.e., the alkane chain) is attached. The bond holding the bonded phase to the silica support is labile to basic pH, and these columns should not be used with basic buffers. However, the columns usually can be used at slightly basic pH (8), as shown in the example in Figure 21, for short periods of time as long as they are returned to acidic conditions at the end of the run. Conditions of high pH or exposure to a slightly basic pH for extended periods of time should be avoided. Be aware that repeated use at basic pH may gradually destroy the column. Some manufacturers are now offering polymer-based supports that are resistant or unaffected by basic pH.

by routinely analyzing all synthetic peptide samples at two different conditions of pH.

The most common bonded phases used for the reverse-phase separation of synthetic peptides consist of alkane chains that range in size from 1 to 18 carbon atoms. The most common are C-4, C-8, and C-18. Because elution from the column is a function of the hydrophobicity of the peptide, and since charge decreases hydrophobicity, very highly charged small peptides tend to be retained better by long-chain bonded supports. Conversely, large hydrophobic peptides tend to exhibit better elution properties from short-chain bonded supports. However, in practice, they can often be used interchangeably with little significant effect on most peptides.

Carbon load, which refers to the amount of bonded phase present per unit of support, also has an effect on the characteristics of a reverse-phase column. For instance, two different C-18 columns can differ significantly in the mass, or amount, of hydrocarbon chain they contain, although they are both C-18 columns. The carbon load not only affects the column's binding capacity, but it can also affect retention time. Unfortunately, carbon load is not a parameter often listed by column manufacturers, and usually it has to be determined by trial and error or learned by word of mouth.

In addition to the type of bonded phase, particle size and pore size are two other parameters that can have significant effects on the resolution of peptides. Particle size or mesh size refers to the actual average size of the particles that make up the support. Modern HPLC columns are generally available in particle size ranges from 3 to 10 microns. Smaller particle sizes produce better resolution, but they are also more expensive and increase operating pressure, referred to as *backpressure*. Commonly, analytical columns (i. e., 4.6 × 250 mm) use 3 to 10 micron particles, and preparative columns (i.e., 22 × 250 mm) utilize 10 micron particles.

The age of a column can have a profound effect on its performance. A column that exhibits an extremely high degree of resolution when it is new may lose its resolving power as it ages. Thus, shoulders will no longer be evident or two similar peptides may co-elute. As a result, heterogeneity may go undetected by a column that is operating at less than optimal efficiency. In this regard, it is helpful to acquire a peptide standard that contains closely eluting peaks that can be used to assess column performance periodically. When a column starts losing resolving power, it should be replaced.

The particles used in reverse-phase chromatography are not solid but rather contain pores (analogous to a honeycomb), which increase the particle's surface area and hence the interaction with the peptide. Pore size refers to the average size of the particle's pores through which solvent and solute pass during interaction with the stationary phase. In order for a molecule to effectively interact with the particle, it must be able to pass through the pores. For large proteins, pore sizes of 300 µ are more effective. This pore size can also be used effectively for most peptides, but for very small peptides (6 residues or less), pore sizes of 80 to 100 µ generally produce sharper peaks and better resolution.

Amino Acid Compositional Analysis

Amino acid analysis reveals the amino acid content in a peptide and quantitates the amount of each. It is routinely the first analysis performed that gives data on the makeup of the peptide. Thus, a mole ratio of the amino acids can be calculated from the analysis and compared to the predicted mole ratio based on the intended sequence of the peptide. With the exception of a few amino acids that do not yield quantitatively (see the following discussion), the procedure gives a fairly accurate picture of the peptide. A general reference for amino acid analysis can be found in *Methods in Enzymology* (Ozols, 1990).

The use of column chromatography for the analysis of amino acids was developed by William Stein and Stanford Moore (Moore and Stein, 1948; Moore et al., 1958; Spackman et al., 1958), for which they won the Nobel Prize in Chemistry in 1972. Their original procedure used ion-exchange chromatography on sulfonated polystyrene resins and post-column detection with ninhydrin. This procedure remained the standard for many years and is still in routine use today. In fact, it is still considered by many to be the method of choice when quantity is not limiting. A limitation of post-column ninhydrin detection is its lowered sensitivity. Even on modern ninhydrin-based instruments, levels below approximately 100 pmol do not quantitate accurately. In addition, post-column ninhydrin analysis requires detection at two wavelengths, 570 and 440 nm, for the detection of both amino and imino (proline and hydroxy-proline) acids, respectively (see Figure 11).

More recently, post-column o-phthalaldehyde detection has been used to increase sensitivity to the low picomole range (Benson and Hare, 1979; Roth and Hampai, 1973). However, the imino acid proline is not detected with o-phthalaldehyde unless it is oxidized with hypochlorite (Bohlen and Mellet, 1979). This procedure also destructively reduces the levels of the other amino acids and, thus, reduces sensitivity. With the development of reverse-phase HPLC, several more sensitive pre-column derivatization procedures have been developed. These procedures employ chemical derivatization of the amino acids after hydrolysis, but before chromatography, with a reagent that reacts quantitatively with the free α-amino group of amino acids and can be detected by fluorescence, ultraviolet, or visible absorbance. These procedures include derivatization with o-phthalaldehyde (Jones, 1986), dansyl chloride (dimethyl-aminonapthalene-1-sulfonyl chloride) (Taphui et al., 1982; Oray et al., 1983; Marquez et al., 1986), PITC (phenylisothiocyanate) (Heinrikson and Meridith, 1984; Cohen and Strydom, 1988), dabsyl chloride (dimethylaminoazobenzene-4-sulfonyl chloride) (Chang et al., 1983; Knecht and Chang, 1986), DABITC (dimethylaminoazobenzene-4-iso-thiocyanate) (Chang, 1983), Fmoc-Cl (9-fluorenylmethyl chloroformate) (Einarsson et al., 1983), and a combination of o-phthalaldehyde and Fmoc-Cl (Einarsson, 1985; Blankenship et al., 1989). Some of these techniques have been automated (Cunico et al., 1986; Gustavsson and Betner, 1990; Simmaco et al., 1990), and today several instrument manufacturers offer one or more chemistries as a package. These packages include Beckman's ninhydrin and DABS (dabsyl chloride) systems, Water's Pico-Tag (PITC) system, Varian's Amino-Tag (Fmoc-Cl) system, and Hewlett Packard's Amino-Quant (OPA/Fmoc-Cl) system. In addition to the references listed, several general reference books also are available that describe the techniques in more detail (Hancock, 1984; Henschen et al., 1985).

Because all amino acid analysis techniques analyze the free amino acids, all polypeptides, including synthetic peptides, must first be broken down into their individual amino acids by hydrolysis of the peptide bonds. This is commonly done by subjecting the peptide to 6N HCl in either the liquid or the gaseous phase. In the original procedure, the peptide was dissolved in 6N HCl, the tube was evacuated and sealed, and hydrolysis was accomplished by incubating the tube at 110 °C for 24 hours (Moore and Stein, 1963). More recently, the technique of vapor-phase hydrolysis has been used (Dupont et al., 1988), in which the peptide does not come into direct contact with the liquid HCl. This procedure tends to produce cleaner analyses because nonvolatile contaminants in the liquid HCl no longer contact the peptide. This factor is more important for high-sensitivity analyses of very low-level samples than it is for

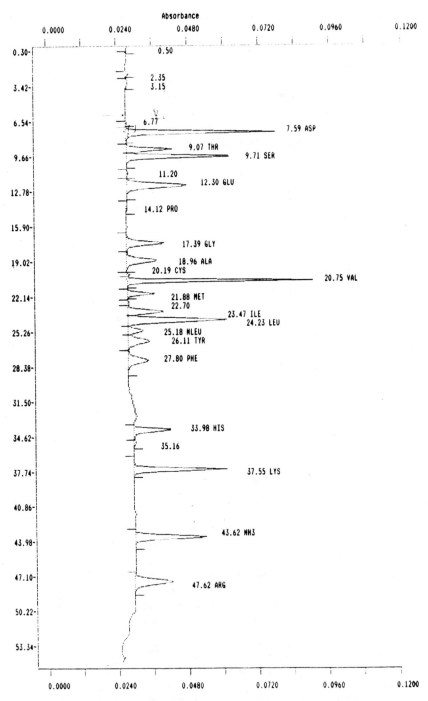

FIGURE 11   Chromatograms of the amino acid analysis of a synthetic peptide. The samples were run on a Beckman 7300 amino acid analyzer equipped with System Gold software. The column was a Beckman amino acid analysis ion-

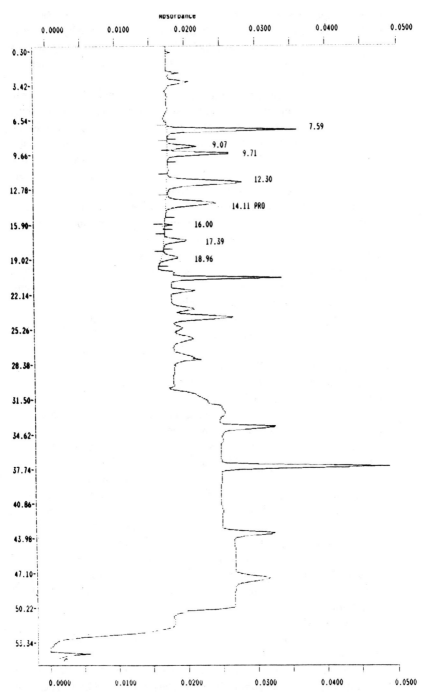

exchange column and detection was post-column with ninhydrin. The left trace was monitored at 570 nm and the right trace at 440 nm. Only the 1-hour hydrolysis is shown.

synthetic peptides, where amounts are usually not a problem, but it is routinely used for both with excellent results. Vapor-phase hydrolysis at 150 to 165 °C for 1 hour (Tarr, 1986; Dupont et al., 1988) is also now used routinely with excellent results and with the obvious advantage of saving time.

As mentioned earlier, several amino acids are not stable to hydrolysis in 6N HCl. Serine, threonine, tyrosine, and methionine are partially destroyed, and glutamine, asparagine, cysteine, and tryptophan are usually completely destroyed. Serine, threonine, and tyrosine can be fairly accurately determined by carrying out a time course of hydrolysis and extrapolating back to zero time (Figure 12). At 110 °C, the times are usually 24, 48, and 72 hours, and at 150 to 165 °C, the times are 1, 2, and 3 hours. Tyrosine destruction can also be decreased by including 0.1% phenol in the hydrolysis mixture. Cysteine/cystine and methionine are usually determined after performic acid oxidation (Hirs, 1967) as cysteic acid and methionine sulfone, respectively. Cysteine can also be determined after modification with iodoacetate or 4-vinylpyridine (Fontana and Gross, 1986; Hawke and Yuan, 1987) as S-carboxymethyl cysteine or S-pyridylethyl cysteine, respectively.

Hydrolysis with methanesulfonic acid (Simpson et al., 1976), toluenesulfonic acid plus tryptamine (Liu and Chang, 1971), mercaptoethanesulfonic acid (Penke et al., 1974), methanesulfonic acid plus 3-(2-aminoethyl) indole (Simpson et al., 1976), or in the presence of thioglycolic acid (Matsubara and Sasaki, 1969) have been reported to preserve tryptophan, but in practice the results have been found to be variable. Alkaline hydrolysis (Hugli and Moore, 1972) also preserves tryptophan but destroys other amino acids in the process. It is probably the method of choice for accurate tryptophan determination, but due to its relative difficulty, it is seldom done on a routine basis.

The integrity of tryptophan in a polypeptide can be evaluated to some extent by its ultraviolet absorbance. Although this process does not generally yield accurate quantification, the absorbance profile of a peptide, scanned from 200 to 320 nm, can provide useful information about the state of the tryptophan in a peptide. An example is shown in Figure 15, which illustrates a UV scan of a 30-residue peptide containing one tryptophan residue and compares it to a scan in which the tryptophan is either missing or destroyed. Intact tryptophan absorbs maximally at 280 nm, while its oxidation products absorb at lower wavelengths and with reduced extinction coefficients. N-formyl tryptophan, which is used in t-Boc synthesis, absorbs maximally at 300 nm, so it would be readily apparent if deformylation was not successful.

Glutamine and asparagine are completely converted to glutamic acid and aspartic acid, respectively, by acid catalyzed deamidation. Therefore, the values for Glu and Asp derived from amino acid analysis of an acid hydrolyzed sample represent the Glu plus Gln and Asp plus Asn content,

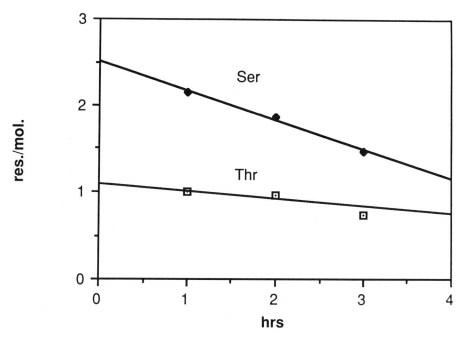

FIGURE 12   Plot of the serine and threonine values from a time-course hydrolysis of a 29-residue synthetic peptide. Hydrolyses were carried out for 1, 2, and 3 hours at 160 °C with vapor phase HCl.

respectively. Asparagine and glutamine can be quantitated if complete enzymatic hydrolysis is performed (Stewart and Young, 1984). However, this procedure is not usually done routinely because of its relative difficulty and potential sequence specific variability.

The rate of peptide bond hydrolysis can also be affected by amino acid sequence. This is most apparent when the two aliphatic side-chain β-branched amino acids, Val or Ile, occur adjacent to one another. In this case, hydrolysis of the peptide bond may not be complete until after 72 hours of hydrolysis (3 hours for high-temperature hydrolysis), so values determined at earlier time points will be low.

Amino acid analysis is particularly well suited for the analysis of synthetic peptides because it is more accurate for smaller molecules, such as peptides, than it is for larger molecules, such as proteins. This situation is a result of the inherent error in the procedure. Recent studies (Crabb et al., 1990; Tarr et al., 1991) indicate an average error in practice of approximately 10 percent for nmol level analyses and as much as 16 to 20 percent for pmol level analyses. For example if a synthetic peptide is analyzed at the nmol level that contained two aspartic acid residues, the expected error would be approximately 0.2 residues. A calculated value of between 1.8 and 2.2 residues would quite accurately indicate that only two residues of aspartic acid were present. If, on the other hand, a protein that contained 20 aspartic acid residues is analyzed, the error would be approximately 2 residues and would yield values between 18 and 22 residues per molecule. Synthetic peptides are usually small enough that the inherent error in the analysis is not sufficiently large to cause such an ambiguity. Thus, amino acid analysis of synthetic peptides is expected to be quite accurate, and any significant deviation from integral values of residues is usually a sign of heterogeneity in the sample. There are, however, two general types of exceptions that must be taken into consideration when evaluating a peptide by amino acid analysis. The first comes from the fact that not all amino acids survive the hydrolysis procedure quantitatively, the second from the fact that the hydrolysis can remove modifying groups that may be present, e.g., side-chain protecting groups, regenerating the free amino acid. The first of these has been discussed already; an example of the second is presented later.

Figure 11 and Table 3 show the results of a typical amino acid analysis of a synthetic peptide performed on an ion-exchange analyzer equipped with post-column ninhydrin detection. In this case, the sample was hydrolyzed with 6N HCl, containing 0.1% phenol in the vapor phase for 1,2, and 3 hours at 160° C. An additional aliquot of the sample was treated with performic acid for the determination of Cys and Met, and the analyses were normalized to each other based on the values for aspartic acid, which is stable to performic acid oxidation, from each analysis. Note the use of a second wavelength in Figure 11 for ninhydrin detection of proline. Time courses, with extrapolation to zero time, are usually necessary only when dealing with larger polypeptides and proteins. The level of destruction of serine and threonine in a typical hydrolysis is usually less than 30 percent and 10 percent, respectively. Thus, the values obtained when these residues are present only a few times in a synthetic peptide are usually sufficient to make a reasonable assessment. If there is any doubt, a time course can be performed. As an example, the data for Ser and Thr from Table 3 are plotted in Figure 12. While the value for Thr does not change, the time-course extrapolation indicates 3 serine residues rather than the 2 residues indicated by the first time point.

Although amino acid analysis provides valuable information in the evaluation of a synthetic peptide, it is dangerous to rely on it as the sole characterization. The analysis shown in Table 4 demonstrates this point. The analysis of this 16-residue peptide appears to be very reasonable when it is compared to the expected values. The peptide contains 1 residue of tryptophan, which is not detected, of course, and the two isoleucines are adjacent to one another so the low Ile value is consistent with the expected slow hydrolysis of the Ile-Ile bond. All other values look good. However, sequence analysis of this peptide revealed that the 2 aspartic acid residues are not entirely present as free aspartic acid.

Table 3   Amino Acid Analysis[a] of a Time Course Hydrolysis of a 29-Residue Synthetic Peptide

| Residue | Mol. ratio | | | Integral value | Expected |
	1 hour	2 hours	3 hours		
Asp	2.92	2.97	2.92	3	3
Thr	1.00	0.96	0.74	1	1
Ser	2.16	1.87	1.48	3	3
Glu	1.94	1.94	1.91	2	2
Pro	0.93	0.92	0.90	1	1
Gly	0.97	0.99	0.97	1	1
Ala	1.07	1.10	1.05	1	1
Cys	0.99	1.00	0.98	1	1
Val	3.01	2.99	2.95	3	3
Met	1.29	1.30	1.27	1	1
Ile	0.93	0.94	0.92	1	1
Leu	2.86	2.84	2.82	3	3
Tyr	0.93	0.81	0.84	1	1
Phe	0.97	0.97	0.97	1	1
His	1.08	0.96	0.95	1	1
Lys	1.92	1.91	1.90	2	2
Arg	1.87	1.77	1.74	2	2
Trp	nd[b]				1

[a] The peptide was vapor hydrolyzed with 6N HCl containing 0.1% phenol for 1, 2, and 3 hours at 160 °C and analyzed on a Beckman 6300 amino acid analyzer. Performic acid oxidation was performed on a separate aliquot.

[b] nd, not determined

Rather, the HPLC chromatograms (Figure 13) of the PTH amino acids indicate that as much as 60 percent of the material at the Asp positions elutes approximately 6 minutes later than aspartic acid, between the positions for alanine and tyrosine. This example clearly illustrates the presence of unintended modification of the aspartic acid residues that was not detected by amino acid analysis, because the adduct was not stable to the hydrolysis procedure and thus regenerated the free amino acid.

Table 4   Amino Acid Analysis[a] of a 16-Residue Peptide

Residue	nmol	Mol. ratio	Expected
Asp	10.9	2.0	2
Thr	6.6	1.2	1
Ser	4.2	0.8	1
Glu	5.7	1.0	1
Pro	5.5	1.0	1
Gly			
Ala	6.0	1.1	1
Cys	4.2	0.8	1
Val	5.9	1.1	1
Met	5.7	1.0	1
Ile	8.7	1.6	2
Leu			
Tyr			
Phe			
His			
Lys	10.3	1.9	2
Arg	5.1	0.9	1

[a] The peptide was vapor hydrolyzed with 6N HCl containing 0.1% phenol for 1 hour at 160 °C and analyzed on a Beckman 6300 amino acid analyzer.

Initial Criteria for Evaluation of Synthetic Peptides

Every peptide, whether it will eventually undergo extensive purification or whether it is intended to be used in a relatively crude form, should be subject to some minimal level of characterization. Generally, the minimal criteria that should be present are (1) a good mass recovery based on a reasonable percentage of the theoretical yield, (2) a predominate peak with reasonable resolution on analytical HPLC or capillary electrophoresis, and (3) an amino acid composition consistent with the presence of the desired peptide. If any of these three criteria are not met, there is reason to suspect a problem, and the peptide should either be evaluated further or resynthesized. The relevant analytical procedures were discussed in detail earlier and are further illustrated in Figures 14 and 15. The top frame of Figure 14 shows the reverse-phase elution profile of a crude 30-residue peptide product that obviously does not meet criter-

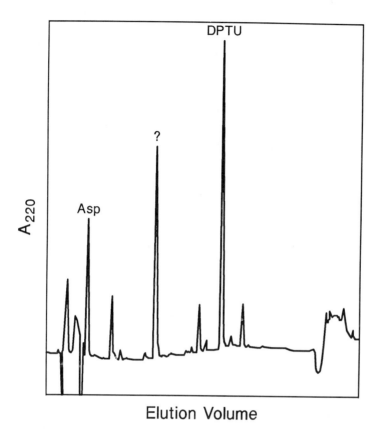

FIGURE 13   Reverse-phase HPLC analysis of PTH amino acids from a cycle of automated Edman degradation on an Applied Biosystems 477A sequencer. Abbreviations: Asp, aspartic acid; DPTU, diphenylthiourea; ?, unknown amino acid adduct.

ion 2. Although the large peak eluting early in the chromatogram appears at first glance to be a well-resolved predominant peak, its broad shape and early elution position are not consistent with that expected for a 30-residue peptide. Subsequent amino acid analysis of separate fractions across the absorbance peaks indicated that this large peak eluting in the first third of the chromatogram was devoid of peptide and that all the peptide was present in the broad, poorly resolved "hump" eluting later. On the other hand, the elution profile shown in the bottom frame of Figure 14 does meet the requirements of criterion 2. Although several species are present, there is obviously a well-resolved predominant peak in this preparation. These two syntheses were, in fact, intended to be the same peptide. The top frame was from the original synthesis and the bottom frame was from a resynthesis. The amino acid analysis, mass

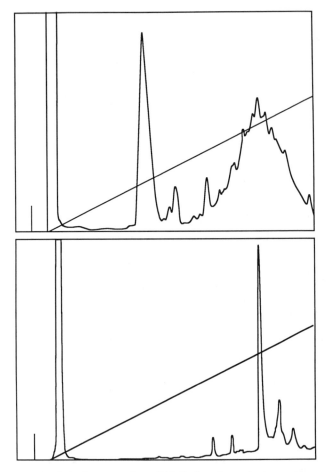

FIGURE 14   Analytical reverse-phase HPLC of crude synthetic peptide
samples. HPLC was on a 4.61 × 250 mm C18 column in 0.1% trifluoroacetic
acid with a 1%/min linear gradient of acetonitrile.

recovery, and UV spectrum of the two syntheses from Figure 14 are
presented in Figure 15. Both the compositional analysis and the mass
yield of the original synthesis indicate inconsistencies. Furthermore, the
ultraviolet absorbance scan of the original synthesis indicates a major
problem with tryptophan while that of the remake, which shows a typical
tryptophan absorption at 280 nm, is what should be expected. Thus, these
data complete the picture and illustrate that the original synthesis failed
to meet criteria 1, 2, and 3, whereas the resynthesis meets all three
criteria and, therefore, appears to be reasonable.

Meeting these three criteria provides complete proof not necessarily
that the desired product is present in good yield, but only that the evalua-

```
        5      10     15     20     25     30
       TWKPYDAADLDPTENPFDLLDFNQTQPERC
```

	Expected	AAA Original	AAA Remake
Asp	7	7.0	7.0
Thr	3	2.9	2.5
Ser	0		
Glu	4	5.2	4.5
Pro	4	3.1	3.4
Gly	0		
Ala	2	1.5	1.9
Cys	1	--	--
Val	0		
Met	0		
Ile	0		
Leu	3	2.0	2.9
Tyr	1	0.4	0.8
Phe	2	1.7	1.9
His	0		
Lys	1	0.7	1.0
Arg	1	1.7	1.2
Trp	1	--	--

Abs.

Abs.

200   240   280   320
nm

30

Yield   409mg   807mg

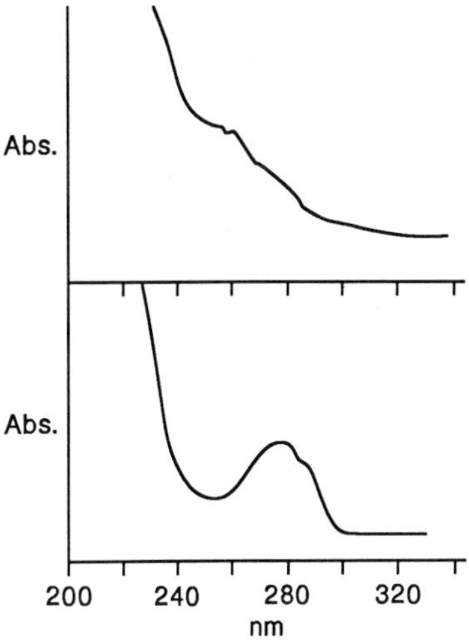

FIGURE 15   Intended sequence, amino acid analysis data, and ultraviolet spectra of the synthetic peptide samples whose HPLC analyses are presented in Figure 14. The column headed AAA original and the top panel of the absorbance spectra correspond to the top panel of Figure 14.

tion is consistent with that conclusion. At this point, a decision, based on the individual situation and the available data, must be made concerning the acceptability of the product. As is usually the case, good judgment and common sense are important ingredients for making that decision. These same criteria can, and should, be applied to all syntheses as part of the initial evaluation, even though the product is slated for additional purification.

PURIFICATION

The crude product that is obtained from stepwise solid-phase peptide synthesis will contain a variety of by-products in addition to the desired product. These products usually will include a family of peptides containing deletion or addition sequences generated during the synthesis itself and a series of peptides containing chemical modifications that are

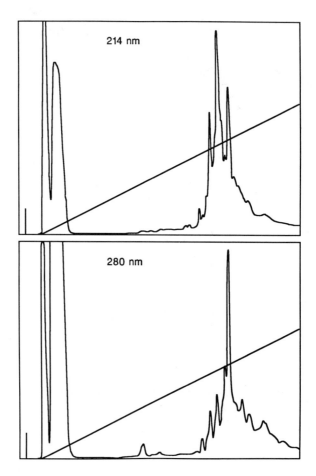

FIGURE 16   Analytical reverse-phase HPLC of a crude synthetic peptide sample monitored at both 214 and 280 nm. HPLC was on a 4.6×250 mm C18 column in 0.1% trifluoroacetic acid with a 1%/min linear gradient of acetonitrile.

usually generated during the cleavage and deprotection steps. The two requirements for obtaining the intended product are (1) that it is present in this mixture in sufficient concentration and (2) that the other components can be successfully removed. In general, using modern synthetic procedures, crude peptides of 30 to 45 residues can be expected to be between 60 and 80 percent target peptide, and those of less than 30 residues will be between 80 and 95 percent target peptide. All that is required then is an effective purification protocol.

    In theory, the purification of a peptide should involve steps that include different chemical or physical principles for the separation of

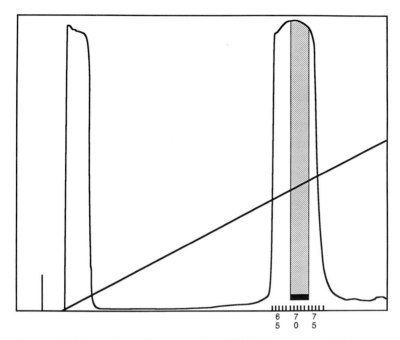

FIGURE 17   Preparative scale reverse-phase HPLC chromatogram of the sample depicted in Figure 16. Approximately 100 mg of peptide was loaded onto a 22 × 250 mm Vydac C18 column equilibrated in 0.1% trifluoroacetic acid. The column was developed with a linear gradient of acetonitrile at 1% min.

heterogeneous species. Historically, the most effective techniques for peptide separation have included ion-exchange chromatography, typically on polystyrene-based resins, partition chromatography, and counter-current distribution (Stewart and Young, 1984). More recently, high-pressure liquid chromatography has been a tremendous addition to this repertoire and has largely surpassed the other techniques for routine use. Today, in practice, most peptides approximately 50 residues or less in length can be purified using reverse-phase HPLC exclusively. In addition, in cases where reverse-phase HPLC is not entirely successful, ion-exchange HPLC can be an effective and valuable complement to reverse-phase techniques. Although seldom reported anymore, low-pressure ion-exchange chromatography using either carboxymethyl (CM) or diethyl-aminoethyl (DEAE) resins with volatile buffer systems can also be useful in some cases (Atherton and Shepard, 1989).

To illustrate the power of reverse-phase HPLC as a purification tool, the following example, illustrating a synthetic product that posed a particularly difficult purification problem, is presented in Figure 16. It shows the reverse-phase HPLC elution profile of the crude product from

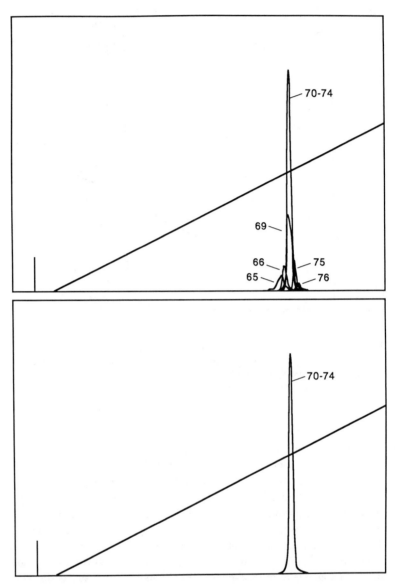

FIGURE 18   Analytical HPLC analyses of the fractions across the preparative
protein peak in Figure 17. HPLC conditions are as described in Figure 16. The
profiles in the top panel are from runs of individual fractions. The profile in the
bottom panel is from the pool of fractions 70–74.

> The process of fraction-by-fraction analysis of complex patterns found in the crude synthetic products by analytical scale HPLC can be a very effective approach to purification. This systematic approach often allows the retrieval of the desired product from a synthesis that would otherwise appear to be too complex.

the synthesis of a 40-residue peptide monitored at both 214 and 280 nm. Based on the 214 nm profile alone, one might conclude that the desired peptide is probably in the major peak, which is the middle peak of the three. However, the 280 nm profile indicated that this is probably not the case, and that the desired peptide is most likely in the peak on the right because it should contain tryptophan, which absorbs strongly at 280 nm. Preparative scale reverse-phase HPLC of 100 mg of crude product on a 2.2 cm × 25 cm C-18 column, under similar buffer and gradient conditions used for the analytical column shown in Figure 16, produced the profile shown in Figure 17. The amount of material loaded produced a profile without any distinguishing features, so analytical scale HPLC was used to analyze each fraction across the peptide peak. These analyses are shown in the top panel of Figure 18 and indicate that fractions 70–74 give similar characteristics with respect to elution time and maximum absorbance at 280 nm. The bottom panel of Figure 18 shows an analytical run of the pooled fractions 70–74. The single symmetrical peak suggests that a homogeneous peptide was obtained from the original sample. This was subsequently verified by amino acid analysis (Table 5), sequence analysis, and mass spectroscopy.

DETAILED CHARACTERIZATION OF COVALENT STRUCTURE

The previous sections dealt mainly with the assessment of peptide homogeneity and purification. With the exception of amino acid compositional analysis, none of these approaches provides any information on the covalent nature of the peptide. The determination that the covalent structure of the peptide is correct is no less important than obtaining a peptide free of by-products. After all, some amount of thought went into the design of the peptide for a particular experiment, and there is much to be lost by performing the experiment with the wrong peptide.

Sequence analysis and mass spectrometry provide the best analysis of the covalent nature of a peptide. Sequence analysis can reveal a great deal of information about a synthetic peptide and, unlike mass spectrometry, it is something that most core laboratories can now do on a routine basis. On the other hand, many peptide chemists now rely on mass analysis as the mainstay of their evaluation of the synthetic product.

Table 5   Amino Acid Analysis[a] of the Synthetic Peptide from Figures 16 through 18

	Mol. ratio		
Residue	Expected	Crude	Fractions 70-74
Asp	3	3.0	3.0
Thr	1	1.2	1.0
Ser	3	3.3	2.5
Glu	5	2.4	4.9
Pro	3	3.7	3.2
Gly	1	0.5	0.9
Ala	2	0.6	1.9
Cys	1	nd	nd
Val	6	5.3	5.4
Met	1	nd	nd
Ile	2	2.6	1.7
Leu	1	1.6	1.1
Tyr	1	0.2	0.8
Phe	1	0.3	0.8
His	2	1.5	1.9
Lys	1	0.7	0.9
Arg	5	5.1	5.1
Trp	1	nd	nd

[a] The peptide was vapor hydrolyzed with 6N HCl containing 0.1% phenol for 1 hour at 160° C and analyzed on a Beckman 6300 amino acid analyzer.

However, both techniques have limitations and they are generally used most effectively when they complement each other.

Sequence Analysis

Automated Edman degradation, more commonly referred to as amino acid sequence analysis, is obviously well suited for determining the order and identity of amino acids in a synthetic peptide. It can also be very helpful in identifying some, although not all, derivatives of amino acids that may be present for a variety of reasons (see Figure 13). The first sequence of a polypeptide was determined in 1951, when Sanger determined the sequence of insulin using fluorodinitrobenzene (Sanger and

Tuppy, 1951). Repetitive degradation of proteins from the amino terminal end with phenylisothiocyanate was first reported by Edman (1950) and remains the chemistry of choice for sequence analysis today (Figure 19). The automation of Edman chemistry in 1967 (Edman and Begg, 1967), with the introduction of the first "spinning cup" sequencer, marked the beginning of the modern age of sequencing. Later, modifications to the spinning cup sequencer greatly improved the sensitivity of sequence analysis (Wittmann-Liebold, 1973; Hunkapiller and Hood, 1978), and the introduction of polybrene (Tarr et al., 1978; Hunkapiller and Hood, 1978), which helped to physically retain the sample in the cup, extended the length of sequence that could be obtained in a single run. Most recently, the development of the "gas-phase" sequencer in 1981 (Hewick et al., 1981) played a major role in advancing the sequential determination of polypeptide structures. Today, several manufacturers offer comparable high sensitivity sequencers.

The original spinning cup sequencer of Edman and Begg required protein amounts in the 100 nmol range or greater (Edman and Begg, 1967). Later, improvements in the spinning cup instrumentation allowed sequence analysis of as little as 1 to 10 nmols of polypeptide, but the introduction of the gas-phase sequencer, which could routinely analyze 100 pmol of polypeptide (Hewick et al., 1981), quickly outpaced the utility of the spinning cup sequencer. Today, the spinning cup sequencer is no longer available and modern sequencers are now capable of providing very clear analyses on as little as 10 pmol of peptide. For a synthetic peptide of 25 residues, this amount constitutes approximately 25 ng of material. Since sensitivity is not a major concern for sequence analysis of synthetic peptides because quantity is seldom a consideration, they are usually analyzed in the 100 to 1000 pmol range. A good current book on sequence analysis is *Protein Sequencing: A Pratical Approach* (Findlay and Geisow, 1989).

Sequence analysis, as the name implies, reveals the sequence, or order, of the amino acids in a polypeptide chain. In this regard, one of the technique's main strengths is in revealing the presence of deletion or addition sequences, which are among the more common and frequent problems encountered in synthetic peptides. It can also be used effectively to reveal the presence of side-chain protecting groups that failed to be removed during the deprotection step. However, it is more effective in this regard for peptides synthesized by t-Boc chemistry rather than by Fmoc chemistry due to the acid lability of the blocking groups used for Fmoc synthesis (see On-Resin Sequencing).

Some significant disadvantages limit the usefulness of sequence analysis. First, unless covalent coupling of the peptide to a solid support is employed, it is not always possible to sequence to the C-terminal residue of a peptide. This problem occurs especially with longer peptides, but it can occur with relatively short peptides as well. It occurs mainly because

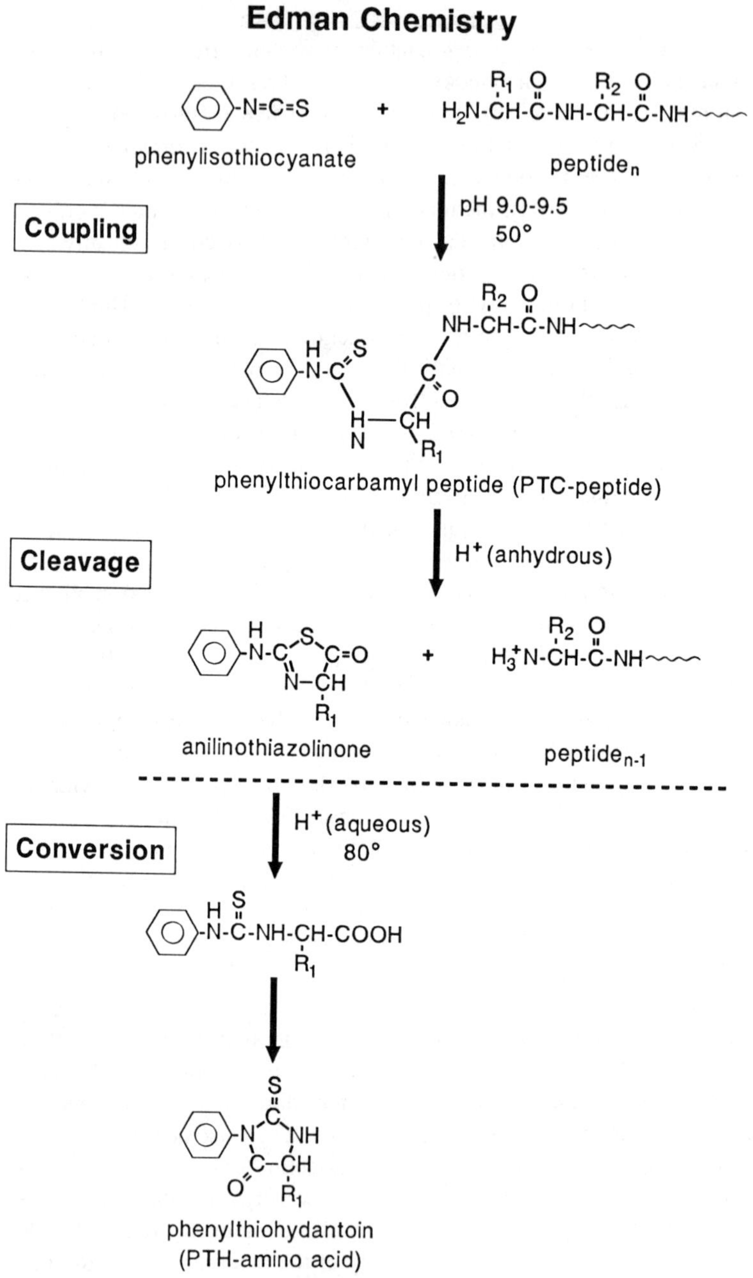
Edman Chemistry
-N=C=S
+
R1 O R2 O
H2N-CH-C-NH-CH-C-NH
phenylisothiocyanate
peptiden
Coupling
pH 9.0-9.5
50°
R2 O
NH-CH-C-NH
H S
-N-C
C
O
H—CH
N
R1
phenylthiocarbamyl peptide (PTC-peptide)
Cleavage
H+ (anhydrous)
H
-N-C
S
C=O
N-CH
R1
+
R2 O
H3+N-CH-C-NH
anilinothiazolinone
peptiden-1
Conversion
H+ (aqueous)
80°
H S
-N-C-NH-CH-COOH
R1
S
C
-N
NH
C-CH
O
R1
phenylthiohydantoin
(PTH-amino acid)

FIGURE 19 Edman chemistry used for the sequential degradation from the amino terminus of proteins and peptides.

of a phenomenon known as *washout,* and it is brought about by the repeated washings, after the coupling and deprotection steps, of the sample in the reaction chamber with solvents. Although these washings are formulated to remove only excess reagents, unwanted reaction by-products, and cleaved ATZ amino acid, and not the peptide itself, some peptide is invariably lost at each cycle. Thus, at some point the level of peptide remaining is below the detection sensitivity of the technique and sequence identification stops. Washout is usually more prone to occur with hydrophobic peptides, and charged residues near the C-terminus tend to reduce washout of the sequentially shortened peptide.

Failure to sequence to the C-terminus of a peptide can also be caused by rising background due to low-level but repeated internal cleavage of the polypeptide chain. This cleavage occurs because the reaction cannot be kept completely anhydrous. At a particular point in the sequence analysis, the signal produced by the sequential degradation of the peptide may still be within the detection range of the HPLC, but the nonspecific background has risen so high that it gives comparable signal intensity, making it impossible to distinguish how much of the signal is caused by sequence and how much is caused by cumulative background. At that point, effective residue identification stops. It usually takes many successive cycles for the background to build up, so this problem usually occurs only with long peptides. In practice, both washout and rising background contribute, but for short peptides, the major cause of cessation of sequencing is washout. For long peptides, sequence identification stops at the point at which the effects of the two converge.

Loss of peptide due to washout can be alleviated to a large extent if the peptide is covalently coupled through its C-terminal residue to an insoluble support so that washout does not occur. Automated solid-phase sequencing was introduced by Laursen (1971), but it has not gained widespread use because of chemical and technical difficulties. Unfortunately, other attempts to provide such chemistry have not been altogether successful either. Today, MilliGen (Pappin et al., 1990) is essentially the only manufacturer of automated sequencers developing supports for covalent attachment or solid-phase sequencing methodologies. However, the techniques and the materials needed to perform the chemistry have been available for only a relatively short period of time, so extensive evaluation in practice has not yet been done. Nonetheless, this approach has potential advantages and bears watching closely. On-resin sequencing (see On-Resin Sequencing) does allow sequencing from an insoluble support, but since it must be followed by the cleavage chemistry, it does not provide the final answer.

It is always desirable, and sometimes necessary, to verify the sequence of a peptide all the way to the C-terminal residue. Occasions where this may be the case occurs when mass spectrometry is not available, when one suspects for some reason that two or more residues may have been transposed due to an operator error in setting up the synthesizer, or when a software error occurs in directing the order of addition. If sequence analysis of the intact peptide terminates before reaching the C-terminal residue, another approach is needed. One approach that has been successful is partial digestion of the synthetic peptide and sequencing the individual peptides after separation on HPLC. The smaller peptides generated by this procedure are usually able to be sequenced to their C-terminus, so that all of the sequence of the larger peptide can be verified by analysis of its parts. The major constraint in this approach is the availability of cleavage sites within the peptide such that reasonably sized subpeptides can be generated and recovered. Table 6 lists the commonly used cleavage agents and the residues or sequences at which they will cleave. General procedures for using these agents can be found in a number of books and references contained therein (Fontana and Gross, 1986; Wilkinson, 1986; Aitken et al., 1989).

A second disadvantage of sequence analysis is that it relies on the availability of standards for identification. That is, identification of the residue at any particular cycle is dependent on comparison, usually of an HPLC peak, to a known compound or standard. This disadvantage is obviously not a limitation for the normal amino acids or even for the side chain protected amino acids because standard compounds are readily identified and available. On the other hand, it can be severely limiting for unexpected or unusual adducts that have formed at some point in the synthesis, such as that presented in Figure 13. Although the presence of an unusual peak may indicate a divergence from normal, it does not provide much useful evidence concerning the identity of the adduct. Furthermore, the situation can be complicated if the unknown adduct co-elutes at the position of a standard amino acid, although it should be immediately obvious that it is not the expected residue at that position.

Another major limitation of sequence analysis is that peptides that are chemically modified at their amino terminus will not be detected. This limitation is particularly important when capping chemistry is employed during the synthesis (see Chapters 3). Although capping can be very helpful in the purification process, any capped peptide that is not removed from the sample will be invisible to sequence analysis.

The sequencing chemistry can also reverse some derivatives of amino acids, regenerating the parent amino acid in the process, so that its presence goes undetected. Examples of partial reversal were seen in Figure 4 with on-resin sequencing of peptides generated by t-Boc chemistry. With the protecting groups used for Fmoc synthesis, the reversal can be complete. Another example of this process is methionine sulfoxide. The sulfur atom of methionine residues can be oxidized during synthesis and handling, usually to the sulfoxide. In fact, methionine sulfoxide is sometimes employed directly as a protecting group in the synthesis of a peptide. After synthesis, the sulfoxide can be converted back to methionine by treatment with thiol reagents such as dithiothreitol and N-methylmercaptoacetamide (Cullwell, 1987; Houghton and Li, 1979; see also Chapter 3). However, if the sulfoxide is present in the peptide and no attempt is made to convert it to methionine prior to sequence analysis, its presence will go undetected in sequencing because DTT present in the sequencing reagents will convert the sulfoxide to methionine.

Finally, sequence analysis is not very reliable for quantitation; some residues are detected only at very low levels or not at all. These residues include serine, threonine, and cysteine, which are degraded to varying degrees by β-elimination during sequencing, and tryptophan, which is sensitive to oxidation either during sequencing or during storage and handling prior to sequence analysis. Although dithiothreitol is often added to the ATZ amino acid extraction solvent to trap the reactive elimination products of serine, threonine, and cysteine, the yields of these amino acid adducts are still low and variable. In fact, cysteine often must be derivatized prior to sequencing in order to detect it at all. The reagents most effective for this purpose are iodoacetic acid, iodoacetamide, and 4-vinylpyridine (Fontana and Gross, 1986). Alkylation with 4-vinylpyridine to the S-pyridylethyl derivative can be performed on the sample after it has been applied to the sequencer cartridge (Andrews and Dixon, 1987; Hawke and Yuan, 1987) and is probably the reagent of choice due to its stability and HPLC elution position. Arginine and histidine also tend to produce low yields due to reduced extraction efficiency of the positively charged amino acids and undesirable interactions with incompletely capped HPLC columns.

In addition to variability in individual PTH amino acid recovery, the overall initial yield of a sequencing run can also vary. This variability can be caused by several factors, such as the operating efficiency of the sequencer and the chemical properties of the sample itself. Thus, even if only stable PTH amino acids are considered, the overall yield of the sequencing run itself can vary severalfold.

In spite of all the limitations listed, sequence analysis is still an indispensable technique for the evaluation of synthetic peptides. What sequence analysis does best, which no other technique can do on a routine or available basis, is to elucidate the actual linear sequence of amino acids as they occur physically in a peptide. In this regard, sequence analysis is outstanding in revealing deletion and addition sequences.

The role of sequence analysis in finding deletion sequences is illustrated in the example presented in Figure 20. The product of this synthesis runs as a single major peak on the standard reverse-phase HPLC analysis system, which is 0.1% TFA with a linear acetonitrile gradient, and the amino acid analysis is entirely consistent with the desired product. As such, it meets the minimal criteria outlined earlier for evaluation of a crude product (see Initial Criteria for the Evaluation of Synthetic Peptides). However, in this case, sequence analysis revealed

Table 6 Common Reagents and Proteases for Cleavage of Peptides[a]

Bond cleaved	Agent	Reference	Source[b]
Arg, ⁻ys—X (X ≠ Pro)	Trypsin	Wilkinson, 1986	Worthington, Pierce, Boehringer Mannheim, Promega
Lys—X (X ≠ Pro)	Endoproteinase Lys-C		Boehringer Mannheim, Promega
	Lysyl endoproteinase	Masaki et al., 1981a, 1981b	Wako Chemicals
Arg—X (X ≠ Pro)	Endoproteinase Arg-C (submaxillary protease)	Levy et al., 1970; Schenkein et al., 1977	Boehringer Mannheim, Pierce, Takara Biochemicals, Promega
Glu—X (X ≠ Pro) or (Asp—X)[c]	Endoproteinase Glu-C *S. aureus* V8 protease	Wilkinson, 1986	Boehringer Mannheim, Pierce, Promega
Asn—X	Asparaginyl endopeptidase	Ishii et al., 1990	Takara Biochemicals
X—Asp	Endoproteinase Asp-N	Maier et al., 1986	Boehringer Mannheim
Trp, Phe, Tyr, Leu—X (X ≠ Pro)	Chymotrypsin	Wilkinson, 1986	Worthington, Boehringer Mannheim
pGlu—X[d]	Pyroglutamate aminopeptidase	Shively, 1986	Boehringer Mannheim, Pierce, Takara Biochemicals

Acetyl—X[e]	Acylamino acid releasing enzyme	Tsunasawa and Narita, 1982	Pierce, Takara Biochemicals
Met—X	Cyanogen bromide	Fontana and Gross, 1986	
Asn—Gly	Hydroxylamine	Fontana and Gross, 1986	
Asp—Pro	Dilute acid	Fontana and Gross, 1986	
Trp—X	Iodosobenzoate, BNPS-Skatole, N-chlorosuccinimide	Fontana and Gross, 1986	

[a] This table is not intended to be exhaustive. Only commonly used agents of high specificity are listed.

[b] For the reader's convenience, a limited number of sources are listed. For some, these are the only source; for others, additional sources exist but no attempt has been made to list them all.

[c] Occasional cleavage produced at these sites.

[d] pGlu, cyclized Gln at the N-terminal position.

[e] Removes amino terminal acetylated residue. Active only on peptides up to 20 to 30 residues.

5 10 15 20
SLPSERILPALSKDKEEEIR

	Expected	AAA
Asp	1	1.1
Thr	0	
Ser	3	2.4
Glu	4	3.9
Pro	2	1.9
Gly	0	
Ala	1	1.0
Cys	0	
Val	0	
Met	0	
Ile	2	1.8
Leu	3	3.0
Tyr	0	
Phe	0	
His	0	
Lys	2	2.0
Arg	2	2.0
Trp	0	
	20	

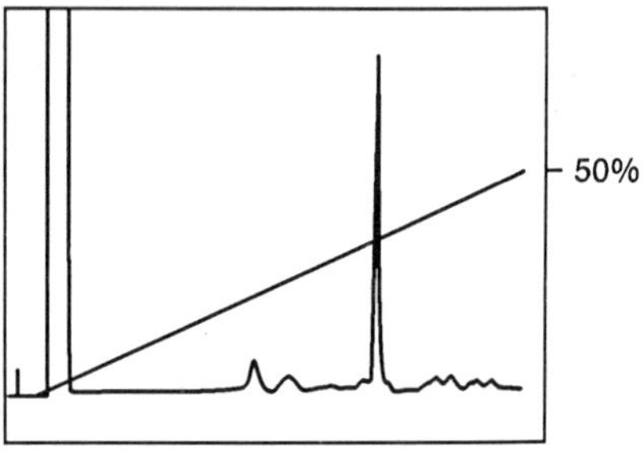

5 10 15 20
SLPSERILPALSKDKEEEIR 40%
SLPSRILPALSKDKEEEIR 60%

FIGURE 20 Evaluation of a 20-residue peptide, indicating the presence of a deletion-sequence peptide. The expected sequence is shown at the top and the experimentally determined sequence of the sample is shown at the bottom. The amino acid analysis and reverse-phase HPLC analysis of the sample are also shown.

the presence of two peptides differing only by a single amino acid. As much as 60 percent of the peptide was present with the deletion of a single glutamic acid residue at position number 5. Quite surprisingly, this occurrence was not well reflected in the amino acid analysis and again illustrates the danger of relying on a single type of analysis. The glutamic acid value may have been more revealing in this case if the peptide contained fewer Glu + Gln residues overall. Subsequent reverse-phase chromatography at basic pH, where the glutamate side chain would now be ionized, easily separated the two peptides (Figure 21). This example also illustrates how reliance on a single set of HPLC conditions can be misleading unless other analytic procedures are brought to bear, and it demonstrates how sequence analysis can complement HPLC as a means of detecting heterogeneity. Presumably, mass spectrometry would also have detected the presence of the two peptides if it had been performed.

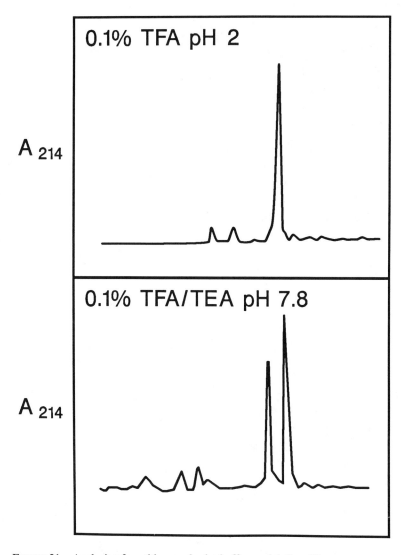

FIGURE 21   Analysis of peptide samples by buffer modulation. Chromatograms are reverse-phase HPLC of the sample depicted in Figure 20 at two different conditions of pH.

Figure 22 illustrates another example of the occurrence of a deletion sequence in a synthetic peptide. In this case, the deletion involves a block of residues rather than just a single residue, which could easily be detected by amino acid analysis. Deletions or additions can occur anywhere

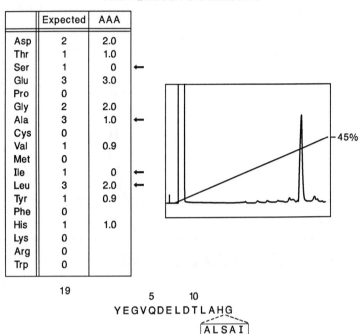

FIGURE 22   Evaluation of a 19-residue peptide, indicating the presence of a total deletion sequence of more than one residue. The expected sequence is shown at the top and the experimentally determined sequence of the sample is shown at the bottom. The amino acid analysis and reverse-phase HPLC analysis of the sample are also shown. Arrows indicate obvious deviations from that expected and the indicated residues correspond to the missing sequence shown in the box.

in the peptide and can be a much more serious problem with longer peptides, where sequencing will not reach to the C-terminus.

Mass Spectrometry

Mass spectrometry is a very valuable tool for the evaluation of synthetic peptides, and many synthetic chemists now rely on it as a routine part of their evaluation process. The determination that the mass of a synthetically produced peptide is correct proves in one step that the synthesis was successful and that the desired product has been obtained. It may, in fact, be the only way to determine the presence of modifications that are unstable or invisible in other analytical procedures. A major problem with mass spectrometry as a routine method for the evaluation of synthetic peptides is that it is still not routine for many laboratories. It requires a great deal of expertise and very expensive and highly sophisticated instrumentation. A recent survey by the Association of Biomolecular Resource Facilities indicated that only approximately 10 percent of member laboratories had routine access to mass analysis. Moreover, much of the technology, particularly for the evaluation of higher mass ranges, is still evolving. However, it is becoming increasingly obvious that mass spectrometry of peptides and proteins has enormous potential and benefits and is becoming the method of choice. The time has not yet arrived, but it may eventually be available to everyone on a routine basis.

Determination of the overall mass of a synthetic peptide is perhaps the most practical and available utilization of mass analysis. While the technology and expertise do exist to produce useful information on the actual sequence of a peptide (Hunt et al., 1988; Griffin et al., 1990), and thus pinpoint the location of any unintended modification, it is generally even less accessible on a routine basis than determination of the total mass. In practice, evaluation of the mass of the molecular ion is usually sufficient, particularly when it is used in conjunction with other analytical methods, to effectively and accurately evaluate a synthetic peptide. Although a detailed discussion of the theory and practice of mass spectrometry in the evaluation of synthetic peptides is beyond the scope of this chapter and is covered in detail elsewhere (McClosky, 1990), a brief general discussion of the main types of methodologies that have proved to be effective for determining the mass of synthetic peptides is appropriate.

The development of particle-induced desorption techniques opened the way for determination of the mass of polar, nonvolatile, thermally unstable molecules, such as peptides, without prior chemical derivatization. Prior to the development of these procedures, samples had to be derivatized to effect vaporization without pyrolytic degradation, and analysis was limited to molecular sizes of 1000 to 1500 daltons. The first particle-induced desorption technique that was effective for other than the smallest of peptides was fast atom bombardment mass analysis (FAB-MS; Barber et al., 1981). This technique is still generally employed today and a good example of a mass spectrum from this technique is shown in Figure 23. The predicted value for the mass ion of this peptide was 4706.4, which is in excellent agreement with the determined value of 4706.1. The peak at 4728.2 is the sodium adduct of the peptide, which often is seen. Although FAB-MS produces very accurate analyses of a wide range of synthetic peptides, a major limitation is its upper mass limit. Moreover, this limit varies somewhat depending on the instrument. While some can produce analyses of peptides in the 5000 to 6000 dalton range, other instruments can handle only those in the 2000 to 3000 dalton range. As modern automated synthesizers and chemistry have improved,

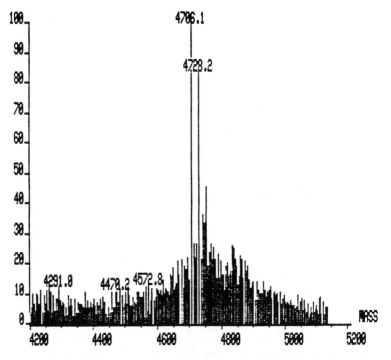

FIGURE 23   Fast atom bombardment mass spectrum (FAB-MS) of a 40-residue synthetic peptide. (Spectra kindly provided by Ken Williams and Walter Mc-Murray of Yale University and the Association of Biomolecular Resource Facilities.)

it has become more and more possible to routinely synthesize peptides with molecular weights above the range of even the best FAB-MS instruments.

Recently, three other types of analysis, which have the advantage of being able to analyze large peptides as well as proteins with masses as high as 30 to 60,000, have become commercially available. These are plasma desorption mass spectrometry (PDMS) (MacFarlane et al., 1974; Cotter, 1988; Fontenot et al., 1991), which is another particle-induced desorption technique, electrospray ionization mass spectrometry (ESMS) (Hail et al., 1990), and laser desorption mass spectrometry (LDMS). ESMS positive ion electrospray ionization produces a series of multiply charged ions, where the charge is carried on the free α-amino group as well as the side chains of arginine and lysine residues. This technique produces a series of ions differing from each other by a single charge, which is then analyzed by a computer to calculate the molecular weight of the polypeptide. It is this multiple charging phenomenon that

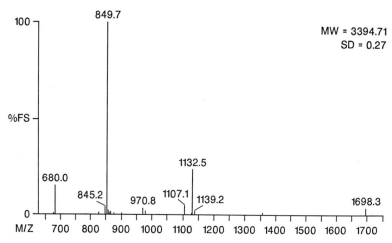

FIGURE 24   Positive ion electrospray mass spectra (ESMS) of a 29-residue
peptide sample. (Analysis kindly provided by M-Scan, Inc.)

has allowed direct molecular weight analysis of large peptides and
proteins at high sensitivity (picomole level) and accuracy. PDMS also
produces multiply charged ions (MH_2+, MH_3+, etc.) in addition to the
mass ion (MH^+) and cluster ions (M_2H^+, M_3H^+). Most plasma desorption analysis of peptides is done on time-of-flight instruments, which
have high mass ranges and allow the simultaneous detection of all ions
without having to scan the spectrum. Examples of electrospray and
PDMS spectra are shown in Figures 24 and 25, respectively. The positive
ion electrospray analysis in Figure 24 is from a peptide whose calculated
mass is 3394.9. Note that a series of multiply charged ions is observed at
m/z 1698 (2+), 1132.5 (3+), 849.7 (4+), and 680.0 (5+). These signals
can be deconvoluted to give a component with a molecular mass of
3394.7, which is in excellent agreement with the calculated mass. The
sample used for this spectrum is, in fact, the 29-residue peptide discussed
in the following section (see An Example). This particular peptide sample was also shown to contain a minor component with a mass of 1106.3,
which corresponds with the weak signal at 1107.1. The other weak signals present remain unassigned. The PDMS analysis shown in Figure 25
illustrates a peptide whose calculated average mass for the singly
charged species is 4542.1. Note that the doubly charged ion is also
present.
    Another recent innovation to high-mass analysis of peptides is
matrix-assisted laser desorption mass spectrometry (LDMS; Gibson,
1991 and references therein; McCloskey, 1990). It employs ultraviolet
absorbing matrices, which absorb energy from a laser pulse and, in turn,

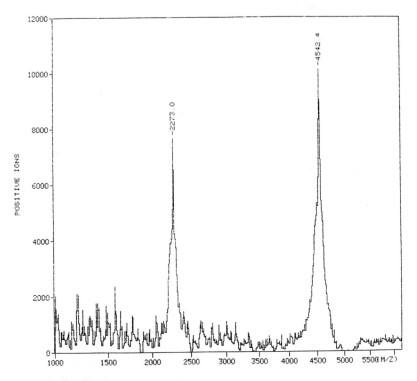

FIGURE 25   Plasma desorption time of flight mass spectra (PDMS) of a 39-residue synthetic peptide. (Spectra kindly provided by Michael Kochersperger of Applied Biosystems, Inc.)

ionize peptides and proteins that have been mixed with the solid matrix. The sensitivity of LDMS is in the picomole range and has the advantage of being unaffected by most components commonly found in peptide solutions.

As this brief discussion of the state of mass spectrometry indicates, the field has come a long way in the last few years. Even with this rapid advance, it is clear that its potential has not yet been fully realized. Even more recent innovations, such as the use of ion-trap mass spectrometers for peptides and proteins, promise even greater capabilities and sensitivity (fmol). Undoubtedly, amazing things will be accomplished with mass spectrometry in the future. Unfortunately, the availability of these new techniques to the general research community has not advanced nearly as rapidly. That situation is slowly changing but it still remains a major problem in the routine evaluation of synthetic peptides.

AN EXAMPLE OF THE COMPLETE EVALUATION OF A SYNTHETIC PEPTIDE FROM CRUDE PRODUCT TO HOMOGENEOUS PEPTIDE

So far in this chapter, examples have been given in individual sections to illustrate particular points that had direct bearing on the subjects being discussed. Those examples were parts of evaluations of various synthetic peptides, all of which underwent complete evaluation protocols, but none of which were presented as a whole. Hopefully, each example was effective in illustrating a particular point, but, taken out of context, it is not always possible to appreciate how each part contributes to the overall picture and how the parts come together to complement each other. In this section, the complete analysis of a synthetic peptide is presented from start to finish to illustrate the evaluation process in a comprehensive and continuous manner.

The peptide selected for this example is 29 residues in length and has the following sequence: GRLKCWLNDLRSYPQEMTHVASIDSV-FKV. It was synthesized with t-Boc chemistry, employing a capping step. Tryptophan was coupled as the N-formyl derivative. After 18 cycles, the synthesis was paused and approximately 25 percent of the resin was removed to reduce the volume of the swollen resin as the synthesis neared completion. This procedure was done to avoid possible mixing and transfer problems near the end of synthesis, because the swollen resin volumes were too large for the reaction vessel. At the conclusion of the synthesis, the dry peptide resin weight of the final product was 1.87 g. Taking into account the amount of resin removed during synthesis, the theoretical peptide resin weight was calculated to be 2.36 g. Thus, the yield of peptide resin was 79 percent of the theoretical yield, which is acceptable. After synthesis, the tryptophan was deformylated, the peptide was cleaved and deprotected with anhydrous HF, it was extracted with 30% acetic acid, and it was lyophilized. The total crude product obtained weighed 617 mg. Based on a molecular weight of 3393 (and taking into account the 25 percent removed), the theoretical yield (100 percent) is 1272 mg. Thus, the peptide weight yield was 49 percent of the theoretical yield, which is a reasonable amount. As discussed earlier (see Peptide Mass Yield), based on practical experience, the rule of thumb is approximately 30 mg per residue for a 0.5 mmol scale synthesis. This yield translates to an expected 650 mg, which is in good agreement with the 617 mg obtained.

The first thing that is always done in the evaluation is to obtain a reverse-phase analytical HPLC profile of the crude material, which is shown in the top left panel of Figure 26. This analysis reveals a significant degree of heterogeneity, but it also shows a single predominant peak. The amino acid analysis of the crude material, presented in Table 7, appears to be consistent with the expected composition of this sample

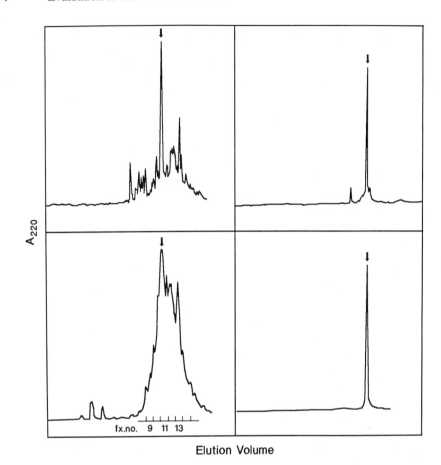

A_{220}

fx.no. 9 11 13

Elution Volume

FIGURE 26   Reverse-phase HPLC chromatograms, depicting the evaluation and purification of a 29-residue peptide. HPLC conditions are as described in Figures 16 and 17. (Upper left) Analytical scale chromatogram of the crude product; (lower left) Preparative scale chromatogram of the crude product; (upper right) Analytical scale chromatogram of fraction 11 from the preparative run; (lower right) Analytical scale chromatogram of the pooled main peak from the analytical run of fraction 11, showing the pure peptide.

with the exception of the values for Arg and Leu. Given the amount of heterogeneity in this sample, these inconsistencies are not surprising at this point. Although the multiple minor peaks seen in Figure 26 have not been specifically characterized, they undoubtedly represent a diverse population of unwanted modified and truncated peptides. This observation is not uncommon for the crude product, especially for larger peptides. All things considered, the overall analysis does not deviate sufficiently from the expected values to cause concern at this stage, and given

Table 7   Amino Acid Analysis[a] of Synthetic Peptides from Figure 26

			Mol. ratio		
Residue	Expected[b]	Crude	Preparative pool	Primary peptide	Secondary peptide
Asp	3	2.8	3.1	3.0	1.1 (1)[c]
Thr	1	1.1	1.1	1.1	
Ser	3	2.4	2.5	2.6	1.5 (2)
Glu	2	2.0	2.0	2.0	
Pro	1	0.9	1.0	1.0	
Gly	1	0.6	1.0	0.9	
Ala	1	1.3	1.3	1.0	1.0 (1)
Cys	1	nd[d]	nd	nd	
Val	3	3.1	3.7	2.6	2.7 (3)
Met	1	nd	nd	nd	
Ile	1	1.2	1.2	1.0	1.0 (1)
Leu	3	2.2	2.8	2.9	
Tyr	1	0.7	0.9	1.0	
Phe	1	1.2	1.2	1.0	1.0 (1)
His	1	1.0	1.0	1.0	
Lys	2	2.0	2.2	2.0	1.0 (1)
Arg	2	1.2	1.9	1.8	
Trp	1	nd	nd	nd	

[a] The peptide was vapor hydrolyzed with 6N HCl containing 0.1% phenol for 1 hour at 160° C and analyzed on a Beckman 6300 amino acid analyzer.

[b] Expected composition for the intended peptide.

[c] Expected composition of secondary peptide based based on mass spectrometry analysis.

[d] nd, not determined.

the presence of a predominant peak in the profile, further purification is well warranted.

Figure 26 also shows a preparative scale run of 100 mg of the crude product in the lower left panel. Elution conditions are similar to the analytical scale run in order to make as direct a comparison as possible for the purpose of pooling fractions. The analytical scale column was a 0.46 × 25 cm Vydac C-18 column, and the preparative scale column was a 2.2 × 25 cm column of the same type from the same manufacturer. Both columns were loaded in 0.1% TFA and eluted with a 1% per minute linear gradient of acetonitrile made 0.1% in TFA. Although the amplitudes of the peaks are greatly increased in the preparative run, the heterogeneity is still evident and there still appears to be a predominant peak eluting at approximately 43% acetonitrile (fraction 11). Based on the similarities in the analytical and preparative runs, fraction 11 was selected for further analysis.

The analytical reverse-phase HPLC profile of fraction 11 from the preparative scale run (upper right panel of Figure 26) appears to be much improved over the crude product, although there is still some evidence of low-level heterogeneity. Automated sequence analysis of this sample indicated good yields of a single sequence that corresponded exactly to the expected sequence. The amino acid analysis of this fraction also appeared to indicate the expected composition with one exception. The value for valine was consistently almost a whole residue too high, even though all the other values fell within the expected tolerances (see Table 7). As mentioned previously, sequence analysis did not show any evidence of a valine addition sequence, although residue identification could not be made past cycle 24, after which there were two valine positions. As it turns out, this additional valine was a definite indication of a problem which was first indicated by a rigorous interpretation of the HPLC profile and confirmed by mass spectrometric analysis.

The FAB-MS analysis of this peptide, shown in Figure 27, indicates the presence of two species with MH+ signals of 1106.6 and 3393.4. The positive ion electrospray analysis shown in Figure 24, which indicates a mass of 3394.7, confirms the FAB-MS data. The calculated mass ion for the intended peptide is 3394.9, which is in excellent agreement with these values. The source of the signal at 1106 was relatively easy to deduce, because it turns out that a capped (acetylated) peptide, starting from the C-terminal residue and ending with the tenth residue into the synthesis (Ac-VASIDSVFKV), has a calculated mass of 1106.3 (remember that the peptide is synthesized from the C-terminus to the N-terminus). Note the difference in signal intensity between the minor component (1106) in the FAB-MS and electrospray spectra. The minor component was verified to be present at approximately 10 percent of the major component by HPLC fractionation and amino acid analysis. The fact that the minor peptide is blocked at the amino terminus explains why it was not detected by sequence analysis, and the fact that three of the ten residues are valine accounts for why valine appeared elevated in relation to the other residues in the amino acid analysis (see Table 7). Further analysis of the fractions from the analytical HPLC of preparative fraction 11 indicated that the predominant peak is the intended peptide and the small peak, which elutes several minutes ahead of the predominant peak is the acetylated ten-residue contaminant. These two peaks were easily separated by fraction collection to yield homogeneous peptides, as reflected in their amino acid compositions (Table 7), HPLC profile (lower right panel of Figure 26), and subsequent mass analysis (not shown).

The FAB-MS spectrum, presented in Figure 27, illustrates a characteristic of this type of analysis that was mentioned previously (see Mass Spectrometry). It cannot be used to quantitate the relative levels of peptides that are present. Although the peak with mass 1106 appears to be the predominant species in the mass analysis, it is, in fact, a minor component, comprising approximately 10 percent of the total sample. This anomaly is a function of the mass range of the instrument and the efficiency of desorption and detection of species at different masses. The situation is reversed in the electrospray data shown in Figure 24.

When all of the data is taken as a whole, there is little doubt that a homogeneous peptide of the correct covalent structure has been obtained. At the same time, it should also be evident that a failure to perform certain of the analyses could have led to incorrect conclusions. Analytical HPLC was indispensable in detecting and assisting in the removal of unwanted by-products. Amino acid analysis indicated the possibility of a problem in the later stages of purification that went unresolved with sequence analysis but was verified by mass analysis. Amino acid analysis and analytical HPLC of individual fractions, with corroboration by mass spectroscopy and sequence analysis, indicated that the predominant species was the target peptide. Sequence analysis, substantiated by mass spectrometry, verified that the sequence was correct. With regard to the last point, however, it must be remembered that the sequence analysis was not able to proceed all the way to the end of the peptide. Verification of the C-terminal sequence by sequence analysis would entail digestion of the peptide with cyanogen bromide or a protease, such as endoproteinase Asp-N, separation of the resulting peptides on reverse-phase HPLC, and sequencing the smaller, individual peptides to their C-terminus.

STORAGE AND HANDLING OF SYNTHETIC PEPTIDES

The manner in which a peptide is handled and stored can be just as important to its integrity and utilization as any of the chemistry used to produce it. In theory, most peptides should be stable at neutral pH in aqueous solution for extended periods of time. However, contamination with microorganisms or metal ions can cause peptide bond cleavage. Moreover, oxygen can adversely affect tryptophan, cysteine, and methionine residues (see Chapter 3), and the presence of metal ions can cause destruction of tryptophan. Aspartic acid-proline bonds are also extremely sensitive to acid cleavage. Storage of peptides containing this bond for extended periods of time in only slightly acidic aqueous solutions can lead to significant hydrolysis. Other peptide bonds with aspartic acid also tend to be more acid labile than most, but not nearly to the degree seen

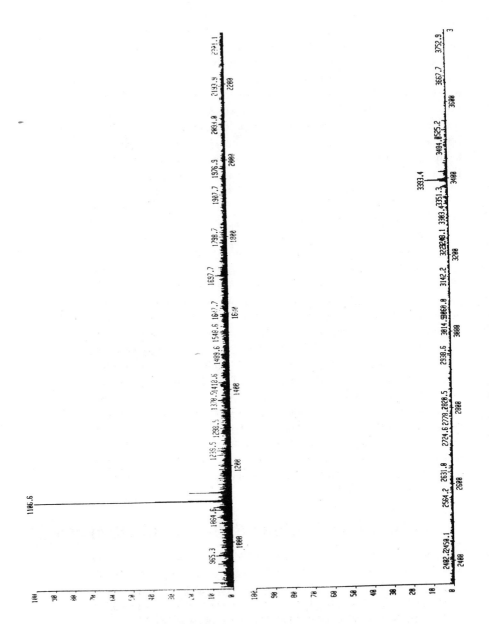

FIGURE 27   FAB-MS spectra of the 29-residue peptide from fraction 11 of the preparative scale chromatogram from Figure 26.

with Asp-Pro bonds. Also in aqueous solution, aspartic acid-glycine bonds, and, in some instances, aspartic acid N-terminal to other short side-chain amino acid residues (Ser, Thr, Ala, Asn), can cyclize to form an aspartimide intermediate that can undergo spontaneous ring opening either to regenerate the original peptide or to produce a β-aspartyl peptide possessing two methylene carbons between the backbone amino and carbonyl group of the residue. When this occurs, the structure of the peptide is drastically altered. The presence of a β-aspartyl bond can be indicated by sequence analysis. Because of the additional carbon in the backbone, the PTC peptide is not able to cyclize efficiently, and sequencing will stop or produce a reduced yield when such a structure is encountered.

For the reasons mentioned already, peptides should be stored in solution only for short periods of time. For long-term storage, the peptide should be lyophilized from a volatile buffer or solvent and stored desiccated at −20° C in either polyethylene, polypropylene, or silanized glass containers. When containers of dry peptides are removed from the cold, they should be allowed to come to room temperature before they are opened to minimize condensation of water vapor on the peptide surface. This process could contribute to hydrolysis when the remainder of the peptide is returned to storage.

If a peptide has been stored for an extended period of time, it is not a bad idea to re-evaluate it before use. Even "dry" peptides, after lyophilization, can contain a significant amount of adsorbed solvent, which can contribute to unwanted reactions. An example is the relatively facile deamidation of Asn and Gln when residual acid is present in the dry peptide. Peptides containing amide side chains should not be dried down from acidic solution or precaution should be taken to assure that all acid is removed from the sample before storage. Minimally, an amino acid composition and an analytical reverse-phase HPLC should be sufficient to indicate possible storage-related problems.

If peptides are going to be stored for any length of time after synthesis, but before they are worked up, they should be desalted before storage. Many salts and organics can be present as a result of the synthesis, cleavage, and deprotection procedures, and these can have an adverse effect upon long-term storage. Desalting can be accomplished either by standard low-pressure gel filtration chromatography or by preparative reverse-phase HPLC.

HOW MUCH IS ENOUGH?

How much evaluation is necessary? That is a question that is always asked, both by investigators who are having peptides made for them by someone else and by the synthetic chemists themselves. In an ideal world, all synthetic peptides would be subjected to all possible evaluation techniques, and there would be no need to ask the question. However, most investigators are restrained by matters of cost, time, and instrument availability, and, therefore, they are often confronted with this question.

Practically speaking, most peptides that can be purified to a single symmetrical peak on HPLC or CZE and yield a good amino acid analysis will be the intended product. Thus, if nothing else is done, the investigator should at least go this far. However, several examples have been given in this chapter where this was not sufficient and where additional characterization indicated problems. Several other examples have been given in which the same problem was indicated in more than one type of analysis. For example, a sample producing an inconsistent amino acid composition was seen to possess two sequences upon sequence analysis or two molecular ions after mass spectrometry. Such a sample will also often show evidence of heterogeneity on analytical HPLC or CZE. It is probably not necessary to have performed all of these analytical procedures on the sample at this stage when any one of them indicates a problem. Conversely, if one procedure indicates a problem while others do not, it is usually not wise to ignore that one in favor of the others. That one procedure is usually trying to tell you something and is sufficient grounds in itself to delve further into the evaluation.

It is also not recommended to rely on a single analytical technique for any evaluation. We have seen examples where apparent single peaks actually contained several species and where an amino acid analysis that appeared to be reasonable was not truly indicative of the complexity of the sample. Evaluation by sequence analysis alone may reveal only one sequence, and although that sequence may be the expected one, other blocked species can also be present. There is a common misconception that mass analysis is all that is necessary if it indicates the correct mass ion. However, for a variety of reasons, it is possible to miss other constituents by using this technique. The contaminant may produce a weaker signal that is not proportional to its actual presence in the sample if it is not desorbed as efficiently or if it is outside the mass range of the instrument. Conversely, the major peptide in regard to the total sample may give a very weak signal or none at all. Comparison of the FAB-MS and electrospray data for the 29-residue peptide shown in Figures 24 and 27 demonstrates how the data can be misleading. In one case (FAB-MS), the lower molecular weight contaminant is dominant. In the other case, it is barely seen and surely would have been overlooked if its presence had not been indicated by other analyses.

A recent study conducted by the peptide synthesis/mass spectrometry subcommittee of the Association of Biomolecular Resource Facilities (Smith et al., 1992) reinforces many of these points. While its original intention was to assess the peptide synthesis capabilities in core facilities, it offers many enlightening observations pertaining to the strengths and weaknesses of the methods of peptide evaluation. It is recommended reading for anyone engaged in peptide synthesis and evaluation.

Many investigators are forced more by circumstance than by choice to forgo mass spectrometric analysis and sometimes sequence analysis on a routine basis. Although the peace of mind that one can get from a spot-on mass result or a perfect sequence is obvious, it may not always be necessary in every case. It is certainly possible to be reasonably confident in the integrity of a synthetic peptide if all of the other analyses presented in this chapter are performed and they are interpreted rigorously.

Unfortunately, no one can say beforehand how much evaluation needs to be done. Each peptide is different and each synthesis is different. The data need to be evaluated as they are collected in order to make the decision. The answer to the question posed at the beginning of this section can only be "as much as is necessary to convince a knowledgeable person that the product is as it should be." You are the ultimate judge. Hopefully, this chapter, as well as the other chapters in this book, has provided a helpful discussion of the means by which to make that decision. How you apply that knowledge is up to you.

SUMMARY

1. Although many peptides can be synthesized on modern automated instruments with predetermined routine packages, many problems can arise for a wide variety of reasons. Therefore, it is never safe to assume that the product of the synthesis is the desired peptide until that has been experimentally verified.

2. Product integrity can be monitored at many stages. These stages include resin sampling during synthesis, on-resin sequencing during or after synthesis and prior to cleavage and deprotection, and extensive evaluation after cleavage and deprotection.

3. The amount and kind of evaluation necessary for a synthetic peptide depends to some extent on the intended use of the peptide. For instance, peptides to be used for structural studies require high levels of purity and rigorous confirmation of chemical composition. Peptides to be used for antibody production do not usually need as rigorous testing since subsequent screening procedures can often assist the selection.

4. In almost all cases, peptide purification can be performed with reverse-phase HPLC or a combination of reverse-phase HPLC and ion exchange HPLC.

5. Amino acid compositional analysis is an absolute necessity for the evaluation of all synthetic peptides. In addition to this, a quantita-

tive separation technique coupled with a mass spectrometric technique provides the most accurate approach for evaluating the purity and authenticity of synthetic peptides.

6. The most common problem encountered is usually the presence of deletion sequences. These can be detected readily by sequence analysis. However, if capping is used during synthesis, a good sequence should not be considered final proof of authenticity. Additional criteria must be applied.

7. Chemical modifications of amino acid side chains are also relatively common. Low levels can usually be removed during purification, but high levels of modification usually require more extensive evaluation for their detection.

8. The ultimate success of the evaluation process relies on experience and good judgment. Don't ignore one piece of data that indicates a problem even though all other evaluations fail to reveal the problem. Also remember that no single technique is foolproof by itself, and all techniques should be used in a complementary fashion.

ACKNOWLEDGMENTS

The author gratefully acknowledges the advice, helpful discussions, and materials provided by Mark Frazier and the staff of the Washington University Protein Chemistry Laboratory, Dan Crimmins and John Gorka of the Howard Hughes Medical Institute, Washington University, Gregg Fields, University of Minnesota, Michael Kochersperger and Anita Hong of Applied Biosystems, Inc., the Association of Biomolecular Resource Facilities, and M-Scan, Inc.

REFERENCES

Alpert, A. J., and Andrews, P. C. 1988. Cation exchange chromatography of peptides on (2-sulfoethyl aspartamide)-silica. J. Chromatogr. 443:85-96.

Andrews, P. C., and Dixon, J. E. 1987. A procedure for *in situ* alkylation of cystine residues on glass fiber prior to protein microsequence analysis. Anal. Biochem. 161:524-528.

Aitken, A., Geisow, M. J., Findlay, J. B. C., Holmes, C., and Yarwood, A. Peptide preparation and characterization. In Protein Sequencing: A Practical Approach, Findlay, J. B. C., and Geisow, M. J., eds.. IRL Press, Oxford, 1989.

Atherton, E., and Sheppard, R. C. Solid Phase Peptide Synthesis: A Practical Approach, IRL Press, Oxford, 1989, pp. 156-161.

Barany, G., Knieb-Cordonier, N., and Mullen, D. G. 1987. Solid phase peptide synthesis: a silver anniversary report. Int. J. Pept. Protein Res. 30:705-739.

Barber, M., Bordoli, R. S., Sedgwick, R. D., and Tyler, A. N. 1981. Fast atom bombardment of solids (F. A. B.): a new ion source for mass spectrometry. J. Chem. Soc. Chem. Commun. 7:325-327.

Benson, J. R., and Hare, P. E. 1979. o-Phthalaldehyde: fluorogenic detection of primary amines in the picomole range, comparison with fluorescamine and ninhydrin. Proc. Natl. Acad. Sci. USA 72:619-622.

Blankenship, D. T., Krivanek, M. A., Ackermann, B. L., and Cardin, A. D. 1989. High sensitivity amino acid analysis by derivatization with o-phthalaldehyde and 9-fluorenylmethylchloroformate using fluoresence detection: applications in protein structure determination. Anal. Biochem. 178:227-232.

Bodanszky, M., Deshmane, S. S., and Martinez, J. 1979. Side reactions in peptide synthesis II. Possible removal of the 9-fluorenylmethylcarbonyl group by the amine components during coupling. J. Org. Chem. 44:1622-1624.

Bohlen, P., and Mellet, M. 1979. Automated fluorometric amino acid analysis: the determination of proline and hydroxyproline. Anal. Biochem. 94:313-321.

Chang, J. -Y. 1983. Manual micro-sequence analysis of polypeptides using dimethylaminoazobenzene isothiocyanate. Methods Enzymol. 91:455-467.

Chang, J. -Y., Knecht, R., and Braun, D. G. 1983. Amino acid analysis in the picomole range by precolumn derivatization and high-performance liquid chromatography. Methods Enzymol. 91:41-48.

Chicz, R. M., and Regnier, F. E. 1990. High performance liquid chromatography: effective protein purification by various chromatographic methods. Methods Enzymol. 182:392-421.

Cohen, S. A., and Strydom, D. J. 1988. Amino acid analysis utilizing phenylisothiocyanate derivatization. Anal. Biochem. 174:1-16.

Cotter, R. J. 1988. Plasma desorption spectrometry comes of age. Anal. Chem. 60:781A-793A.

Crabb, J. W., Ericsson, L., Atherton, D., Smith, A. J., and Kutny, R. A collaborative amino acid analysis study from the Association of Biomolecular Resource Facilities. In Current Research in Protein Chemistry: Techniques, Structure, and Function, Villafranca, J. J., ed., Academic Press, San Diego,1990, pp. 49-61.

Crimmins, D. L., Gorka, J., Toma, R. S., and Schwartz, B. D. 1988. Peptide characterization with a sulfoethyl aspartimide column. J. Chromatogr. 443:63-71.

Cullwell, A. 1987. Reduction of methionine sulfoxide in peptides using N-methylmercaptoacetamide. Applied Biosystems User Bulletin, Model 430, No. 17.

Cunico, R., Mayer, A. G., Wehr, C. T., and Sheehan, T. 1986. Bio-Chromatography 1:6-14.

Darbre, A., ed. Practical Protein Chemistry, A Handbook. John Wiley and Sons, New York, 1986.

Dupont, D., Keim, P., Chui, A., Bozzini, M., and Wilson, K. J. 1988. Gas-phase hydrolysis for PTC-amino acid analysis. Applied Biosystems User Bulletin, Model 420A Issue No. 2.

Edman, P. 1950. Method for determination of amino acid sequence in peptides. Acta. Chem. Scand. 4:283-293.

Edman, P., and Begg, G. 1967. A protein sequenator. Eur. J. Biochem. 1:80-91.

Einarsson, S. 1985. Selective determination of secondary amino acids using precolumn derivatization with 9-fluorenylmethylchloroformate and reversed-phase high-performance liquid chromatography. J. Chromatogr. 348:213-220.

Einarsson, S., Josefsson, B., and Lagenkvist, S. 1983. Determination of amino acids with 9-fluorenylmethylchloroformate and revered-phase high-performance liquid chromatography. J. Chromatogr. 282:609-618.

Ewing, A. G., Wallingford, R. A., and Olefirowicz, T. M. 1989. Capillary electrophoresis. Anal. Chem. 61:292A-303A.

Feldhoff, R. 1991. Why not ethanol-based solvents for RP-HPLC of peptides and proteins? In Techniques in Protein Chemistry II, Villafranca, J. J., ed., Academic Press, San Diego, 1991, pp. 55-63.

Findlay, J. B. C., and Geison, M. J. Protein Sequencing: A Pratical Approach, IRL Press, Oxford, 1989.

Fontana, A., and Gross, E. Fragmentation of polypeptides by chemical methods. In Practical Protein Chemistry, A Handbook, Darbre, A., ed., John Wiley and Sons, New York, 1986, pp. 67-120.

Fontenot, J. D., Ball, J. M., Miller, M. A., David, C. M., and Montelaro, R. C. 1991. A survey of potential problems and quality control in peptide synthesis by the fluorenylmethylcarbonyl procedure. Peptide Res. 4:19-25.

Fujii, N., Otaka, A., Funakoshi, S., Bessho, K., Watanabe, T., Akaji, K., and Yajim, H. 1987. Studies on peptides CLI: Synthesis of cystine-peptides by oxidation of S-protected cysteine peptides with thallium (III) trifluoroacetate. Chem. Pharm. Bull. 26:539-548.

Futaki, S., Yajami, T., Taike, T., Akita, T., and Kitagawa, K. 1990. Sulphur trioxide/thiol: a novel system for the reduction of methionine sulphoxide. J. Chem. Soc. Perkin Trans. 1:653-658.

Gibson, B. W. A brief overview of mass spectrometric methods for the analysis of peptides and proteins. In Techniques in Protein Chemistry II, Villafranca, J. J., ed., Academic Press, San Diego, 1991, pp. 419-425.

Griffin, P. R., Martino, P. A., McCormack, A. L., Shabanowitz, J., and Hunt, D. F. Protein and oligopeptide sequence analysis on the TSQ-70 triple quadrupole mass spectrometer. In Current Research in Protein Chemistry:Techniques, Structure, and Function, Villafranca, J. J., ed., Academic Press, San Diego, 1990, pp. 117-126.

Gustavsson, B., and Betner, I. 1990. Fully automated amino acid analysis for protein and peptide hydrolysates by pre-column derivatization with 9-fluorenylmethylchloroformate and 1-aminoadamantane. J. Chromatogr. 507:67-77.

Hail, M., Lewis, S., Zhou, J., Jardine, I., and Whitehouse, C. Analysis of peptides and proteins by tandem quadrupole mass spectrometry. In Current Research in Protein Chemistry: Techniques, Structure, and Function, Villafranca, J. J., ed., Academic Press, San Diego, 1990, pp. 105-116.

Hancock, W. S., ed. CRC Handbook of HPLC for the separation of Amino Acids, Peptides, and Proteins, Vol. I and II, CRC Press, Boca Raton, Florida, 1984.

Hawke, D., and Yuan, P. 1987. S-Pyridylethylation of cystine residues. Applied Biosystems User Bulletin for 470A/477A-120A, No. 28.

Heinrikson, R. L., and Meridith, S. C. 1984. Amino acid analysis by reverse-phase high-performance liquid chromatography: Pre-column derivatization with phenylisothiocyanate. Anal. Biochem. 136:65-74.

Henschen, A., Hupe, K. P., Lottspeich, F., and Voelter, W., eds. High Performance Liquid Chromatography in Biochemistry, VCH, Weinheim, Germany, 1985.

Hewick, R. M., Hunkapiller, M. W., Hood, L. E., and Dreyer, W. J. 1981. A gas-liquid solid phase peptide and protein sequenator. J. Biol. Chem. 256:7990-7997.

Hirs, C. H. W. 1967. Determining cysteine as cysteic acid. Methods Enzymol. 11:59-62.

Houghton, R. A., and Li, C. H. 1979. Reduction of sulfoxides in peptides and proteins. Anal. Biochem. 98:36-46.

Hugli, T. E., and Moore, S. 1972. Determination of the trypyophan content of proteins by ion exchange chromatography of alkaline hydrolysates. J. Biol. Chem. 247:2828-2834.

Hunkapiller, M. W., and Hood, L. E. 1978. Direct microsequence analysis of polypeptides using an improved sequenator, a non-protein carrier (polybrene), and high pressure liquid chromatography. Biochemistry 17:2124-2133.

Hunt, D. F., Shabanowitz, J., Yates, J. R., Griffin, P. R., and Zhu, N. Z. In Analysis of Peptides and Proteins, McNeil, C., ed., John Wiley and Sons, New York, 1988, pp. 151-165.

Ishii, S., Abe, Y., Matsushita, H., and Kato, I. 1990. An asparaginyl endopeptidase purified from jack bean seeds. J. Protein Chem. 9:294- 295.

Jones, B. N. Amino acid analysis by o-phthaldialdehyde precolumn derivatization and reverse Phase HPLC. In Methods of Protein Microcharacterization, A Practical Handbook, Shively, J. E., ed., Humana Press, Clifton, N. J., 1986, pp. 121-151.

Jorgenson, J. W., and Lukacs, K. D. 1981. Zone electrophoresis in open-tubular glass capillaries. Anal. Chem. 53:1298-1302.

Kaiser, E., Colescott, R. L., Bossinger, C. D., and Cook, P. I. 1970. Color test for the detection of free terminal groups in solid-phase synthesis of peptides. Anal. Biochem. 34:595-598.

Kaiser, E., Bossinger, C. D., Colescott, R. L., and Olsen, D. B. 1980. Color test for terminal prolyl residues in the solid phase synthesis of peptides. Anal. Chim. Acta 118:149-151.

Kent, S. B. H. 1988. Chemical synthesis of peptides and proteins. Ann. Rev. Biochem. 57:957-989.

Kent, S. B. H., Riemen, M., LeDoux, M., and Merrifield, R. B. A study of the Edman degradation in the assessment of the purity of synthetic peptides. In Methods in Protein Sequence Analysis, Elzinga, M., ed., Humana Press, Clifton, New Jersey, 1982, pp. 205-213.

Knecht, R., and Chang, J. -Y. 1986. Liquid chromatographic determination of amino acids after gas-phase hydrolysis and derivatization with (dimethylamino) azobenzenesulfonylchloride. Anal. Chem. 58:2375-2379.

Kochersperger, M. L., Blacher, R., Kelly, P., Pierce, L., and Hawke, D. H. 1989. Sequencing of peptides on solid phase supports. Am. Biotech. Lab. 7:26-37.

Laursen, R. A. 1971. Solid phase Edman degradation, an automated peptide sequencer. Eur. J. Biochem. 20:89-102.

Levy, M., Fishman, L., and Schenkein, I. 1970. Mouse submaxillary gland proteases. Meth Enzymol. 19:672-681.

Liu, T. Y., and Chang, Y. H. 1971. Hydrolysis of proteins with p-toluenesulfonic acid. J. Biol. Chem. 246:2842-2848.

MacFarlane, R. D., Skowronski, R. P., and Torgenson, D. F. 1974. New approach to the mass spectroscopy of non-volatile compounds. Biochem. Biophys. Res. Commun. 60:616-621.

Mahoney, W. C., and Hermodson, M. A. 1980. Separation of large denatured peptides by reverse phase high performance liquid chromatography. J. Biol. Chem. 255:11199-11203.

Maier, G., Drapeau, G. R., Doenges, K. H., and Ponstingl, H. Generation of starting points for microsequencing with a protease specific for the amino side of aspartyl residues. In Methods of Protein Sequence Analysis, Walsh, K. A., ed., Humana Press, Clifton, New Jersey, 1986.

Mant, C. T., and Hodges, R. S. HPLC of peptides. In HPLC of Biological Macromolecules, Chromatographic Science Series, Vol. 51, Gooding, K. M., and Regnier, F. E., eds., Marcel Dekker, New York, 1990, pp. 301-332.

Masaki, T., Tanabe, M., Nakamura, K., and Soejima, M. 1981. Studies on a new proteolytic enzyme from Achromobacter lyticus M497-1 I. Purification and some enzymatic properties. Biochem. Biophys. Acta 660:44- 50.

Masaki, T., Fujihashi, T., Nakamura, K., and Soejima, M. 1981. Studies on a new proteolytic enzyme from Achromobacter lyticus M497-1 II. Specificity and inhibition studies of Achromobacter protease I. Biochem. Biophys. Acta 660:51- 55.

Marquez, F. G., Quesada, A. R., Sanchez-Jimines, F., and Nunez de Castro, I. 1986. Determination of 27 dansyl amino acid derivatives in biological fluids by reverse-phase high-performance liquid chromatography. J. Chromatogr. 380:275-283.

Matsubara, H., and Sasaki, R. M. 1969. High recovery of tryptophan from acid hydrolysates of proteins. Biochem. Biophys. Res. Commun. 35:175-181.

McCloskey, J. A., ed. 1990. Mass spectrometry. Methods Enzymol. 193:1-960.

Moore, S., Spackman, D. H., and Stein, W. F. 1958. Chromatography of amino acids on sulfonated polystyrene resins. Anal. Chem. 30:1185-1190.

Moore, S., and Stein, W. H. 1958. Photometric ninhydrin method for use in the chromatography of amino acids. J. Biol Chem. 176:367-388.

Moore, S., and Stein, W. H. 1963. Chromatographic determination of amino acids by the use of automatic recording equipment. Methods Enzymol. 6:819-831.

Oray, B., Lu, H. S., and Gray, R. W. 1983. High-performance liquid chromatographic separation of Dns-amino acid derivatives and application to protein and peptide structural studies. J. Chromatogr. 270:253-266.

Ozols, J. 1990. Amino acid analysis. Methods Enzymol. 182:587-601.

Pappin, D. J. C., Coull, J. M., and Koester, H. New approaches to covalent sequence analysis. In Current Research in Protein Chemistry: Techniques, Structure, and Function, Villafranca, J. J., ed., Academic Press, San Diego, 1990, pp. 191-202.

Penke, B., Ferenczi, R., and Kovacs, K. 1974. A new acid hydrolysis method for determining tryptophan in peptides and proteins. Anal. Biochem. 60:45-50.

Regnier, F. E. 1987. The role of protein structure in chromatographic behavior. Science 238:319-323.

Rose, D. J., and Jorgenson, J. W. 1988. Fraction collector for capillary zone electrophoresis. J. Chromatogr. 438:23-24.

Roth, M., and Hampai, A. 1973. Column chromatography of amino acids with fluoresence detection. J. Chromatogr. 83:353-356.

Rubinstein, M., 1979. Preparative high performance liquid partition chromatography of proteins. Anal. Biochem. 98:1-7.

Sanger, F., and Tuppy, H. 1951. The amino acid sequence in the phenylalanyl chain of insulin. Biochem. J. 49:463-481.

Sarin, V. K., Kent, S. B. H., Tam, J. P., and Merrifield, R. B. 1981. Quantitative monitoring of solid phase peptide synthesis by the ninhydrin reaction. Anal. Biochem. 117:147-157.

Schenkein, J., Levy, M., Franklin, E. C., and Frangione, B. 1977. Proteolytic enzymes from the mouse submaxillary gland. Arch. Biochem. Biophys. 182:64-70.

Schroll, A. L., and Barany, G. 1989. A new protecting group for the sulphhydryl groups of cysteine. J. Org. Chem. 54:244-247.

Shively, J. E. Reverse phase HPLC isolation and microsequence analysis. In Methods of Protein Microcharacterization, A Practical Handbook, Shively, J. E., ed., Humana Press, Clifton, New Jersey, 1986, pp. 41-87.

Simmaco, M., DeBiase, D., Barra, D., and Bossa, F. 1990. Automated amino acid analysis using pre-column derivatization with dansyl chloride and reversed-phase high-performance liquid chromatography. J. Chromatogr. 504:129-138.

Simpson, R. J., Neuberger, M. R., and Liu, T. -Y. 1976. Complete amino acid analysis of proteins from a single hydrolysate. J. Biol. Chem. 251:1936-1940.

Smith, A. J., Young, J. D., Carr, S. A., Marshak, D. R., Williams, L. C., and Williams, K. R. State of the art peptide synthesis: comparative characterization of a 16-mer synthesized in 31 different laboratories. In Techniques in Protein Chemistry III, Angeletti, R. H., ed., Academic Press, San Diego, 1992, pp. 219-232.

Snyder, L. R., and Kirkland, J. J. Introduction to Modern Liquid Chromatography, John Wiley and Sons, New York, 1979.

Spackman, D. H., Stein W. H., and Moore, S. 1958. Automatic recording apparatus for use in chromatography of amino acids. Anal. Chem. 30:1190-1206.

Stewart, J. M., and Young, J. D. Solid Phase Peptide Synthesis, Pierce Chemical Co., Rockford, Illinois, 1984.

Tapuhi, Y., Schmidt, D. E., Linder, W., and Karger, B. L. 1982. Analysis of dansyl amino acids by reverse-phase high performance liquid chromatography. Anal. Biochem. 127:49-54.

Tarr, G. E., Beecher, J. F., Bell, M., and McKean, D. J. 1978. Polyquaternary amines prevent peptide loss from sequenators. Anal. Biochem. 84:622-627.

Tarr, G. E. Manual Edman sequencing system. In Methods of Protein Microcharacterization, A Practical Handbook, Shively, J. E., ed., Humana Press, Clifton, New Jersey, 1986, pp. 155-194.

Tarr, G. E., Paxton, R. J., Pan, Y-C. E., Ericsson, L. H., and Crabb, J. W. Amino acid analysis 1990: The third collaborative study from the Association of Biomolecular Resource Facilities (ABRF). In Techniques in Protein

Chemistry II, Villafranca, J. J., ed., Academic Press, San Diego, 1991, pp. 155-194.

Tregear, G. W., van Rietschoten, J., Sauer, R., Niall, H. D., Keutmann, H. T., and Potts, J. T., Jr. 1977. Synthesis, purification, and chemical characterization of the amino-terminal 1-34 fragment of bovine parathyroid hormone synthesized with the solid phase procedure. Biochemistry 16:2817-2823.

Tsunasawa, S., and Narita, K. 1982. Micro-identification of amino-terminal acetyl amino acids in protein. J. Biochem. 92:607-613.

Wilkinson, J. M. Fragmentation of polypeptides by enzymic methods. In Practical Protein Chemistry, A Handbook, John Wiley and Sons, New York, 1986, pp. 121-148.

Wittman-Liebold, B. 1973. Amino acid sequence studies on ten ribosomal proteins of *Escherichia coli* with an improved sequenator equipped with an automatic conversion device. Hoppe-Seyler's Z Physiol. Chem. 354:1415-1431.

Young, P. M., and Merion, M. Capillary electrophoresis analysis of species variations in the tryptic maps of cytochrome C. In Current Research in Protein Chemistry: Techniques, Structure, and Function, Villafranca, J. J., ed., Academic Press, San Diego, 1990, pp. 217-232.

5

Applications of
Synthetic Peptides

*Victor J. Hruby, Shubh D. Sharma,
Nathan Collins, Terry O. Matsunaga, and
K. C. Russel*

The tremendous advances in the development of methods for the synthesis of peptides, pseudo-peptides, and related methods as well as the corresponding advances in our understanding of peptide and protein structure, conformation, and dynamics provides unique opportunities to apply designed synthetic peptides for an enormous variety of problems in biology and medicine. In addition, if these advances can be coupled to the advances in molecular biology and cloning, on the one hand, and asymmetric synthesis and catalysis, on the other, it should be possible to provide hitherto unavailable, indeed unthinkable, approaches to diverse areas of drug design, chemotherapy, molecular immunology, and a wide variety of other uses.

Already it is clear that peptide therapy has enormous potential in such diverse areas as growth control, blood pressure, neurotransmission, hormone action, digestion, reproduction, and so forth. Nature has "discovered" that it can control nearly all biological processes by various kinds of molecular recognition and that peptides and proteins are uniquely suited for this control because of their enormous potential for diversity. This finding may, perhaps, be most readily understood if one recognizes that, considering only the 20 normal eucaryotic amino acids, the number of unique chemical entities for a pentapeptide is 3,200,000 (20^5); for a hexapeptide, it is 64,000,000 (20^6). Considered from this perspective, perhaps it is not unexpected that nature has "discovered" that peptides and proteins can do it all, from providing structure and motion, to catalysis, to growth and maturation. The ability of the immune system in higher animals, including humans, to recognize literally millions of foreign materials made by nature as well as humans, and to get rid of them as part of a survival strategy, is just one example that illustrates the potential of peptide-based drugs, therapeutics, and modulators of biological function.

Despite the enormous potential of peptides and small proteins for these areas, surprisingly little advantage has been taken of the potential of these molecules as drugs and tools for use in basic and clinical research. The reasons given for this are many and include such factors as (1) the lack of knowledge of most chemists regarding the structures, conformations, and synthesis of peptides and proteins; (2) the putative instability of these compounds in biological systems; (3) the lack of bioavailability of these compounds; (4) their high potency and very small concentrations in most biological systems; and (5) the enormous diversity that is possible. In fact, many of these problems can be overcome or are irrelevant to a particular biological or medical problem. In the case of diversity, this can, in fact, serve as an advantage, since it is now possible to construct very large libraries of peptides (10^6 to 10^7 and more) by both biological (Scott and Smith, 1990; Devlin et al., 1990; Cwirla et al., 1990) and chemical (Lam et al., 1992a, 1992b) methods and to screen them for a very wide variety of functions. These new approaches can be expected to accelerate the use of peptides and peptide-derived compounds for medical and biological purposes.

In this chapter, we provide a brief summary of some of the medical and biological applications for synthetic peptides. Quite a bit already has been accomplished, but more importantly, current studies are providing a framework for much more comprehensive success in the future. Indeed, it can be anticipated that peptide and peptide-related compounds will probably be among the most numerous drugs of the future. If this chapter helps in this process in some small way, we will have accomplished our purpose. (For other recent reviews or overviews of this area, see, for example, Emmett, 1990; Tam and Kaiser, 1988; Hruby et al., 1990a, 1990b; Ward, 1990.)

ANTIGENIC AND IMMUNOGENIC USES OF SYNTHETIC PEPTIDES

The use of synthetic peptides in the study of the immune response stems from the ability of peptides to mimic similar antigenic sequences in proteins. Antibodies raised against an antigen in a native protein can bind to the same antigenic sequence in the absence of the rest of the protein. With the advent of the Merrifield solid-phase procedure (1963), rapid and selective synthesis of pure peptides has become routine, and it is now widely used in the study and prediction of protein antigenicity. Primarily, synthetic peptides are used to identify key antigenic epitope regions of proteins. This identification is achieved by initiating the immune response against the native protein and then testing the binding of the resulting antisera to synthetic fragments of the protein in an appropriate immunoassay, such as the enzyme-linked immunosorbent assay (ELISA).

ELISA

Identification of various protein antigens has far-reaching applications in the fields of biochemistry, immunology, and molecular biology. Short peptide epitopes can be used in the elucidation of the mechanism of the immune response (Smith, 1989), and they allow investigation of the molecular recognition process of the antibody/antigen interaction complex (Hinds et al., 1991). The cross-reactivity of antibodies raised against synthetic epitopes for the native protein has facilitated the development of a new generation of synthetic peptide vaccines, which are potentially useful against viral diseases, such as foot and mouth disease, hepatitis B, and HIV. In addition, the cross-selectivity of antipeptide antibodies for the native protein has been utilized in the isolation and characterization of gene products (see the following discussion). Given the ease of synthesis of short peptides, the major difficulty in any study of this nature is deciding which antigenic fragments to synthesize. The minimum epitope length of a peptide fragment has been determined to be about six to ten residues, and that length has been used to determine which regions of a macromolecular structure, such as a protein, will be antigenic. Two protocols have been developed to this end. The first of these *epitope mapping* procedures involves the synthesis and testing of all possible peptide fragments in the protein of interest. The second protocol is based on prediction of the antigenic regions from the primary sequence of the protein, and, hence, it involves the synthesis of fewer peptides.

Elucidation of Antigenic Peptide Sequences in Proteins

With the introduction of immunological testing of peptides that were still bound to the solid support used in their synthesis (Smith et al., 1977), investigators realized the value of producing all the overlapping peptide

1   213

H$_2$N–T –T –S –A –G –E –S –A –D –P –(201 amino acids)–T–L–COOH

(i)   –T –T –S –A –G –E

   (ii) –T –S –A –G –E –S

      (iii) –S –A –G –E –S –A

         (iv) –A –G –E –S –A –D

            (v) –G –E –S –A –D –P

FIGURE 1   An example of systematic synthesis of all possible overlapping hexapeptides of the foot and mouth disease virus protein VP1, as utilized by Geyson and coworkers (1986). (Only the first five fragments are indicated.)

fragments in a protein to locate the highly antigenic sequences (Figure 1). This systematic protocol was greatly enhanced by the alternative synthetic procedure of Geysen et al. (1984). They prepared peptides on polyethylene rods utilizing standard Merrifield solid-phase chemistry. The linker between the peptide and the support was a polyacrylic acid derivative that was compatible with both water and the organic solvent. The polyethylene rod, with attached peptide, could then be used as the solid phase in an ELISA immunoassay. The feasibility of this procedure was demonstrated by Rodda and coworkers (1986), who synthesized all possible hepta -, octa -, nona-, and decapeptides of the 153-residue protein myoglobin, and they assessed the binding of these to polyclonal antibodies that were raised against the native protein. In this way, they identified the continuous epitope LKTEAE, comprising residues 49 to 54 of myoglobin as its strongest antibody-binding fragment.

   Although this technique yields a complete, systematic epitope mapping, it can be applied only to small proteins due to the prohibitive expense and time required for such a procedure. However, an example in which a nonselective, systematic, synthetic approach is essential can be found in the study of the recognition of T helper cells for processed foreign protein fragments, which are located on the surface of macrophages (Unanue and Allen, 1987). These fragments are produced by enzymatic degradation or denaturation of the native protein, and they do not necessarily correspond to the regions of high antigenicity that are recognized by antibodies. Thus, when studying the immune response mechanism of T helper cell activation, one cannot easily predict sequences that are T cell epitopes (DeLisi and Berzofsky 1986; Rothbard and Taylor, 1988), and the most efficient method of finding the antigenic specificity of T cell recognition is the synthesis of an overlapping series of fragments of the protein. T cells actually bind to relatively few antigens due to major histocompatibility complex restriction, which is thought to suppress autoimmunity. Finnegan and coworkers (1986) investigated this theory by synthesizing fourteen 20-residue overlapping peptides of staphylococcal nuclease, which is a 149 amino acid bacterial protein. Binding studies of the overlapping fragments to mouse T cell clones raised against staphylococcal nuclease indicated only two strongly antigenic sequences, corresponding to segments 81 to 100 and 91 to 110. The observation of so few epitopes for mouse T cells on a phylogenetically distant bacterial protein suggested that inhibition of autoimmunity is not a major component in the limited repertoire of T cell antigens.

Prediction of Protein Antigenic Regions for Antibodies

In predicting an epitope region of a protein, the primary consideration is that an antibody will bind only to the surface without disrupting its overall conformation. Hence, a reasonable assumption is that all highly antigenic regions of a protein are on its surface and are well exposed, for example, secondary structural elements, such as amphipathic helices, loops, and random coil regions containing high densities of hydrophilic amino acids. When three-dimensional structural information is available from X-ray or NMR, the surface accessibility of the protein can be immediately determined (Thornton et al., 1986; Fanning et al., 1986). However, in the absence of such tertiary structural data, accessible regions can be predicted from our knowledge about primary structure according to expected surface properties. The range of predictive algorithms has been reviewed (van Regenmortel et al., 1988) and is briefly described in the following section. (Also see Chapter 2.)

Use of Tertiary Structural Information

One of the foremost X-ray studies on antigenic interactions was the determination, to 2.8 Å resolution, of the three-dimensional structure of an antibody/antigen complex (Amit, 1986). This structure comprised the Fab fragment from a monoclonal antibody bound to its antigenic region on lysozyme. One of the key findings of this work was that two continuous epitopes of lysozyme were in contact with the antibody fragment. These epitopes were chain residues 18 to 27 and 116 to 129 (see Figure 3 of Amit et al., 1986). Although this study graphically demonstrated the key epitope binding sites of an antibody, it also exemplified a major drawback in the use of synthetic peptides as antigens. It has been calculated that the average contact surface area between an antigenic globular protein and antibody is 16 to 20 Å in diameter (Barlow et al., 1986), and, indeed, in the aforementioned X-ray structure study the entire antibody/antigen contact area was shown to be ca. 20×30 Å. This area constitutes a protein surface that is too large to contain only one continuous chain of residues, which indicates that all protein antigens are, in fact, made up of a set of " discontinuous" epitopes. Solid-phase peptide synthesis allows the preparation of linear, or "continuous," peptide epitopes only. Thus, only partial binding of antibody to linear synthetic peptide epitopes of the native polypeptide can ever be observed relative to the native protein. Despite this drawback, binding between antibodies and continuous synthetic epitopes is substantial enough for most uses and, at least, for those linear sequences corresponding to the regions of highest antigenicity.

Prediction of Antigenicity Based on Primary Structure

Any prediction about regions of high antigenicity from primary structure is concomitant with a prediction about the characteristics of surface regions of a protein. Of the surface characteristics of a protein, the most obvious is hydrophilicity. Globular proteins generally fold, so many of the hydrophobic residues are buried in the protein core, while many of the hydrophilic residues are located at the protein surface in order to maximize good energetic interactions with the aqueous environment. In order to predict protein tertiary structure, relative hydrophilicity (or hydrophobicity) scales have been calculated for the 20 naturally occurring amino acids; thus, the average hydrophilicity over six or seven residues in a chain can be estimated. Those segments with the highest average hydrophilicity (lowest hydrophobicity) are then very likely to correspond with exposed surface regions, and, hence, they are potentially highly antigenic. Several relative hydrophilicity scales have been calculated and compared (Hopp, 1986; van Regenmortel et al., 1988), and, generally, they have given investigators a success rate of 65 to 75 percent in locating continuous peptide fragments of high antigenicity. An example of the Hopp and Woods' scale (1981) is plotted for myoglobin in Figure 2(a), together with the known epitopes, and reasonable agreement is observed. A special case of hydrophilicity in proteins is that of amphipathicity. Secondary structural regions are said to be amphipathic if one side of the structure is appreciably more polar than the other. Amphipathic α-helices are generally observed on the surface of proteins, where one-half of the helix, as viewed down its axis, contains hydrophilic residues on the protein exterior, and the other half comprises hydrophobic residues that interact with the protein core. (This finding is particularly well represented using a helical wheel diagram; see Figure 3). As such, these regions may be exceptionally antigenic. When the hydrophilic residues are presented to the surface in a secondary structural motif such as an amphipathic α helix, the hydrophilicity scale described previously will predict only a region of intermediate antigenicity, since about half of the residues in the primary sequence are hydrophobic. Hence, based on primary sequence alone, these potentially highly antigenic sites would be

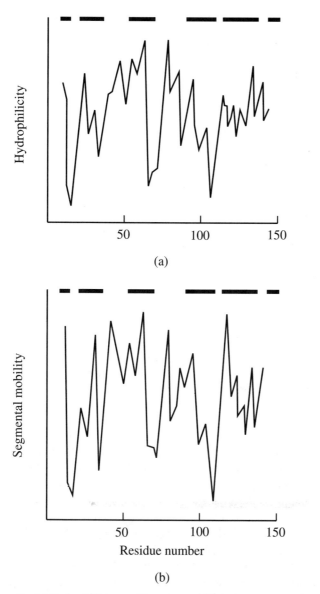

FIGURE 2   (a) Hydrophilicity profiles of myoglobin calculated with the scale of Hopp and Woods (1981) and plotted against residue number. The black bars above the profile indicate regions of greatest antigenicity. (b) Segmental mobility profile of myoglobin calculated from the scale of Karplus and Schulz (1985). Figures adapted from van Regenmortel et al. (1988).

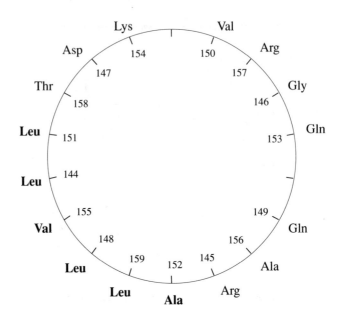

$$L^{144}RGDLQVLAQKVARTL^{159}$$

FIGURE 3   Primary sequence and helical wheel representation of the amphipathic α-helix in VP1 protein of foot and mouth disease virus. The residues constituting the hydrophobic section of the helix are shown in boldface type.

missed. This error has been corrected in some hydrophilicity scales, which incorporate prediction of amphipathic helices from the primary sequence (DeLisi et al., 1986; Sette et al., 1986).

The antigenicity of amphipathic α helices was demonstrated by Pfaff and coworkers (1982). They predicted the presence of seven helices in the foot and mouth disease virus (FMDV) viral protein VP1, and assessed the potential antigenicity of these on their amphipathic character. The fifth helix in the sequence appeared to have the most amphipathic nature (Figure 3) and, indeed, a synthetic hexadecapeptide fragment generated from it induced antibodies that neutralized the intact virus. Interestingly, a standard Hopp and Woods (1981) hydrophilicity plot of this hexadecapeptide predicted that it would possess limited antigenicity.

Another surface characteristic that can be predicted from primary structure is segmental mobility. The mobility of chains of amino acids can be observed by X-ray or NMR. It allows one to calculate an atomic temperature factor (B), which is proportional to the mean square atomic displacement from the equilibrium crystal position. When calculated for

segment
mobility
determined by
temp. factor

the α carbons of a protein chain and averaged over segments of six to ten residues, a plot of B versus residue number can be made. Since the segments at the surface will be relatively unhindered by the rest of the protein, these will generally have a high B factor and will correspond to relatively accessible antigenic regions. A B factor scale for the 20 naturally occurring amino acids, parameterized from 31 protein X-ray structures from the Brookhaven Protein Data Bank, has been calculated (Karplus and Schultz, 1985), and an example profile for myoglobin is shown in Figure 2(b). Notably, the regions of high segmental mobility correspond well with those predicted to be hydrophilic, and a fair agreement between predicted and experimentally observed antigenicity is seen.

Analogues of Antigenic Peptides

The real potential of synthetic peptides lies in the researcher's ability to design novel analogues that can mimic or accentuate certain properties of the original biological sequence, using conformational information from solution NMR and/or computer-aided molecular modeling (Hruby et al., 1990). This approach is widely used in the synthesis of potent and selective analogues of peptide hormones and neurotransmitters. It has only recently been applied to the conformational analysis of antigenic peptides in studying the molecular recognition processes between an antibody and an antigen.

Dyson and coworkers (1988) studied the immunodominant nonapeptide (YPYDVPDYA) from influenza virus hemagglutinin, noting that the first four residues of this epitope formed a β-turn in aqueous solution, as indicated by ^{1}H NMR. Assuming that this conformational motif is important for antigenic recognition, analogues of the pentapeptide YPYDV were prepared, in which positions 3 and 4 were substituted for the other 19 natural amino acids. The β-turn population about Pro^2 and position 3 was assessed by 1 and 2 D NMR in water. In this study, the nature of the residues in position 3 was found to be most influential in stabilizing the turn conformation, and, in most cases, a type II turn was favored. Consistent with this finding, YPGDV was observed to produce the highest turn population, as indicated by the low temperature dependence of the Asp^4 amide proton resonance (Table 1). The Pro-Gly dipeptide is well known to favor a type II reverse turn (Gierasch et al., 1981). The dependence of antibody binding to this nonapeptide on the type of β-turn conformation about Pro^2 Tyr^3 was further investigated by Hinds and coworkers (1991). In these studies, α-alkyl derivatives of Pro^2 were synthesized. In particular, the γ-spirolactam (1) and α-Me Pro (2) (Figure 4) analogues were studied to determine their ability to stabilize turns when compared to their binding affinity for two antisera, for which YPYDV was proved to be the minimum antigenic sequence. The cyclic constraint in (1) forces the backbone to adopt a type II reverse turn, as

Table 1   Correlation of the relative turn formation of YPYDV and its analogues with the dissociation constant for equilibrium binding to antibody

Sequence	$^1H\ NMR$ $\Delta\delta$ (-p.p.b.)	Kd (nM)[+]
YPYDV	4.6	180
YPGDV	3.3	—
1	1.5	No binding observed
2	0.0	3.6

Turns about positions 2 and 3 were determined by observation of the relevant nOes by ^{1}H NMR. The relative population of the turn is given here by the degree of intramolecular hydrogen bonding between residues 1 and 4 as indicated by the D^4 NH chemical shift temperature dependence. The lower NH temperature dependencies indicate more stable - turns. Hence, equilibrium binding increases with turn stability, as long as the amide NH in position 3 remains intact. (+) Measured against antibody DB 19/1 in Hinds et al., 1991.

indicated by NMR (nOe, nuclear Overhauser effect) and the Asp^4 NH temperature dependence (Table 1). The total lack of binding of (1) to either of the antibodies studied appeared to suggest that a type II turn is not favorable for recognition. However, YP[N-Me]YDV also showed no binding affinity, which indicates that the Tyr^3 NH is critical for formation of the antibody/antigen complex. α-Methylation of Pro^2 in (2) was shown to favor a type I β-turn by gas-phase molecular modeling, while ^{1}H NMR indicated the presence of a highly stabilized turn (Table 1), which was a mixture of both type I and type II turns in solution. The enhanced binding of (2) relative to the native pentapeptide sequence was thus attributed to the increased stability of the turn about positions 2 and 3, and this epitope most probably binds in a type I β-turn conformation. The information gathered in such molecular recognition studies not only aids our understanding of the antibody/antigen complex, but it also provides general data on receptor-ligand theory.

Antigenic Peptides in the Isolation of Gene Products

Rapid advances in gene cloning and nucleotide sequencing have made routine the location of genes encoding proteins that have not been isolated and for which a function has not been assigned. Since the non-processed primary sequence of such a protein is determined from the gene, antigenic regions can be predicted, as described previously. Hence, raising antipeptide antibodies that are selective for the native protein facilitates a very accurate method of isolating the putative gene product.

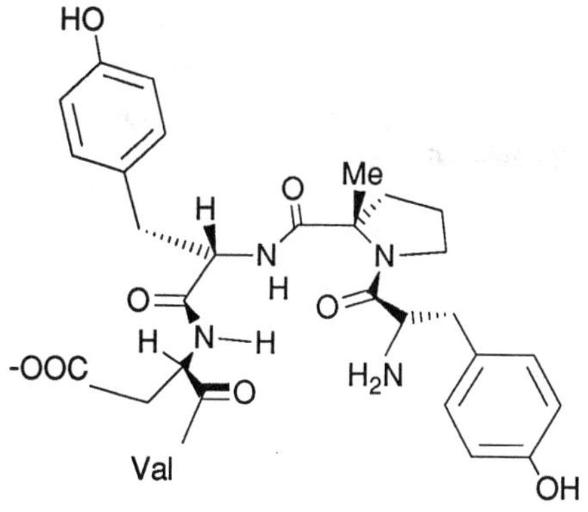

2

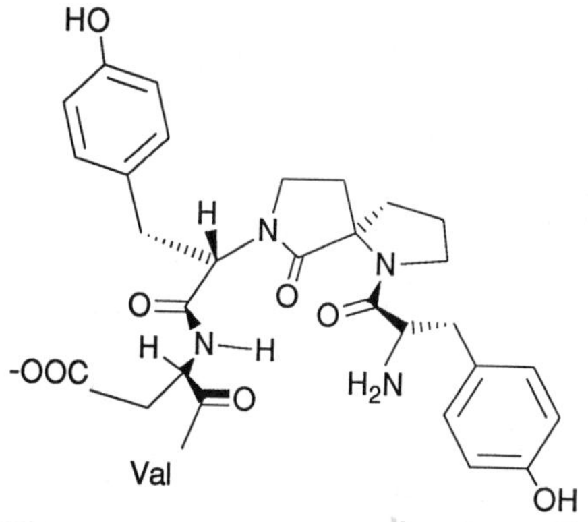

1

FIGURE 4   Analogues of YPYDV tested by Hinds et al. (1991). γ-spirolactam (1) shown in the constrained type II β-turn conformation; α-MePro derivative (2) shown in a type I turn.

The use of this technique has become widespread in the field of protein purification and characterization and the literature abounds with many variations of its application, especially in the detection of viral proteins and protein intermediates.

Indeed, the first example of the characterization of a putative gene product was the detection and isolation of the R protein in cells infected with Moloney murine leukemia virus (Sutcliff et al., 1980). After sequencing of the retroviral *env* gene DNA, three gene products were detected. Two of these were already known to code for the gp 70 and p15E polypeptides, while the remaining codons at the 3' end coded for the newly identified R protein. Several C-terminal fragments of the putative R protein were synthesized, and antibodies were raised against them. All of the resulting antisera were found to be selective for an 80-kilodalton protein located in the cell lysate. This large polypeptide was shown to correspond to the entire unprocessed *env* gene product by its affinity for cross-selective antisera raised against the gp 70 5' end polypeptide. A 12-kilodalton protein was also isolated, which bound only anti-R protein peptide antibodies, and this was assumed to be the processed R protein cleaved from the other *env* gene products. The key evidence that a single gene product has been located is that the putative protein is recognized by antibodies raised against *more than one peptide fragment* of its predicted protein sequence. This is important because some epitopes may not be as highly antigenic as predicted, or they may not be present after processing of the initial gene product (Lerner, 1984).

As mentioned, this is one of many possible examples in an ever-expanding field of application. (For more detailed reviews, see Lerner, 1982, 1984; Sutcliff et al., 1983; Walter, 1986; van Regenmortel et al., 1988.)

Mimotopes, Peptide Libraries, and the Future

All the examples cited so far have been based on protein/antigens of known sequence, and the synthesis of the peptide fragments have been limited to continuous epitopes. An exciting alternative protocol in the search for synthetic peptide antigens, which circumvents these restrictions, was proposed by Geysen and coworkers in 1986. They introduced the concept of the "mimotope," which is defined as a molecule that is able to bind to the combining site of an antibody, but which is not necessarily identical to the native epitope that induced that antibody. A mimotope thus mimics the essential binding properties of the native epitope. Moreover, this definition may be extended to the mimicking of discontinuous epitopes, since a linear mimotope may, in theory, contain the major binding residues from more than one continuous epitope in the correct topographical configuration for binding.

If one shifts the emphasis from searching for the native epitope of a protein to finding mimotopes of the native epitope, a totally different approach becomes apparent. Searching for a suitable mimotope requires no previous knowledge of the protein sequence or the native epitopes. Instead, having raised antibodies to the native protein, potent mimotopes may be located by their binding to a library of randomized peptides that incorporate all the 20 natural amino acids in all positions. However, for a library of all possible hexapeptides (the approximate minimum antigenic length), this procedure would require the synthesis and testing of 20^6 (that is, 64,000,000) peptides, which, until recently, was totally impractical. Recently, however, two synthetic procedures for creating such peptide libraries have been presented, which offer exciting possibilities for the future. Chronologically, the first of these is "light directed, spatially addressable parallel chemical synthesis" (Fodor et al., 1991). This procedure combines N_α photolabile protection with photolithography to allow simultaneous multiple peptide synthesis on a glass microscope slide, using solid-phase techniques. This procedure facilitates selective deprotection of the N terminus by simply masking one region of the slide during light irradiation. Using a computer-controlled binary masking system, 2^n compounds can be synthesized in n steps on the same slide. Since the position of each peptide in the library is known (that is, the synthesis is spatially addressed), potent mimotopes are easily found in a fluorescence immunoassay simply by the location of the fluorescing regions of the slide.

Although this procedure represents an efficient means of preparing peptide libraries, it relies heavily on a new, unproved, and expensive technology, and it has only so far been applied to half of the naturally occurring amino acids. The second method for preparing peptide libraries depends entirely on standard solid-phase chemistry. Thus, it is more attractive to the peptide chemist. This method relies on "split synthesis" for randomization in the library (Lam et al., 1991, 1992) (Figure 5). A somewhat modified standard beaded solid support is split into an equal number of aliquots, and a single protected amino acid is coupled to each aliquot. After completing the couplings the aliquots are thoroughly mixed together and split again for deprotection and coupling of the next selection of protected amino acids. This process can be repeated as many times as required, and it allows for the synthesis of larger libraries than could be obtained using the previously described technique. Having completed the library, the mixed peptide resin will contain beads that bear only one peptide sequence each. Location of potent mimotopes is achieved by using assay conditions in which the bead bearing the binding peptide will be stained. This bead then can be removed with the aid of a microscope and a micromanipulator, and the peptide on the isolated bead can be microsequenced.

Both of these synthetic technologies provide efficient ways to map antigenic peptides. Use of these approaches will greatly enhance our understanding of the nature of discontinuous epitopes and the molecular mechanisms of antibody/antigen recognition. Possibly of more importance, peptide libraries may facilitate the discovery of mimotope peptide immunogens that can be used as vaccines. With these methodologies, vaccine candidates can be directly mapped from monoclonal antibodies that are known to neutralize biologically important molecules or pathogens. In either case, peptide libraries will become invaluable tools in the elucidation of receptors of all types and in the general immunochemistry of the future.

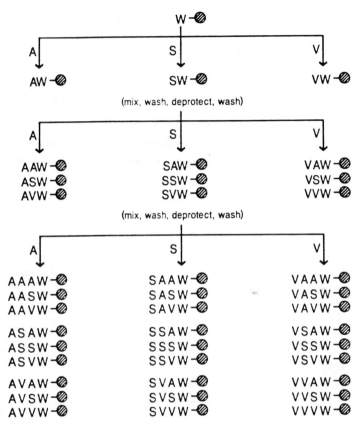

FIGURE 5   Scheme for random peptide synthesis using the split synthesis method for a random tripeptide containing S, A, or V with a terminal tryptophan added. Diagram from Lam et al. (1991).

USE OF PEPTIDES AS ENZYME INHIBITORS

Inhibition of enzymatic systems remains a prime target for the regulation and control of different disease states. Currently, numerous pharmaceuticals exist that are targeted toward the inhibition of enzymes in (1) cancer cells, (2) bacteria, (3) the systemic system of humans, and (4) many other systems. For example, 5-fluorouracil and methotrexate are commonly used as chemotherapeutic agents for the inhibition of dihydrofolate reductase and thymidylate synthetase. Penicillins, native and semisynthetic, which can be viewed as peptidic in nature, are still considered first line antibacterials. Similarly, lovastatin, isolated from *Aspergillus tereus,* is a cholesterol-lowering agent that works by inhibiting the con-

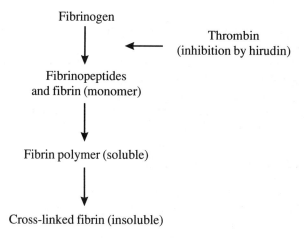

FIGURE 6   Scheme describing the mechanism of fibrin clot formation and inhibition of thrombin activity by hirudin.

version of 3-hydroxy-3 methyl-glutaryl Co-A to mevalonate (Stryer, 1975).

More recently, peptides have become attractive as potential therapeutic agents for targeting specific enzymes or enzyme systems. For example, hirudin is a 65 amino acid (sulfated on Tyr[63]) polypeptide inhibitor of the serine protease thrombin, isolated from the salivary gland of the medicinal leech *Hirudo medicinalis* (Stone et al., 1990). Interestingly, peptides that are related to hirudin are already being tried experimentally in clinical trials for the prevention of blood clots in patients with rejoined limbs. Mechanistically, hirudin acts as a tight-binding inhibitor of thrombin. Thrombin, being a serine protease enzyme, acts primarily by cleaving fibrinogen to fibrin monomers, which later cascade into a fibrin clot-forming polymer. Once hirudin is bound to thrombin, thrombin is no longer able to bind to fibrinogen, thus preventing the formation of fibrin monomers, which, in turn, would polymerize to form the fibrin (blood) clot (see Figure 6) (Mann, 1987). For a review of the clotting cascade pathway, the reader is referred to a review by Kayser (1983). Kinetically, hiridun is a slow, tight-binding inhibitor that interacts with thrombin in an apparent two-step fashion. The first step is a slow complexation with thrombin at a remote site, while the second step is essentially diffusion controlled (4.7×10^8 M^{-1} sec^{-1}) and involves irreversible complexation to the active site (Stone and Hofsteenge, 1986). Since thrombin is a serine protease, it (as do most enzymes of this class) has a preference for cleavage at the carboxy termini of amino acid residues, with thrombin being specific for basic residues, particularly arginine (Stone et al., 1990). The peptide affinity label D-Phe-Pro-Arg-CH_2-Cl acts as an irreversible inhibitor of α thrombin (Sonder and Fenton, 1984; Walker, Wikstrom, and Shaw, 1985). This peptide was shown to bind covalently to a histidine residue in thrombin and consequently blocked the interaction of thrombin with hirudin.

Based upon the proposal that the C terminus of hirudin may be important in binding to thrombin, Krstenansky and Mao (1987) synthesized, by solid-phase methods, the unsulfated C-terminal fragment N-α-acetyl hirudin$_{45-65}$. They found that this peptide bound to thrombin at a single site with an association rate constant of 10^5 M^{-1}. In addition, they observed that the synthetic peptide D-Phe-Pip-Arg-p-nitroanilide was still hydrolyzed by the thrombin-hirudin$_{45-65}$ complex. At the same time though, the complex was able to inhibit fibrin clot formation. This occurrence led to the suggestion that the C-terminal portion of hirudin contained the binding domain needed to complex with the α thrombin, yet it did not alter the region near the active catalytic site. This work was followed up later by Mao and coworkers (1988), who synthesized smaller C-terminal fragments and determined that the last ten amino acid residues were sufficient for binding and inhibition of thrombin-mediated clotting. However, once the Phe^{56} was removed from the C terminus, no further anticlotting activity was observed. In addition, replacement of Phe^{56} by either acidic (Glu), hydrophobic (Leu), or enantiomeric (D-Phe) amino acids resulted in complete loss of activity. Therefore, the ten amino acid peptides represented a unique class of noncatalytic site thrombin inhibitors. It was also shown in these studies that binding of the C-terminal fragment induced a conformational change in the thrombin when complexed.

Recent work by Dennis and coworkers (1990), who used site-specific mutagenesis to replace Asn^{52} with methionine in hirudin followed by cyanogen bromide cleavage at this single methionine site, found, as did Mao and coworkers, that a conformational change in the complex did occur that affects the active site of thrombin. This occurrence was shown by a change in the Michaelis constant (K_m) for the substrate D-Phe-Ala-Pip-Arg-p-nitroanilide.

Thus, the work on the hirudin-thrombin complex via synthesis of peptide fragments yielded important information about the mechanism of thrombin-induced coagulation pathways. Similar studies in many areas are likely to lead to many clinically valuable peptide pharmaceuticals in coming years.

STRUCTURE-FUNCTION STUDIES

One of the principal goals of peptide and protein research is the rational design of peptide ligands whose chemical, physical, and biological properties can be predicted a priori. It is clearly impossible to come up with a general set of rules that would apply to structure-function studies of all receptors. However, the goal here is to discuss some general approaches in the design of synthetic peptides that can be used to better understand ligand-receptor interactions through structure-function studies.

It now seems clear that there are three separate states of the ligand-receptor complex occurring during the ligand-receptor interaction. First, the ligand must be "recognized" by the receptor for binding to take place. In a second step, a change in conformation of the hormone-receptor complex that results in the transduction of the biological response through a second messenger occurs. Finally, the hormone is released from the receptor. The differentiation between binding and transduction (or inhibitory) states relating to hormone interactions have been well established as being the result of the hormone exhibiting different conformational and topological features (Rudinger et al., 1971; Meraldi et al., 1975, 1977; Schywzer, 1977; Walter, 1977; Hruby et al., 1981a, b).

Ideally, one would like to be able to "know" what a receptor is "looking for" for binding, transduction, and release of its complementary binding ligand. In some cases, remarkable progress has been made in these areas. But even with the ability to isolate receptors, there is currently no method to design ligands with specific properties a priori. The use of structure-function studies is the classical method for the design of synthetic peptides with specific activities. The ability to perform structure-function studies relies heavily upon the availability of a well-defined ligand-receptor system with sensitive binding assays and bioassays that can answer questions related to selectivity, binding, and transduction for synthetic peptide analogues.

Structure-function studies usually begin with simple modifications of the native ligand for a receptor or receptor class. The hormone-receptor system is complex so that, in general, only by individual modifications can one ensure that a single change is providing the change in effect. Thus, a large number of single modifications built upon one another can result in quite sophisticated analogues. In some cases, the modifications in the analogue can be so extensive that it is difficult for an "outside observer" to see the relation of its structure to that of the endogenous hormone.

In a somewhat related approach, it has been found that many hormones have a minimum sequence that may have full biological potency. Since most peptide analogues are synthesized by either solid-phase or solution-phase synthesis, if a minimum active sequence can be found, it can greatly simplify the synthesis of all subsequent analogues prepared. A great deal of information can often be gleaned from the kinds of studies mentioned, and, in some cases, the separation of peptide sequences required for binding and transduction can be determined. This, in turn, can lead to the design of agonist and antagonist ligands. Often, assays that relate modes of action through primary and secondary messengers can be examined. In this way, it sometimes is possible to separate an endogenous ligand's affinity for subtypes of multiple receptors. A few examples of some of these approaches will now be discussed.

Deletion Peptides

Deletion peptides of melanin concentrating hormone (MCH), whose native structure is

Asp-Thr-Met-Arg-Cys-Met-Val-Gly-Arg-Val-Tyr-Arg-Pro-Cys-Trp-Glu-Val,

showed that only residues 5 to 15 were necessary for full biological activity (Table 2) (Matsunaga et al., 1989). Removal of the tryptophan in position 15 lowers the biological activity by more than tenfold, and the potency continues to decrease, with fragments sequentially deleting residues 1 to 5. These studies suggested the importance of Trp^{15} for interaction with the MCH receptor, with the 5 to 14 fragment having only 1/300 the potency of the native hormone.

Stereochemical Substitutions

Once a minimum sequence has been determined, it is often convenient to search for stereochemical requirements of the receptor with the systematic substitution of D for L amino acids. The incorporation of D amino acids also can provide clues about the importance of certain secondary structures (for example, β-turns, α-helix) and may also yield further information about specific residues necessary for peptide-receptor interaction as well as contributing stability to analogues against enzymatic breakdown (Hruby, 1982).

In one example, the incorporation of D-Phe at position 7 of $[Nle^4]\alpha$-MSH (melanotropin stimulating hormone) resulted in a melanotropin superagonist (Sawyer et al., 1980). The native linear peptide has the active sequence Ac-Ser-Tyr-Ser-Met-Glu-His-Phe-Arg-Trp-Gly-Lys-Pro-Val-NH$_2$. The substitution was made based on the evidence that heat-alkaline treated α-MSH and $[Nle^4]\alpha$-MSH led to prolonged activity with an unusual amount of racemization at the Phe^7 position. The $[Nle^4,$ D-$Phe^7]\alpha$-MSH analogue was shown to be 60 times as potent as the native hormone, with extremely prolonged activity and greatly reduced biodegradation. Other examples of the use of D amino acids include LHRH agonists and antagonists (Coy et al., 1982; Freidinger et al., 1985), substance P antagonists (Piercy et al., 1981), oxytocin antagonists (Melin et al., 1983; Lebl et al., 1985), enkephalin (Morgan et al., 1977), and somatostatin analogues (Freidinger and Veber, 1984; Torchiana et al., 1978) among others.

Table 2   Relative Potencies of MCH Fragment Analogues as Determined by the In Vitro Fish (*Synbranchus marmoratus*) Skin Bioassay.

Peptide	Potency Relative to MCH
MCH	1.0
MCH(2-17)	1.0
MCH(3-17)	1.0
MCH(4-17)	1.0
MCH(5-17)	1.0
MCH(1-16)	1.0
MCH(2-16)	1.0
MCH(3-16)	1.0
MCH(4-16)	1.0
MCH(5-16)	1.0
MCH(1-15)	1.0
MCH(2-15)	1.0
MCH(3-15)	1.0
MCH(4-15)	1.0
MCH(5-15)	1.0
MCH(1-14)	0.1
MCH(2-14)	0.07
MCH(3-14)	0.05
MCH(4-14)	0.023
MCH(5-14)	0.014

Isosteric and Other Substitutions

Pseudo-isosteric replacements are also quite useful for determining peptide-receptor interactions. Amino acids can be replaced with other amino acids of similar size but different electronic properties. Such substitutions would include the use of Nle for the oxidizable methionine, where a methylene unit is used as a replacement for sulfur. This substitution was carried out for α-MSH. The resulting [Nle^4]α-MSH peptide was shown to be 1.5 to 2.2 times more potent than the native hormone having methionine in the fourth position. This slight increase in potency may, in part, be the result of a lack of methionine oxidation during the bioassays

(Hruby et al., 1980). Substitutions could include changes in the aromatic amino acids tyrosine, tryptophan, phenylalanine, and histidine. Switching these amino acids might allow study of spatial requirements and/or hydrogen-bonding interactions. In oxytocin, the importance of the phenolic OH in the Tyr^2 position was illustrated by the synthesis of $[Phe^2]$-oxytocin (Bodanszky and du Vigneaud, 1959) and $[Trp^2]$oxytocin (Kaurov et al., 1972), which show weak agonist activity, while $[p\text{-}MePhe^2]$oxytocin (Rudinger et al., 1972) is an antagonist. In other cases (but not oxytocin), acidic residues aspartic acid and glutamic acid might be exchanged for their amide counterparts asparagine and glutamine and visa versa. These types of substitutions might have a pronounced impact on the analogues' chemical nature because the net charge on the ligand is changed. The basic residues lysine and arginine are also often interconverted to examine how well the amine functional group fits the receptor relative to the guanidine functional group. These types of replacements have been very effective in the design of analogues of enkephalins (Kruszynski et al., 1980; Mosberg et al., 1983), somatostatins (Freidinger and Veber, 1984), and μ-opioid selective peptides (Kazmierski and Hruby, 1988).

Constraint

In addition to these more classical approaches, the use of conformational and topographical constraint has been a very powerful approach to peptide ligand design (Hruby, 1982; Hruby et al., 1990), including increasing potency, receptor selectivity, providing antagonists, and increasing stability against biodegradation. We now provide a few examples of these approaches.

The incorporation of N-methyl amino acids has a tendency to constrain the χ_1 angles of amino acids and provide *cis-trans* isomerism around the amide bond of which they are part. One excellent example is in the case of certain $[N\text{-}MeNle^3]$CCK-8 (CCK = cholecystokinin) analogues, which showed a significant increase in the specificity for the CCK-B (brain) receptor over the CCK-A (peripheral) receptor. It has been suggested from NMR studies that the enhanced selectivity is the result of the brain receptor's preference for the *cis* amide bond which is not seen in related $[Nle^3]$CCK-8 analogues (Hruby et al., 1990). Another example includes the substitution of N-methyl-D-phenylalanine in the one position of Phe-Cys-Tyr-D-Trp-Orn-Thr-Pen-Thr-NH_2 (CTOP). This analogue was shown to be 200 times less potent than $[D\text{-}Phe^1]$CTOP at the μ-opioid receptor, simply because of the change in the χ_1 angle from $-60°$ to $180°$ (Kazmierski et al., 1988). N-Methylalanine was used in place of proline in some bradykinin analogues, decreasing their potency (Filatova et al., 1986). A review of the literature reveals that many similar examples could be cited.

Amino acids that are α-alkylated have also been widely used to examine a receptor's conformational requirements. These amino acids, besides adding potential additional steric bulk to the peptide, tend to induce β-turn structures (De Grado, 1988; Toniolo, 1990). One amino acid that has proved to be extremely useful for this purpose is amino-isobutyric acid (Aib) in the place of glycine.

To study the steric and stereoelectronic nature of a receptor, bulky nonproteinogenic amino acids are often incorporated into peptide analogues. One of the most widely used replacements is the substitution of penicillamine (Pen, 3-mercapto-D-valine) for cysteine. This substitution has led to the differentiation between agonist and antagonist analogues in oxytocin (Table 3). A series of analogues that have even larger side chains at the β-carbon were also examined. In these analogues, antagonists were obtained only when both hydrogens on the β-carbon were replaced with alkyl groups. Replacement of only one methyl group was not sufficient for antagonism at the uterine receptor. There was little difference in the inhibitory properties observed for β,β-dimethyl or ethyl analogues and bicyclic [Cpp^1]oxytocin. It has been suggested that this antagonism is caused, in part, by the restriction of disulfide interchange between R and S chiralities and to disposition of the χ_1 and χ_2 angles of the Tyr^2 side chain group. The substitution of phenylalanine for cysteine and other larger β,β-dialkyl cysteine derivatives has also been examined in the development of potent V$_1$ and V$_2$ vasopressin antagonists (Bankowski et al., 1978; Dyckes et al., 1974), δ-selective cyclic en-kephalin analogues (Mosberg et al., 1983a,b), and μ-selective somato-statin analogues (Pelton et al., 1985; Kazmierski and Hruby, 1988). The possibility of using other sterically demanding amino acid side chains, such as t-butyl, adamantyl, napthyl, and so on, are almost endless, and many such amino acids have been incorporated into peptides.

Ring Size

For cyclic peptides, the optimization of ring size is of great use. The effects of ring size and the necessity of the sulfur atoms in disulfide-con-taining peptides have probably been most completely studied in oxytocin and vasopressin analogues. Replacement of either or both sulfur atoms in oxytocin resulted in analogues with agonist activity (Hruby and Smith, 1987; Jost, 1987). Attempts to modify the 20-membered ring size by addition or deletion of single residues or single side-chain carbons or sulfurs led to a 100-fold decrease in potency in all cases. Similar experiments with vasopressin generally resulted in less potent analogues, and the relative effect was largely dependent on the hormone receptor ex-amined. One quite successful exploration of ring size was in the develop-ment of somatostatin agonists. Removal of six residues from the ring of somatostatin

Table 3   Activities of Some 1-Substituted Oxytocin Analogues

Analogue[a]	Activity U/mg	Reference
Oxytocin	546	Chan and Kelly, 1967
[Mpa¹]oxytocin	803	Ferrier et al., 1965
[Dmp¹]oxytocin	$pA_2 = 6.94$	Vavrek et al., 1972
[D,L-β-MeMpa¹]oxytocin	91	Schulz and du Vigneaud, 1966
[Dep¹]oxytocin	$pA_2 = 7.24$	Vavrek et al., 1972
[Cpp¹]oxytocin	$pA_2 = 7.61$	Lowbridge et al., 1979
[Pen¹]oxytocin	$pA_2 = 6.86$	Vavrek et al., 1972

[a]Mpa: β-mercaptopropanoic acid; Dmp: β,β-dimethyl-β-mercaptopropanoic acid; Dep: β,β-diethyl-β-mercaptopropanoic acid; Cpp: β-cyclopentamethylene-β-mercaptopropanoic acid; Pen: penicillamine.

Definition: pA_2 is defined as the negative logarithm of the molar concentration of a competitive antagonist that reduces the response of the agonist by 50 percent of what it would be in the absence of the antagonist.

Mathematically:

$$E = \frac{E_{max}}{\dfrac{D_2}{a}\left(1 + \dfrac{i}{A_2}\right) + 1} \quad \text{and } pA_2 = -\log A_2$$

E = observed response

E_{max} = maximum response in absence of antagonist

D_2 = half-maximum concentration

i = antagonist concentration

a = concentration of agonist

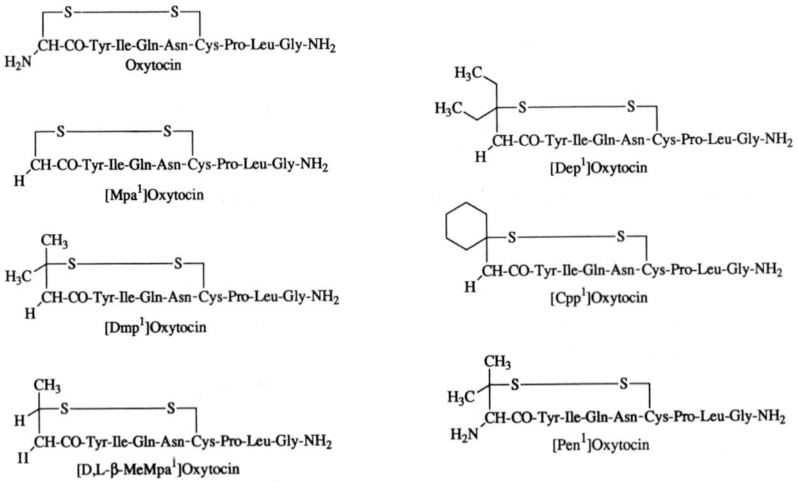

Ala-Gly-Cys-Lys-Asn-Phe-Phe-Trp-Lys-Thr-Phe-Thr-Cys

gave rise to the 20-membered ring analogue

D-Phe-Cys-D-Trp-Lys-Thr-Cys-Thr-ol,

which is superactive in inhibiting growth hormone release (Bauer et al., 1983). The ring was subsequently reduced to the 18-membered cyclic hexapeptide

Pro-Phe-D-Trp-Lys-Thr-Phe,

which is equipotent to somatostatin in the inhibition of growth hormone, insulin, and glucagon release (Veber et al., 1981).

Cyclic Constraints

Conformational features in peptides that are important for receptor interaction often can be stabilized by introducing cyclic constraints upon the peptide ligands. One excellent example is in the development of α-MSH superagonists. A proposed β turn around Phe[7] was stabilized by the replacement of Met[4] and Gly[10] with cysteine residues and cyclization to afford the cyclic

$[Cys^4,Cys^{10}]\alpha$-MSH,

an analogue with much greater potency than the native hormone, which also exhibits prolonged activity (Sawyer et al., 1982). In a similar manner, substitution of the Met[5] and Gly[2] positions of enkephalins with D-Pen, followed by cyclization, led to the highly potent and δ-opioid selective cyclic peptide

$[D-Pen^2,D-Pen^5]$enkephalin

(Mosberg et al., 1983).

Side-chain cyclizations not involving disulfides generally are not found in nature, but they can be important tools in structure-function studies with synthetic peptides. There are many possibilities for side-chain–side-chain cyclizations, including, but not limited to, amides (lactams), esters (lactones), ethers, ketones, and so on. Many possibilities have already been examined. Perhaps the most common cyclization, due to its convenience and stability, is the amide bond formed between free carboxylate and free amine groups found on some amino acid residue side chains. This type of cyclization has been used quite successfully in a series of oxytocin analogues. The weak monocyclic agonist

$$[\overline{Mpa^1,Glu^5,Cys^6}, Lys^8]oxytocin$$

was converted to the highly potent bicyclic antagonist

$$[Mpa^1,\overline{Glu^5,Cys^6},Lys^8]oxytocin$$

by amide bond formation between Glu^5 and Lys^8 (Hill et al., 1989). Not only did this cyclization cause antagonist activity, but it also increased the binding to the receptor over 1000-fold. Cyclizations of a similar manner were also studied in CCK-8. The monocyclic lactam analogue

$$Boc-\overline{Asp-Tyr(SO_3H)-Nle-D-Lys}-Trp-Nle-Asp-Phe-NH_2$$

exhibited a significant decrease in potency at the CCK-A (peripheral) receptor (Charpentier et al., 1987). A monocyclic bis-lactam analogue of CCK-8 was prepared by replacement of $Met^{28,31}$ residues with lysine and cyclizing the N^ε amino groups with succinic acid. This procedure resulted in the highly CCK-B (brain) selective analogue (Rodriguez et al., 1990). Based on computer modeling, α-MSH analogues were prepared incorporating amide cyclizations (Al-Obeidi et al., 1989a,b). This led to the synthesis of the highly potent

$$Ac-[Nle^4, \overline{Asp^5,D-Phe^7,Lys^{10}}]\alpha-MSH-(4-10)-NH_2$$

analogue. Amide cyclizations have also been studied in the glucagon

(Lin et al., 1992), enkephalin (Schiller et al., 1985), and GnRH systems (Struthers et al., 1984).

Other cyclizations have been reported, including biphenyl type cyclizations used in an approach to the synthesis of the antitumor agent bovardin (Bates et al., 1984) and diazobridges in enkephalin analogues (Siemeon et al., 1980).

Peptidomimetics

More recently, an increasing emphasis of structure-function studies has been directed toward the synthesis and use of ligands containing pseudopeptides and peptidomimetics (Hruby et al., 1990). While it is at present possible to stabilize certain conformational features in peptides, the use of peptidomimetics as rigid templates for these features should provide even better analogues in the future. There is also interest in peptides containing unusual amino acids, such as β-methylphenylalanine, which allow peptides to maintain their conformation but change the topology of the ligand. The use of synthesis and the use of these types of amino acids should continue to provide greater insight to many ligand-receptor systems in the future. Additional discussion on the use of peptidomimetics can be found in Chapter 2.

PEPTIDE-BASED VACCINES

The eradication of most life-threatening bacterial diseases from the western world has magnified interest in the prevention of infection by more untreatable pathogens, such as viruses, using vaccines. Vaccines often consist of killed or genetically mutated whole pathogens, which are injected into the patient to induce immunity against the active pathogen. With the accumulation of information on the location and antigenicity of continuous epitopes in the constituent proteins of these pathogens and their toxins, considerable interest has arisen in the induction of cross-selective immunity from short synthetic peptide vaccines (Lerner, 1982).

The use of synthetic peptide antigens as vaccines has several potential advantages over whole viral or bacterial preparations. In the inactivation of whole organism vaccines, problems can arise when the antigenic proteins are too denatured to elicit adequate cross-selective antibodies for the active pathogen. Alternatively, inefficient inactivation can produce vaccines contaminated with live pathogens, which may promote infection. When using live attenuated nonpathological viruses, genetic reversion to the pathological state is a potential hazard. The economic cost and dangers of growing the pathogen in a suitable medium prior to inactivation can also be daunting. A more serious difficulty with whole organism vaccines is antigenic variability. Most of the present-day life-threatening or debilitating diseases employ a survival strategy of frequently altering their antigenic regions by genetic or phenotypic mechanisms (Birkbeck and Penn, 1986). Thus, new vaccines to the same pathogen must be continuously developed.

Synthetic peptide vaccines overcome these problems because they are nontoxic and relatively inexpensive. They also provide two possible methods for dealing with antigenic variation. First, since linear synthetic epitopes are fairly easily prepared, mutations, once located, can be readily incorporated into the synthesis. Second, and perhaps of greater importance, invariant antigenic sequences of the pathogen can be utilized to promote longer-lasting immunity.

Examples of Synthetic Peptide Vaccines

Investigation of synthetic peptide vaccines has largely centered on antiviral agents, such as for foot and mouth disease virus, hepatitis B, influenza virus, polio virus, and, of enormous recent impact, human immunodeficiency virus (HIV). For the sake of brevity, only two of these examples are described, which, however, highlight the major considerations in peptide-based vaccine design. It should be noted that antibacterial peptide vaccines are also of current interest, such as for diphtheria and cholera toxins (Audibert et al., 1981; Jacobs et al., 1983), as well as antiparasitic immunogens for prevention of malaria (Richman et al., 1989).

Foot and Mouth Disease Virus

Foot and mouth disease virus (FMDV) terminally afflicts domestic livestock and has had devastating economic effects on the agricultural industry. FMDV consists of four protein subunits, VP1–4, of which only VP1 showed any immunogenic activity when the individual proteins were tested (Brown, 1985). However, the antigenicity of VP1 was noted to be considerably reduced relative to the whole intact virus, which suggests that the other subunits are required for the correct folding of VP1. Pfaff and coworkers (1982) predicted that residues 144 to 159 of VP1 adopted a highly antigenic amphipathic helix conformation, and they proved their hypothesis by raising cross-selective antibodies to this sequence, which neutralized the whole virus. Other studies by Strohmaier et al. (1982) and Bittle et al. (1982) indicated that the most effective immunogen was the slightly longer sequence 141 to 160. It is intriguing that this segment is actually more antigenic than whole protein VP1. However, it is still two orders of magnitude less efficient as a vaccine than the intact inactivated virus, despite inoculation when conjugated to a variety of carrier proteins (Brown, 1989).

The diversity of antigenic variability within a species is well illustrated by FMDV. The virus has seven different serotypes throughout the world, and vaccination with one inactivated serotype isolate does not protect against the others. Also, a great deal of variation exists within each serotype such that intact virus vaccination does not necessarily prevent reinfection with the same serotype. However, despite fears that peptide-based vaccines may be highly strain selective, the segment 141 to 160 has been found to be cross-selective for several of the serotypes of FMDV (Brown, 1989). This occurrence is most probably the result of the relatively invariant sequence homology of residues 145 to 151 of VP1 across the various serotypes (Brown, 1989).

Hence, if invariant regions in the various subclasses of a pathogen can be located that are immunogenic, then the number of synthetic peptide vaccines required to protect against an antigenically diverse species can be reduced.

Human Immunodeficiency Virus

The spread of human immunodeficiency virus (HIV), the causative agent of acquired immunodeficiency syndrome (AIDS), is now reaching worldwide epidemic proportions and so has become a great concern in the search of neutralizing vaccines. HIV is a retrovirus that bears a reverse transcriptase, so that viral RNA can be encoded into the host cell's DNA. The viral particles are enclosed in a lipid bilayer envelope into which two glycoproteins are embedded. The major extracellular protein involved in recognition and binding to the T helper cell host is gp 120 (Allan et al. 1985), which is attached to the transmembrane protein gp 41 (Veronese et al., 1985). These proteins contain all the potential immunogenic sites on HIV (Barin et al., 1985; Kennedy et al., 1986).

The primary structures of gp 120 and gp 41 were found by sequencing of the *env* gene coding the unprocessed protein gp 160, which comprises both smaller proteins. Using the hydrophilicity scale profile of Hopp and Woods (1981), Kennedy and coworkers (1986) predicted a highly antigenic segment between residues 735 and 752 of gp 160 corresponding to a region in gp 41. This sequence was synthesized and coupled to keyhole limpet hemocyanin by means of a disulfide bridge, and it was shown to invoke neutralizing antibodies for HIV in rabbits (Chanh et al., 1986). Similarly a fragment of the C terminal of gp 120 (corresponding to residues 508 to 537 of gp 160) was located as a surface epitope and shown to induce an immune response in rabbits (Chanh et al., 1986). Since then, several recombinant vaccines derived from these glycoproteins have been proposed and patented (for example, Anderson et al., 1989; Putney et al., 1989).

Despite the relative speed with which antigens have been located on HIV, a sobering drawback in the prevention of infection with this and other retroviruses has recently been discovered. In studies on dengue (Halstead, 1979) and yellow fever virus (Barrett and Gould, 1986), it was observed that the antiviral antibodies actually enhanced the rate of infection. This activity also was shown for HIV in a recent study, where 3 out of 16 human monoclonal test antibodies were found to increase infection and proliferation of the virus in vitro (Robinson et al., 1990). This recurrence has been indicated to be caused by a complement-mediated antibody-dependent enhancement mechanism (Robinson et al., 1988). Depending on the prevalence of such "enhancing" antibodies, the protective effects of neutralizing antibodies can be completely overridden (Robinson, 1988).

This observation raises doubts about the use of recombinant whole protein vaccines based on gp 120, gp 41, or gp 160 since these proteins must contain a mixture of enhancing and neutralizing epitopes. Thus, one must be very cautious in selecting a neutralizing antigen for use as a vaccine. Carefully chosen synthetic peptide-based vaccines may, hence, prove be to the most effective weaponry in curbing the spread of the HIV epidemic.

Peptide-Carrier Conjugation

To elicit the maximum immunogenic response from synthetic antigens of molecular weight lower than about 1500, it is generally necessary to couple the peptide to a carrier protein. The reasons for this procedure have, until recently, been unclear (see the following discussion).

Several choices of carrier protein now exist, the most popular of which are bovine serum albumin (Dayhoff, 1976), tetanus toxoid (Bizzini et al., 1970), ovalbumin (Nizbet et al., 1981), and keyhole limpet hemocyanin (Malley et al., 1965). Antigenic peptides may be coupled to these proteins by a variety of chemical procedures (for a review, see van Regenmortel, 1988; also see Chapter 3). Coupling of the carboxyl terminus of the antigen to the carrier protein can occur primarily at α- and ε-NH_2 groups using glutaraldehyde (Habeeb et al., 1968), carbodiimide (Goodfriend et al., 1964), or various active ester procedures (Means and Feeney, 1971). Secondary acylation sites on the carrier are the Tyr OH or His imidazole side-chain groups. Alternatively, if cysteine is added to the antigenic sequence, coupling may be achieved by means of a disulfide bridge linkage (Gilliland, et al. 1980) or with a heterobifunctional reagent (see Table 6 and Chapter 3).

The carrier and coupling procedure must be chosen so as to maximize linkage to the protein without affecting its solubility or significantly altering the structure/conformation of the coupled antigen. Also, since most synthetic antigens are coupled in their fully deprotected state, the conditions must be carefully controlled to minimize intramolecular or polymerization reactions of the peptide.

Carrier-Free Peptide Vaccines

One of the obstacles in developing and administering peptide-based vaccines is the necessity of a carrier protein. In conjugating the peptide to the carrier, it is unclear what conformational or topographical constraints are placed on the antigen or whether the peptide is uniformly coupled. Also, the protein carrier can induce hypersensitive, allergenic side reactions in the patient after repeated inoculations. These difficulties can lead to irreproducibility in the immune response (Kobbs-Conrad, 1991). To circumvent these problems, two new protocols in vaccine design are now emerging that are based entirely on synthetic peptides.

The Multiple Antigen Peptide System

The multiple antigenic peptide (MAP) concept was recently introduced by Tam (1989). This protocol avoids the chemically undefined entity of an antigen-carrier system by utilizing a small peptidyl core consisting of seven radially branching lysyl residues covalently connected to a conventional solid support (Figure 7a). Antigen sequences can then be built onto the octa-branching core using standard solid-phase chemistry. Cleavage from the resin then produces the MAP, bearing eight copies of the antigenic sequence, on an inner core that generally accounts for less than 10 percent of the total molecular weight (Figure 7b).

The dense packing of so many copies of a highly antigenic epitope has been shown to produce a strong immunogenic response. Tam (1989) tested 14 different MAPs in which the antigenic sequences varied from 7 to 21 residues in length. All produced high titers of antibodies in mice and rabbits. Eleven of the 14 MAP examples also induced polyclonal antibodies that were cross-selective for the native protein. Notably, none of these antisera showed any affinity for the hepta-lysyl core (Tam, 1989). The MAP system has not yet been proved to be an effective vaccination vehicle, although it does offer exciting possibilities for the future.

The Importance of T Cell Epitopes to Increased Immunogenic Response—Multivalent B and T Cell Epitope Vaccines

As discussed in a previous section, T helper cells recognize and bind to processed epitopes, by means of the mediation of antigen presenting cells (Unanue et al., 1987). These processed epitopes are not necessarily the same as those that promote and bind to antibodies. Antigens that induce B cells to produce antibodies are labeled B cell epitopes. In the presence of T cell antigens, one of the functions of T helper cells is to stimulate B cells into boosting their antibody production. Although B cells can and do respond to foreign antigens directly, the response is often minimal without T cell intervention. Hence, to elicit full immunogenic activity, current thinking suggests that a vaccine should consist of B *and* T cell epitopes to be most effective. This theory was first supported by Francis and coworkers (1987), who noted that the carrier-free FMDV vaccine derived from residues 141 to 160 of VP1 did not protectively immunize

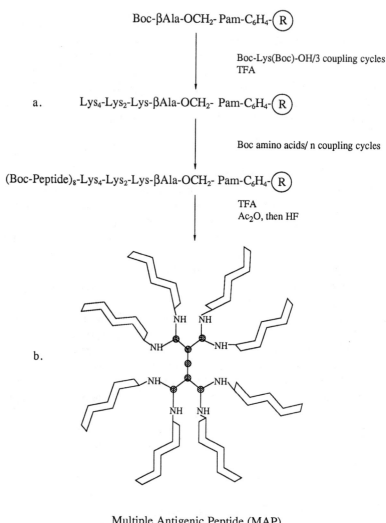

Boc-βAla-OCH$_2$- Pam-C$_6$H$_4$-(R)

Boc-Lys(Boc)-OH/3 coupling cycles
TFA

a.   Lys$_4$-Lys$_2$-Lys-βAla-OCH$_2$- Pam-C$_6$H$_4$-(R)

Boc amino acids/ n coupling cycles

(Boc-Peptide)$_8$-Lys$_4$-Lys$_2$-Lys-βAla-OCH$_2$- Pam-C$_6$H$_4$-(R)

TFA
Ac$_2$O, then HF

b.

Multiple Antigenic Peptide (MAP)

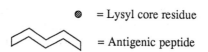

        = Lysyl core residue

        = Antigenic peptide

FIGURE 7   (a) Stepwise solid-phase synthesis of a MAP. (b) The final oc-
tabranching MAP core, with eight copies of the antigenic peptide.

the H-2^d strain of mice. By coupling this B cell antigen to known T cell
epitopes of ovalbumin and sperm whale myoglobin, high levels of cross-
selective antibodies were invoked, which neutralized FMDV in subse-

quent challenge experiments. What this example also graphically demonstrated was the reason for the use of carrier proteins in the majority of past immunogenic studies. Coupling of B cell epitopes to small carrier proteins includes the T cell epitopes of the carrier into the vaccines, and hence it serves to boost an otherwise limited immune response. This observation has led to the search for potent T cell antigens on some of the more effective carriers, such as tetanus toxoid (Ho et al., 1990), with a view to incorporating the epitope fragment into future vaccines rather than the whole proteins.

The synthetic potential of multivalent B and T cell epitope vaccines was recently extended to allow more flexibility in the choice of antigens incorporated. Utilizing the template-assembled synthetic protein (TASP) engineering techniques of Mutter (1988), Kobbs-Conrad and coworkers (1991) designed a totally synthetic vaccine consisting of single or multiple copies of B and T cell epitopes built onto a β-sheet template peptide (Figure 8). They observed high titers of antibodies in response to this template-assembled vaccine bearing B cell epitopes of LDH-C$_4$ and T cell antigens of tetanus toxoid. Following this methodology, any combination of B and T cell epitopes could, in theory, be assembled on a small peptide template. In principle, this method is similar to the MAP system of Tam (1989), although it is interesting that in all the examples he considered, only B cell epitopes were built onto the hepta-lysyl core. The large immunogenic response observed for the MAP protocol is presumably due to the density (or effective concentration) of antigen(s) on the vaccine, which may be sufficient to stimulate significant antibody production. One could, of course, use an orthogonal protection procedure to build MAP systems containing a mixture of B and T cell epitopes to produce greater immunogenicity, if required.

Both methodologies provide significant steps forward in the design of synthetic peptide-based vaccines. Such vaccines may now be prepared entirely by efficient solid-phase procedures, and thus structure-immunogen correlation of antigenic sequences may be more easily assessed.

ANTISENSE PEPTIDES

What? . . . How? . . . Why? . . . Why not? These are the types of questions persons with inquisitive minds, in quest of an understanding of the laws of nature, have always asked. A question of similar nature, seeking the structural or functional relationship between two peptides coded individually by each of the two opposing strands of a double-stranded DNA stretch, is leading to the development of a new area referred to as antisense peptides. The term *antisense peptide* or *complementary peptide,* therefore, refers to a peptide sequence that could be *hypothetically* deduced from the nucleotide sequence that is complementary to the nucleotide sequence coding for a naturally occurring peptide, peptide hormone,

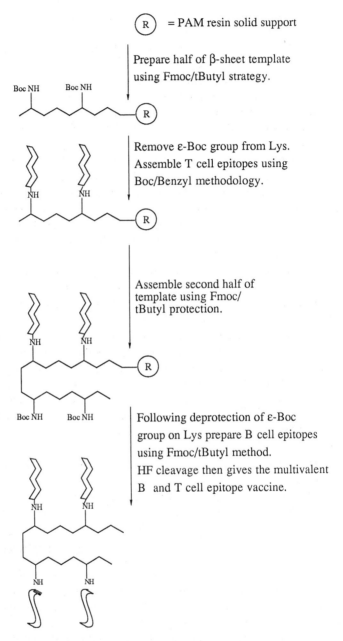

R = PAM resin solid support

Prepare half of β-sheet template using Fmoc/tButyl strategy.

Remove ε-Boc group from Lys. Assemble T cell epitopes using Boc/Benzyl methodology.

Assemble second half of template using Fmoc/tButyl protection.

Following deprotection of ε-Boc group on Lys prepare B cell epitopes using Fmoc/tButyl method.
HF cleavage then gives the multivalent B and T cell epitope vaccine.

FIGURE 8   Construction of multivalent B and T cell epitope vaccines on a peptide template.

neurotransmitter, or the *sense peptide*. Although not studied extensively, antisense peptides, in general, have shown some selectivity in binding with their sense counterparts. In certain cases, the antisense peptides have even exhibited surprising biological activity profiles, mediated through the biological receptors for the sense peptide.

It appears that, originally, this concept was proposed by Mekler (1969). Jones (1972) independently proposed and tested this concept by synthesizing peptides that were antisense to the C-terminal tetra-peptide of gastrins. Because the gene for this tetrapeptide was not known at that time, these authors had to synthesize various permuta-tions and combinations of the possible antisense peptide sequences. It was observed that one of these peptides exhibited inhibitory activity in a gastric acid release assay. A theoretical treatment to this concept was later proposed by Biro (1981a, 1981b, 1981c). A thorough inves-tigation into this concept was initiated by Blalock and coworkers (Bost et al., 1985) in a series of studies aimed at investigating the interaction between the antisense and the sense peptide in solid-phase binding assays. Later on, Chaiken and coworkers (Shai et al., 1987, 1989) developed high-performance affinity chromatography (HPAC) techniques to investigate the binding between antisense and sense peptides. These two techniques, discussed in the next section, remain as the only methods to study the interactions between a pair of sense and antisense peptides. This section is aimed at providing an overview of this concept and highlighting the potential applications of this con-cept in biomedical research as well as various issues concerning the nature of the binding exhibited by a set of antisense-sense peptides.

Antisense Peptides as Internal Images of the Sense Peptides

It was observed by Blalock and coworkers (Blalock and Smith, 1984; Blalock and Bost, 1986) that, in general, the codons for hydrophobic and hydrophilic amino acids on one strand of DNA may be complemented on the complementary strand by the codons for hydrophilic and hydropho-bic amino acids, respectively. Uncharged (slightly hydrophilic) amino acids, on the other hand, are complemented by uncharged amino acids (Table 4). Interestingly, either $5' \rightarrow 3'$ or $3' \rightarrow 5'$ reading of the comple-mentary nucleotide sequence displays the same pattern in hydropathicity of the coded amino acid sequence with respect to the same sense peptide sequence. This similarity in the hydropathic profile of the sequences, obtained in either way, stems from the fact that the middle base of the triplet codon specifies the hydropathic nature of the coded amino acid, and this base remains second in both $5' \rightarrow 3'$ and $3' \rightarrow 5'$ readings. This remarkable pattern in genetic code, termed *anticomplementarity,* appears to form the central theme of the concept of antisense peptides. Table 5 depicts the peptide systems in which this concept has been applied and

Table 4   List of Amino Acids, Their Codons, Corresponding Complementary Codons, and Antisense Amino Acids with Their Respective Hydropathic Scores, as Given by Kyte and Doolittle (1982)

Amino Acid	Hydropathic Value	Codon [5'→3']	Complementary Codon [3'→5']	Amino Acid Encoded by the Complementary Codon			
				[5'→3'] Direction	Hydropathic Score	[3'→5'] Direction	Hydropathic Score
Arg	−4.5	AGA	UCU	Ser	−0.9	Ser	−0.9
		AGG	UCC	Pro	−1.6	Ser	−0.9
		CGA	GCU	Ser	−0.9	Ala	+1.8
		CGC	GCG	Ala	+1.8	Ala	+1.8
		CGG	GCC	Pro	−1.6	Ala	+1.8
		CGU	GCA	Thr	−0.7	Ala	+1.8
Lys	−3.9	AAA	UUU	Phe	+2.7	Phe	+2.7
		AAG	UUC	Leu	+3.7	Phe	+2.7
Asp	−3.5	GAC	CUG	Val	+4.2	Leu	+3.7
		GAU	CUA	Ile	+4.5	Leu	+3.7
Glu	−3.5	GAA	CUU	Phe	+2.7	Leu	+3.7
		GAG	CUC	Leu	+3.7	Leu	+3.7

Asn	−3.5	AAC	UUG	Val	+4.2	Leu	+3.7
		AAU	UUA	Ile	+4.5	Leu	+3.7
Gln	−3.5	CAA	GUU	Leu	+3.7	Val	+4.2
		CAG	GUC	Leu	+3.7	Val	+4.2
His	−3.2	CAC	GUG	Val	+4.2	Val	+4.2
		CAU	GUA	Met	+1.9	Val	+4.2
Pro	−1.6	CCA	GGU	Trp	−0.9	Gly	−0.4
		CCC	GGG	Gly	−0.4	Gly	−0.4
		CCG	GGC	Arg	−4.5	Gly	−0.4
		CCU	GGA	Arg	−4.5	Gly	−0.4
Tyr	−1.3	UAC	AUG	Val	+4.2	Met	+1.9
		UAU	AUA	Ile	+4.5	Ile	+4.5
Ser	−0.9	UCA	AGU	Stop	—	Ser	−0.9
		UCC	AGG	Gly	−0.4	Arg	−4.5
		UCG	AGC	Arg	−4.5	Ser	−0.9
		UCU	AGA	Arg	−4.5	Arg	−4.5
		AGC	UCG	Ala	+1.8	Ser	−0.9
		AGU	UCA	Thr	−0.7	Ser	−0.9

Table 4 (continued)

Amino Acid	Hydropathic Value	Codon [5'→3']	Complementary Codon [3'→5']	Amino Acid Encoded by the Complementary Codon			
				[5'→3'] Direction	Hydropathic Score	[3'→5'] Direction	Hydropathic Score
Trp	-0.9	UGG	ACC	Pro	-1.6	Thr	-0.7
Thr	-0.7	ACA	UGU	Cys	+2.5	Cys	+2.5
		ACC	UGG	Gly	-0.4	Trp	-0.9
		ACG	UGC	Arg	-4.5	Cys	+2.5
		ACU	UGA	Ser	-0.9	Stop	—
Gly	-0.4	GGA	CCU	Ser	-0.9	Pro	-1.6
		GGC	CCG	Ala	+1.8	Pro	-1.6
		GGG	CCC	Pro	-1.6	Pro	-1.6
		GGU	CCA	Thr	-0.7	Pro	-1.6
Ala	+1.8	GCA	CGU	Cys	+2.5	Arg	-4.5
		GCC	CGG	Gly	-0.4	Arg	-4.5
		GCG	CGC	Arg	-4.5	Arg	-4.5
		GCU	CGA	Ser	-0.9	Arg	-4.5
Met	+1.9	AUG	UAC	His	-3.2	Tyr	-1.3

Cys	+2.5	UGC	ACG	Ala	+1.8	Thr	-0.7
		UGU	ACA	Thr	-0.7	Thr	-0.7
Phe	+2.7	UUC	AAG	Glu	-3.5	Lys	-3.9
		UUU	AAA	Lys	-3.9	Lys	-3.9
Leu	+3.7	CUA	GAU	Stop	—	Asp	-3.5
		CUC	GAG	Glu	-3.5	Glu	-3.5
		CUG	GAC	Gln	-3.5	Asp	-3.5
		CUU	GAA	Lys	-3.9	Glu	-3.5
		UUA	AAU	Stop	—	Asn	-3.5
		UUG	AAC	Gln	-3.5	Asn	-3.5
Val	+4.2	GUA	CAU	Tyr	-1.3	His	-3.2
		GUC	CAG	Asp	-3.5	Gln	-3.5
		GUG	CAC	His	-3.2	His	-3.2
		GUU	CAA	Asn	-3.5	Gln	-3.5
Ile	+4.5	AUA	UAU	Tyr	-1.3	Tyr	-1.3
		AUC	UAG	Asp	-3.5	Stop	—
		AUU	UAA	Asn	-3.5	Stop	—

Table 5 Binding Constants for Various Antisense-Sense Peptide Interactions

Sense Peptide	Antisense Peptide	Reading Direction	Dissociation Constant (M)	Assay	Reference
SYSMEHFRWGKPVGKKRRPVKVYP (ACTH)	GVHLHRAPLLAHRLAPAEVFHGVR (HTCA)	5'→3'	0.3×10^{-9}	SPBA[a]	Blalock et al., 1986
			1.9×10^{-6}	SPBA	Bost et al., 1985
	RMRYLVKATPFGHPFFAAGHFHMG	3'→5'	0.3×10^{-9}	SPBA	Blalock et al., 1986
YGGFMTSEKSQTPLVTL (γ-Endorphin)	QRDKGRLALLGGHEPAV	5'→3'	2×10^{-5}	SPBA	Carr et al., 1986
QHwSYGLRPG (LHRH)	SRAQSIGPVL	5'→3'	$\sim 1 \times 10^{-4}$	Inhibition of a biological assay	Mulcahey et al., 1986; also see Bost and Blalock, 1989
RPKPQQFFGLM (Substance P)	AGFGVKKPNY	3'→5'	6×10^{-6}	SPBA	Bost and Blalock, 1989
KETAAAKFERQHMDSSTSAA (Bovine pancreatic RNase S-peptide)	RGGSRSVHVLPLKLSRRSFL	5'→3'	4.0×10^{-7}	ZE-HPAC[b]	Shai et al., 1987; Shai et al., 1989
	SRSVHVLPLKLSRRSFL		3.5×10^{-6}		
	RGGSRSVHVLPLK		1.8×10^{-5}		
	VHVLPLKLSRRSFL		8.3×10^{-6}		
	PLKLSRRSFL		2.3×10^{-5}		
	RSVHVLPLK		$>5.9 \times 10^{-5}$		
	VHVLPLKLSRRSFL		1.2×10^{-5}		

Sequence	Description	K_d (M)	Method	Reference
LFSRRSLKLPLVHVSRSGGR	N→C terminal inversion of the above	8.0×10^{-7}		
SRRSLKLPLVHVSRSGGR		1.0×10^{-6}		
RSLKLPLVHVSRSGGR		6.5×10^{-6}		
KLPLVHVSRSGGR		2.3×10^{-5}		
LVHVSRSGGR		$>5.9 \times 10^{-5}$		
KESRRSLKLPLVHVSRSGGR	Further substitution of underlined amino acids	1.5×10^{-6}		
EKSRRSLKLPLVHVSRSGGR		1.5×10^{-6}		
EESRRSLKLPLVHVSRSGGR		3.7×10^{-6}		
LFPRRLKLPLVPVRSGGR		1.2×10^{-5}		
AAQHGGGLGVQVE	$5' \to 3'$ B.G.S.[c]	920×10^{-6}	ZE-HPAC	Omichinski et al., 1989
AAQHGRGFGIQIE	$5' \to 3'$ B.G.S.R[d]	460×10^{-6}		
TAKDSGSLSVKVE	Computer optimized sequences	460×10^{-6}		
AGKDSGSFSVKVE		160×10^{-6}		
PAKDSGSFSVKVK		59×10^{-6}		
TAKDSWSFSVKVK		22×10^{-6}		
AAKDSRGLGIEIE		40×10^{-6}		
AGKYSWRLRIKIK		0.7×10^{-6}		

FNLDAEAFAVLSG
(Rat liver glycoprotein fragment)

Peptide	Sequence	Direction	K_d	Method	Reference
GSGSFGTVYKGKWHGDVAVK (c-RAF$_{356-375}$ sequence)	LYSNISMPLALVHSAKGARA SAKGARA	$5' \rightarrow 3'$	380×10^{-6} 1500×10^{-6}	ZE-HPAC	Fassina et al., 1989a
	All D-isomers of the above: lysnismplalvhsakgara sakgara		380×10^{-6} 1500×10^{-6}		
	FHGHISMPFAFVYSSKAAAA	Computer optimized sequence	8×10^{-6}		
CYFQNCPRGGKR (Arg^8-Vasopressin-Gly-Lys-Arg)	SQLQVGHGPLAAPWAVLEVA	$5' \rightarrow 3'$	3.5×10^{-3} 5.9×10^{-6} [e]	ZE-HPAC	Fassina et al., 1989b
	PLAAPWAVLEVA		7.3×10^{-3}		
	SQLQVGHG		$>30 \times 10^{-3}$		

[a] SPBA = solid-phase binding assay.

[b] ZE-HPAC = zonal elution — high-performance affinity chromatography in which sense peptide has been immobilized on the column matrix.

[c] B.G.S. = best guess sequence.

[d] E.G.S.R. = best guess sequence for rabbit.

[e] Binding constant for soluble sense and immobilized antisense peptide.

the interactions have been studied either by a solid-phase binding assay or HPAC. As is evident from these results, a remarkable binding affinity (in general, from 10^{-3}M to 10^{-6}) exists between a sense peptide and an antisense peptide. The ability of an antisense peptide to recognize its sense counterpart and their anticomplementarity has given rise to the suggestion that an antisense peptide may represent an "internal image" of its sense counterpart.

Blalock and coworkers further tested this suggestion by raising antibodies against HTCA, an antisense peptide for adrenocorticotropin hormone (ACTH), and testing their ability to bind the receptors for ACTH (Bost and Blalock, 1986). Indeed, it was found that these antibodies, being an image of an image, behaved like ACTH in the receptor binding assays and were successfully used to purify an ACTH binding protein or receptor. This concept of receptor purification has since been applied successfully to the purification of binding proteins for ACTH (Bost and Blalock, 1986), fibronectin (Brentani et al., 1988), vasopressin receptor (Abood et al., 1989), γ endorphin (Carr et al., 1986), luteinizing hormone releasing hormone, LHRH (Mulchahey et al., 1986), and angiotensin-II (Elton et al., 1988a, 1988b). It has also been shown that antibodies against a pair of complementary peptides exhibit idiotypic–anti-idiotypic relationships (Smith et al., 1987). The antibodies against synthetic ACTH(1-13) bound the antibodies against "HTCA," the peptide antisense to ACTH.

The selective interactions discussed previously gave rise to a suggestion that the DNA strand that is complementary to the strand coding a peptide hormone might be coding for its receptor. Sequence search analysis performed by Bost et al. (1985) revealed that regions of complementarity exist between mRNAs for epidermal growth factor (EGF), transferrin (TF), interleukin-2 (IL-2), and their receptors. These complementary sequences, however, were small (six amino acids) and were obtained upon $3' {\rightarrow} 5'$ reading of the nucleotide sequence, which is not the usual transcriptional or translational direction during protein synthesis. A similar six amino acid homology has been found (Weigent et al., 1986) in a region of human interleukin-2 and coat protein of human T-lymphotropic retrovirus (HTLV-III). Ghiso and coworkers (1990) have reported another instance where such a homology between two naturally occurring protein sequences has been observed. It has been found that a peptide, Ser-Tyr-Asp-Leu, which would be antisense to Gln-Ile-Val-Ala-Gly (residues 55 to 59) in cystatin-C (an inhibitor of cysteine protease), is located at positions 611 to 614 of β-chain human C4, the fourth component of complement. These two fragments have been found to exhibit binding to each other as studied by enzyme-linked immunosorbent assay (ELISA) and affinity chromatography. The occurrence of such small complementary fragments in nature appear not to be common. For example, the two complementary hexapeptide stretches found in EGF and its receptor had no other match among the known sequences of 3060 proteins (Bost et al., 1985). Similarly, the comparison of the nucleotide sequences of human insulin mRNA with insulin receptor mRNA revealed the presence of 16 complementary nucleotide sequences that are of 23 to 26 base lengths (Segerstéen et al., 1986). However, a comparison of an insulin mRNA sequence with mRNA sequences of ten different proteins, such as corticotropin, progrowth hormone, endorphin, albumin, presomatotropin, thyrotropin, and so on, revealed as many as 81 short sequences that were complementary to human insulin mRNA. Further, it has been argued that if an antisense DNA strand codes for the receptor for the ligand, there should be a deficiency of stop codons in the antisense strand (Goldstein and Brutlag, 1989). The anticodons for two leucine codons and one codon for serine would be stop codons, thereby terminating the transcription of the antisense strand. Interestingly, antisense strands of bovine pro-opiomelanocortin was found to have an open reading frame for its full extent without stop codons. These authors, however, have cautioned that this case might be just unique. Needless to say, a more exhaustive study is required to shed light on this aspect of antisense-sense peptides.

Only one case where an antisense mRNA is produced and blocks the translation of its complementary mRNA is known (Mizuno et al., 1984). The gene for an osmoregulatory membrane protein, OmpC, in *Escherichia coli* was found to be transcribed under conditions of high osmolarity. A large portion of its 174 base transcript is complementary to the 5′ end region of OmpF mRNA, which codes for another osmoregulatory protein. The interaction of these complementary mRNAs has been suggested to be the cause for premature termination of the transcription of OmpF gene.

There are instances where no measurable interactions between antisense-sense peptide pairs have been observed (Rasmussen and Hesch, 1987; Eberle et al., 1989; and Eberle and Huber, 1991). For example, Eberle and coworkers also failed to raise antibodies against a peptide antisense to human ACTH. Negative results on the studies with antisense angiotensin II peptide have also been reported (Guillemette et al., 1989; De Gasparo et al., 1989). Such suggestions as nonuniversality of this phenomenon, weak unmeasurable interactions, or difficulties in generating antiself antibodies by an animal (antisense peptide making antireceptor antibodies) have been made to explain these cases. It can, therefore, be concluded that, at present, the concept of antisense-sense peptides is in its incipient stage. There are important questions to answer. The biological relevance and the mechanistic aspects of the interactions exhibited by these peptide pairs remain as the main issues to be understood.

Mechanistic Basis for Selective Interactions between Antisense-Sense Peptide Pairs

The interactions observed between an antisense-sense peptide pair can be highly selective, allowing discrimination among two closely related peptides. For example, an immobilized peptide antisense to Arg^8-vasopressin-related peptide was able to efficiently distinguish between two closely related peptides,

Cys-Tyr-Ile-Gln-Asn-Cys-Pro-Leu-Gly-NH$_2$ (oxytocin)

and

Cys-Tyr-Phe-Gln-Asn-Cys-Pro-Arg-Gly-NH$_2$ (Arg^8-vasopressin)

in an affinity binding experiment (Fassina et al., 1989b). However, the structural and mechanistic basis of these interactions has remained an enigma. The lack of any apparent structural similarity between a pair of antisense-sense peptides, the preservation of binding ability of an antisense peptide after sequence variation, and substitution of D amino acids and even on inverting the sequence from N terminus to C terminus (Shai et al., 1987, 1989; Fassina et al., 1989a; Table 5) suggest that these interactions are degenerate in terms of both sequence and conformation.

It has been hypothesized that these interactions occur as a result of recognition of a pattern of hydrophobic and hydrophilic residues in an antisense-sense peptide pair (Bost and Blalock, 1989; Shai et al., 1987, 1989; Fassina et al., 1989a, 1989b). Further, shortening or simplification of the sequence of an antisense peptide leads to a gradual decrease in binding affinity, suggesting a multiplicity of contacts at multiple sites along the whole sequence of the antisense and sense peptides. In addition, the observation of Shai and coworkers (1987, 1989) that a stoichiometric ratio of greater than 1:1 existed in the interaction of antisense and sense RNase S peptide is also consistent with the suggestion of multiple contacts between two complementary peptides.

Recent studies by Chaiken and coworkers (Lu et al., 1991) have added yet another dimension to this already complicated and hard to explain scenario. They have purified a vasopressin receptor–Arg^8-vasopressin (AVP) complex from a rat liver membrane preparation by affinity chromatography. The immobilized ligand on the column matrix was a 20 amino acid peptide that is antisense to the N terminus of the common biosynthetic precursor to AVP and bovine neurophysin II. Based on these results it has been hypothesized that a three-way interaction of the antisense peptide–sense peptide–sense peptide receptor might exist as a result of multiple-point group interactions between a pair of sense and antisense peptides. In a receptor-sense peptide complex, a part of the sense peptide surface may still be available for establishing multipoint interactions with the antisense peptide.

Approaches for the Generation of Antisense Peptides

The generation of an antisense sequence is straightforward in cases where DNA sequence information on a sense peptide is available. A sequence of complementary bases is written opposite the nucleotide sequence of the sense peptide. The entire complementary or antisense base sequence is then read in both $5' \rightarrow 3'$ and $3' \rightarrow 5'$ directions and translated to the peptide sequence according to the genetic code. It has been observed that antisense peptides generated in both the natural $5' \rightarrow 3'$ reading of the nucleotide sequence frame as well as the opposite $3' \rightarrow 5'$ direction exhibit the ability to selectively bind the sense peptide. The possible reason for these interactions has been discussed elsewhere in this section. Table 4 describes the amino acids and their corresponding complementary or antisense amino acids when translated in both $5' \rightarrow 3'$ and $3' \rightarrow 5'$ directions.

As is evident from Table 4, in the absence of knowledge of the nucleotide sequence of a sense peptide, one would be required to test all the various combinations of antisense amino acid sequences resulting from the degeneracy of the genetic code. Nonetheless, few approaches to generate antisense peptides in such instances have been proposed. One such approach makes a series of educated guesses based on the use of preferred codon usage tables (Granthan et al., 1980; Aota et al., 1988; Hartfield and Rice, 1986; Hartfield et al., 1979), which allows one to assess the probability of a particular codon to be used for each amino acid for a given peptide. In certain instances, the frequency of encoding a particular antisense amino acid among a combination of antisense amino acids may also help in making a choice. For example, in $3' \rightarrow 5'$ reading, the arginine codon may give rise to six antisense codons, two of which code for serine; the remaining four code for alanine. Alanine may, therefore, be preferred in designating an amino acid antisense to arginine. Similarly, in the $5' \rightarrow 3'$ reading frame, arginine may serve as an antisense replacement for proline. The combined use of these approaches may help one to obtain a "best guess," or "consensus," antisense peptide sequence. Examples of this approach have been discussed in the generation of correct antisense peptides to substance P, read in the $3' \rightarrow 5'$ direction (Bost and Blalock, 1989), and design of a 13-mer peptide antisense to a rat glycoprotein sequence (Omichinski et al., 1989). As is evident from Table 4, certain anticodons, in either 5′→3′ or 3′→5′ direction, are stop codons. To overcome this problem in generating antisense peptides, Bost and Blalock (1989) have proposed to substitute this codon by an amino acid that is suggested by the best guess or consensus amino acid for that position. Similarly, Shai et al. (1989) replaced the half-cysteine residues in a deduced antisense peptide by serine residues.

Omichinski and coworkers (1989) and Fassina and coworkers (1989a) have successfully demonstrated another approach to the design of antisense peptides. Their approach stems from the observed hydropathy anticomplementarity in the sense-antisense amino acid pairs and consequent suggestion that a hydropathic complementarity may be responsible for the selective interactions between a pair of sense-antisense peptides (Blalock and Smith, 1984; Bost et al., 1985). As described in Table 4, hydrophobic and hydrophilic amino acids are, respectively, complemented by hydrophilic and hydrophobic amino acids. Uncharged (slightly hydrophilic) amino acids are complemented by uncharged amino acids. Using this approach, the consensus, or best-guess, sequence obtained earlier, or the actual antisense peptide sequence, if known, is subjected to further refinements in order to maximize the existing hydropathic complementarity. The amino acid substitutions are made in the antisense peptide to maximize a defined "moving average hydropathic score" for it with respect to the sense peptide. In one of the examples, these authors (Omichinski et al., 1989) demonstrated the use of this approach in refining a best-guess antisense sequence for a rat glycoprotein fragment (Table 5). One of the optimized sequences, AGKYSWRLRIKIK, exhibited 1310-fold higher binding affinity for the 13-mer sense peptide, FNLDAEAPAVLSG, than the previous best-guess sequence, AAQHGGGLGVQVE. In the other example, Fassina and coworkers (1989a) have shown that an optimized sequence antisense to the 356 to 375 peptide sequence of c-raf protein exhibited a 47-fold higher binding affinity to immobilized c-$raf_{356-375}$ peptide than the actual c-$raf_{356-375}$ antisense peptide (Table 5). Although not applied yet, these examples strongly suggest that this approach, in principle, may allow a de novo design of antisense peptides that are highly selective to a given sense peptide.

Assays for Studying the Interactions between an Antisense Peptide and Its Sense Counterpart

In general, two types of assays, solid-phase binding assay and high-performance affinity chromatography (HPAC), are being used to study antisense-sense peptide interactions. Both these assays are solid-phase assays designed to overcome the main difficulty that one would face in applying the standard techniques of receptor-ligand interactions, where the unbound ligand is removed by filtration, centrifugation, or selective adsorption on an inert insoluble additive, such as charcoal.

Solid-Phase Binding Assay

Well-established techniques of the enzyme-linked immunosorbent assay (ELISA) were used by Blalock and coworkers (Bost et al., 1985) to develop the binding assay for studying the interactions between a sense and an antisense peptide. Typically, multiwell microtiter plates are coated with varying amounts of an antisense peptide (10 to 50 µg), another peptide or protein serving as control, for 12 to 18 hours. The wells are then washed and incubated with a fixed concentration of the radiolabeled sense peptide for 1 to 2 hours. The plates are then washed with appropriate buffer to remove the unbound radiolabeled sense peptide, and the amount of bound radiolabeled sense peptide is measured. In case antibodies against the sense peptide are available, "cold" sense peptide may be substituted for radiolabeled sense peptide, and the amounts of the bound sense peptide may be estimated by the standard techniques of ELISA using antibodies. The use of this technique has been demonstrated by Blalock and coworkers (1985) in their study of the binding of ACTH, γ endorphin, LHRH, and substance P to their respective antisense peptides (Table 5).

High-Performance Affinity Chromatography (HPAC) Method

Chaiken and coworkers (Shai et al., 1987) have developed the methods of high-performance affinity chromatography (HPAC) as a direct, powerful tool to study the interactions between immobilized sense peptide and a mobile antisense peptide. The affinity matrix is prepared using the standard techniques and reacting the sense peptide with the activated matrix. Chaiken and coworkers (Shai et al., 1987, 1989) have employed as matrix Accell 78 (Waters Chromatography Division, Milford, Mass.), which is a silica-based support containing a six-carbon spacer with terminal N-hydroxysuccinimide active ester for immobilizing molecules through their amino functionalities. The matrix is reacted with the peptide, and then it is end-capped with aminoethanol treatment and packed in a glass column. An identical column containing blank end-capped matrix also is prepared. Alternatively, Sharma and coworkers (S.D. Sharma, F. Al-Obeidi, M.E. Hadley, and V.J. Hruby, unpublished data) have used a prepacked activated tresyl column (SelectiSphere 10, Pierce Chemical Company, Rockford, Illinois) for immobilizing [Nle[4]]-α-melanotropin (α-MSH), a tridecapeptide analogue of α-MSH. This was achieved by pumping a solution containing a determined amount of α-melanotropin through the column at pH 7.2. Spectrophotometric evaluations of the eluant indicated the immobilization of the peptide on the column. The column was subsequently end-capped with aminoethanol to block all the remaining reactive tresyl groups.

The affinity column thus prepared is then used to study the interactions of antisense peptide with the immobilized sense peptide according to the experimental and theoretical techniques defined by Chaiken and coworkers. Typically, the extent of retardation of different concentrations of a mobile antisense peptide, under binding conditions on this column, can then be used to calculate an equilibrium binding constant for the mobile antisense–immobile sense peptide complex according to Eq. 1.

$$ 1 / V - V_o = K_{M/P} / [M]_T (V_o - V_m) + [P] / [M]_T (V_o - V_m), \qquad \text{(Eq. 1)} $$

where V = the experimental elution volume of mobile antisense peptide, V_o = the unretarded elution volume, V_m = the mobile phase volume, $[M]_T$ = the total concentration of matrix-bound sense peptide, $[P]$ = the concentration of mobile antisense peptide in the chromatographed zone, and $K_{M/P}$ = the dissociation constant of the complex of mobile antisense peptide (P) and matrix-bound sense peptide (M).

Alternatively, a fixed concentration of antisense peptide can be made to compete on this column with different concentrations of soluble sense peptide. This procedure affords the binding constants for a mobile antisense–mobile sense peptide complex (Eq. 2). In both these methods, called *zonal elution methods,* the antisense peptide is applied to the column as a fixed zone of different concentrations.

$$ 1 / V - V_o = K_{M/P} / [M]_T (V_o - V_m) + $$
$$ K_{M/P}[L]_T / K_{L/P}[M]_T (V_o - V_m), \qquad \text{(Eq. 2)} $$

where $[L]_T$ = the total concentration of competing mobile sense peptide and $K_{L/P}$ = the dissociation constant between mobile antisense peptide and mobile competing sense peptide. The rest of the parameters are the same as in Eq. 1.

In yet another method, referred to as the *continuous elution method,* the solutions containing different concentrations of an antisense peptide are passed through the affinity column, and the plateau of maximum absorbance (UV) is observed. The volume of the eluant at which the affinity matrix is half saturated is used to calculate the binding constant of the mobile antisense–matrix-bound sense peptide complex, according to Eq. 3.

$$ 1 / \overline{V} - V_o = K_{M/P} / M_T + [1 / M_T] / [P]_o, \qquad \text{(Eq. 3)} $$

where $\overline{V}$ = the volume at which the affinity matrix is half saturated, V_o = the void volume, $K_{M/P}$ = the dissociation constant of the complex of the immobilized sense peptide and antisense peptide, M_T = the total amount of immobilized sense peptide, and $[P_o]$ = the initial concentration of mobile antisense peptide.

Biological Activity of Antisense Peptides

Although not studied extensively, the antisense peptides have been found to exert biological responses through the biological receptors for the sense peptides. Al-Obeidi and coworkers (1990) assayed a series of antisense peptides corresponding to the β-melanotropin (β-MSH) sequence for melanotropic (MSH-like) activity in frog skin bioassay and for melanin concentrating hormone (MCH-like) activity in frog skin and fish skin bioassays. Surprisingly, it was found that some of the antisense peptides exhibited either agonistic or antagonistic activity in these in vitro bioassays, suggesting the interaction of these peptides with MSH or MCH receptors.

The biological response exerted by an antisense peptide in intact animals has not been studied extensively. Jones (1972) has reported that after administration to conscious dogs of a peptide antisense to the C-terminal tetrapeptide of gastrins, some inhibition of gastric juice volume and pepsin output was observed. Bost and Blalock (1989) have also observed a partial inhibition of stress-induced steroid production after in vivo injection of a peptide antisense to ACTH. It could be argued that these inhibitions may result either by the antagonistic action of the antisense peptide on the sense peptide receptors or as a result of interaction between administered antisense peptide and endogenous sense peptide, thereby leaving low levels of sense peptide to elicit their normal biological responses.

Antibodies against antisense peptides may also exert biological responses when administered in whole animals. As shown by Görcs and coworkers (1986), intravenous administration of antiserum against a peptide antisense to human placental progonadotropin-releasing hormone, when injected in ovariectomized rats, caused a significant decrease in plasma levels of luteinizing hormone. Because the antibodies against an antisense peptide will behave as the sense peptide in recognition aspects (being an image of an image), this effect has been interpreted in terms of an interference of this antiserum with the actions of endogenous gonadotropin releasing hormone. Immunocytochemical staining of receptors for gonadotropin releasing hormone on female rat pituitary cells by this antiserum further supported this inference.

Potential Applications of Antisense Peptides

Because biomolecular recognition is the hallmark of antisense-sense peptide systems, the most important use of antisense peptides lies in the area of biomedical research. As demonstrated by various studies, described previously, the antisense peptide, or antibodies against it, may help in the purification of both endogenous ligands as well as biological receptors. Anti-antisense peptide antibodies, because of their relatively higher binding affinities, may score better than the sense peptide-based ligands and may, therefore, replace the sense peptides in various diagnostic assays. Highly selective antisense peptides that bind sense peptides with high affinity may, in turn, replace antibodies in various radioimmunoassays for the sense peptides. Likewise, the development of highly selective antisense peptides against tumor cell markers may also aid in diagnosis and therapeutic modalities of the disease state. Peptides that are antisense to the peptide sequences of coat protein of a virus or a retrovirus may also aid in the management of disorders caused by a virus or retrovirus by interfering with the virus-host interactions at the molecular level.

The in vivo administration of antisense peptides may aid in better modulation of biological responses exerted by their endogenous sense counterparts. An administered antisense peptide may act as a sink for an elevated level of an endogenous sense peptide, thereby managing a disease or disorder caused due to elevated levels of the sense peptide. Bost and Blalock (1989) have also made the suggestion that this phenomenon of antisense peptides may make it possible to develop the peptide equivalent of the nonpeptidic ligands for biological receptors. Undoubtedly, the success of such applications will depend on the unequivocal demonstration of a high degree of specificity of interactions between a pair of antisense and sense peptides. This will be essential in the light of observed degeneracy in sequence and conformational features for such interactions (Shai et al., 1989; and Fassina et al., 1989a, 1989b).

THE USE OF PEPTIDES FOR AFFINITY LABELING OF RECEPTORS

Establishment of some type of irreversible, but specific, interaction between a ligand and its biological acceptor, termed *affinity labeling,* is central in attempts to identify, characterize, and even isolate the receptor molecules for hormones and neurotransmitters. In the case of enzyme-substrate interactions, this approach helps to identify the chemical constitution (amino acid residues) of the binding and/or catalytic site(s). The most general approach for establishing irreversible binding between a ligand and its acceptor relies on the establishment of a covalent bond between the two entities. Two main methods for achieving this goal are chemical affinity labeling and photoaffinity labeling. While both techniques have found their applicability in numerous peptide-hormone receptor systems, photoaffinity labeling generally is preferred because of its nondependence on the proper juxtaposition of reactive groups as well

Table 6   Reagents for the Introduction of Maleimido Groups in Peptides

Reagent	Reference
	Yochitake et al., 1979
	Kitagawa et al., 1976
	Martin and Papahadjopoulos, 1982

as the presence or absence of certain specific chemical functionalities at the receptor binding site.

Another technique for affinity labeling of acceptor molecules involves the use of a class of homo- as well as hetero-bifunctional cross-linking reagents. These reagents are added to the incubation medium after the binding of a ligand with its acceptor has been established. This technique, therefore, also causes a covalent cross-linkage between the ligand and the acceptor. A number of these reagents, which utilize essentially the chemical affinity and photoaffinity approaches, have been developed and are available commercially (see Tables 6–9). The photoreactive and/or chemically reactive groups used in these bifunctional cross-linking reagents are, in principle, the same as those that are described in the following discussion on the development and use of chemical affinity and photoaffinity peptide ligands.

Chemical Affinity Labels

Chemical affinity labeling involves the establishment of a covalent bond between a ligand and its biological acceptor during their specific binding to each other. Usually, the ligand used in this operation is first modified to bear a chemically reactive group that is capable of reacting with any of the chemical functionalities that are usually on the acceptor molecule. Most importantly, the nature of this reactive group should be one that allows this reaction to take place under physiologically compatible conditions, that is, aqueous medium, ambient temperature, and pH 6–8 values. These requirements limit the type of chemical reactions one can efficiently perform under these conditions with functionalities, such as amino, carboxyl, thiol, imidazole, phenol, hydroxyl, etc., that are usually found on the acceptor molecule. Among these, the amino and thiol groups, because of their chemical reactivities under the prescribed conditions, are the best-suited targets for chemical affinity labeling. A variety of derivatives capable of reacting with thiol and amino groups are now known and can be efficiently incorporated in the ligand molecule. The following section provides a general discussion about these chemical affinity groups and their incorporation in the peptide-based ligands. The main point while designing these chemical affinity ligands, however, is the consideration of the site in the ligand molecule at which one can incorporate this chemically reactive group. Because specific biomolecular recognition and binding is of utmost importance for any labeling experiment, the site of derivatization must be one that causes little or no change in the recognition and the binding characteristics of the ligand.

Thiol-Reacting Derivatives

The use of affinity labels reactive toward thiol is rather limited, because the acceptor molecules, in general, lack free thiol groups. Most of the cysteine residues present in a biomolecule usually exist in their respective oxidative state, forming a disulfide linkage among a pair of cysteine residues. In spite of this, the chemistry for thiol-sensitive derivatives is well evolved and can be performed under mild biocompatible conditions. As a result, it is the most favored one for conjugating a biomolecule either with an insoluble matrix for use in affinity chromatography or with another biomolecule for use in diagnostic assays, immunization, antibody, and vaccine production. Two main strategies followed for reactions with a thiol group are S alkylation and disulfide bond formation.

S-Alkylating Reactions

Thiols undergo reaction with α-halo ketones and maleimide groups under physiological conditions. These ketones, particularly the iodo- and bromo-acetyl groups, have been used in many instances for alkylating an

Table 7   Heterobifunctional Reagents for the Synthesis of Photoaffinity Peptides

Reagent	Reference
Reactive for amino group (may also react with SH groups):	
N_3—⟨benzene⟩—COBr	Hixson and Hixson, 1975
N_3—⟨benzene⟩—CH_2-COO—⟨benzene⟩—NO_2	Eberle and de Graan, 1985
N_3—⟨benzene, NO_2⟩—CH_2-COO—⟨benzene⟩—NO_2	Eberle and de Graan, 1985
N_3—⟨benzene, NO_2⟩—F N_3—⟨benzene, F⟩—NO_2	Levy, 1973 Smith and Knowles, 1974 Fleet et al., 1969
N_3—⟨benzene⟩—N=C=S	Sigrist and Zahler, 1980
N_3—⟨benzene⟩—$\overset{\overset{+}{NH_2}\ \overset{-}{Cl}}{\underset{\parallel}{C}}$-$OCH_3$	Ji, 1977
N_3—⟨benzene⟩—CH_2—$\overset{\overset{+}{NH_2}\ \overset{-}{Cl}}{\underset{\parallel}{C}}$-$OCH_3$	Ji, 1977
N_3—⟨benzene⟩—COO-N⟨succinimide⟩	Yip et al., 1980 Galardy et al., 1974

Table 7   (continued)

Reagent	Reference
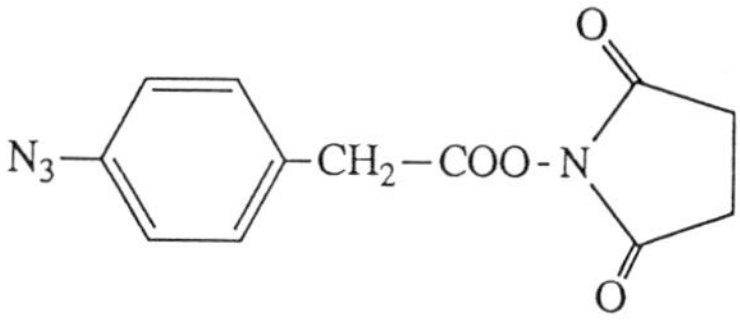	Eberle and de Graan, 1985

N_3—⟨benzene ring⟩—CH_2—COO—N(succinimidyl)

| 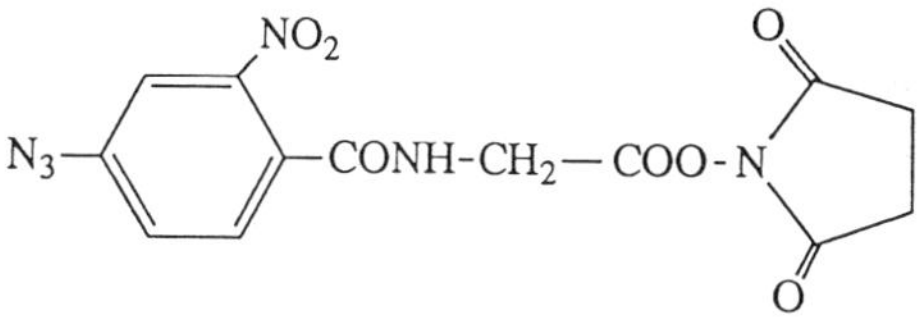 | Yip et al., 1980 |

N_3—⟨benzene ring, NO_2⟩—$CONH$–CH_2—COO–N(succinimidyl)

| 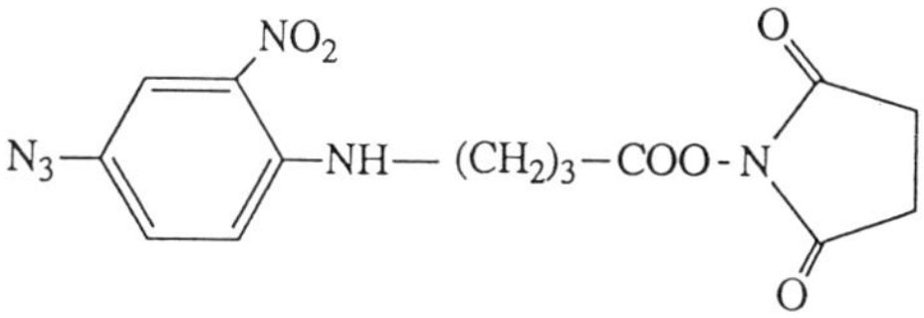 | Guire et al., 1977 |

N_3—⟨benzene ring, NO_2⟩—NH—$(CH_2)_3$—COO–N(succinimidyl)

| 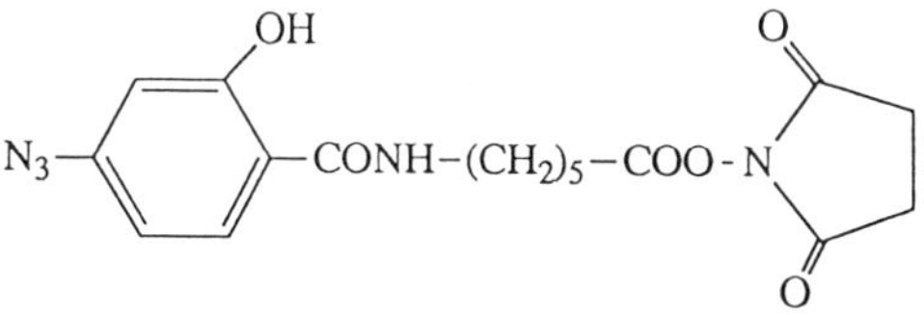 | Ji and Ji, 1982 |

N_3—⟨benzene ring, OH⟩—$CONH$–$(CH_2)_5$—COO–N(succinimidyl)

Reactive for carboxy group:

| 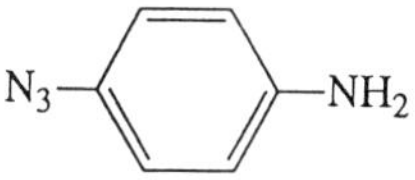 | Escher et al., 1979 |

N_3—⟨benzene ring⟩—NH_2

| 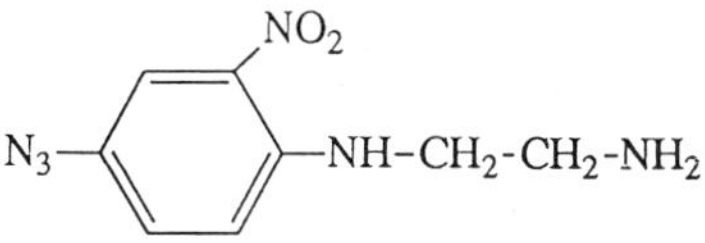 | Das and Fox, 1978 |

N_3—⟨benzene ring, NO_2⟩—NH–CH_2-CH_2-NH_2

| 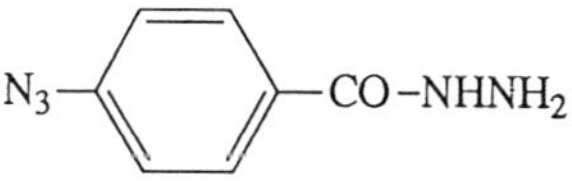 | Eberle and de Graan, 1985 |

N_3—⟨benzene ring⟩—CO–$NHNH_2$

Table 7   (continued)

Reagent	Reference

Reactive for arginine side-chain functionality:

Vanin et al., 1981
Ngo et al., 1981

Reactive for cysteine side-chain functionality:

Hixson and Hixson, 1975
Schwartz and Offengand, 1974

Seela and Rosemeyer, 1977

Budker et al., 1974

Trommer and Hendrick, 1973

Reactive for tryptophan side-chain functionality:

Muramoto and Ramachandran, 1980
Demoliou and Epand, 1980

Table 7 (continued)

Reagent	Reference
Reactive for tyrosine/histidine side-chain functionality	
N_3—⟨phenyl⟩—N_2^+ Cl^-	Escher et al., 1979
Carbene generating reagents reactive for amino groups:	
H_3C—⟨phenyl⟩—SO_2-$\overset{N_2}{\underset{\Vert}{C}}$-$COCl$	Chowdhry and Westheimer, 1978
$N_2{=}\overset{CF_3}{\underset{\vert}{C}}{—}COCl$	Chowdhry et al., 1976
$N_2{=}\overset{CF_3}{\underset{\vert}{C}}{—}COO$—⟨phenyl⟩—$NO_2$	Chowdhry et al., 1976
$\overset{CF_3}{\underset{}{}}$ diazirine—⟨phenyl⟩—$COOH$	Burgermeister et al., 1983 Nassal, 1983

SH group in a biomolecule. The order of reactivity among α-halo ketones is as follows: iodo >> bromo > chloro > fluoro. The reactivity of α-iodo as well as α-bromo haloketones toward -SH functionality is not entirely specific. While a variety of other functionalities, such as water, OH, NH_2, imidazole ring of histidine, and so on, react with α-iodo as well as α-bromo ketones, the reaction of α-chloroketone with thiols is selective, but slow, under physiologically compatible conditions. The α-fluoroketones are practically inert. Consequently, α-bromo ketones are

Table 8   Radioactive Heterobifunctional Reagents for Specific Derivatization of Amino Groups in the Peptides for Photoaffinity Labeling of Their Respective Receptors

Reagent	*Reference*
	Ji and Ji, 1982
	Ji and Ji, 1982
	Ji and Ji, 1982

* Radioactive iodine atoms may easily be incorporated in these molecules.

the most favored chemical affinity alkylating reagents. Bromoacetic acid can be coupled using the usual methods of peptide synthesis to either N-terminal or side-chain amino functions in a peptide ligand to afford this type of chemical label. For example, such derivatives of α-melanotropin have been prepared and used for conjugation with other large molecular weight proteins (Eberle et al., 1977).

Unlike α-haloketones, the reaction between a thiol and a maleimido group is highly specific. Maleimido groups are fully compatible with the methods of peptide synthesis and purification. A variety of reagents containing maleimido groups are available commercially as free carboxylic acids or as active esters, such as N-hydroxysuccinimide or p-nitrophenyl ester (Table 6), for facile introduction into peptide ligands.

Table 9   Cleavable Heterobifunctional Reagents for Derivatization of the Peptides for Reversible Labeling of Their Respective Receptors

Reagent	Reference

Amine reactive groups:

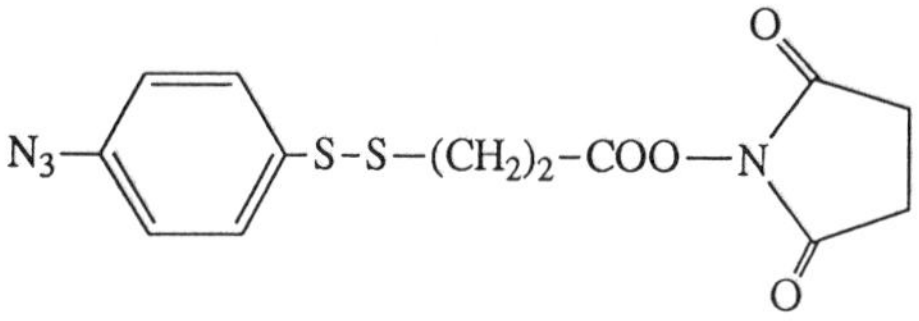

Ji, 1979
Vanin and Ji, 1981

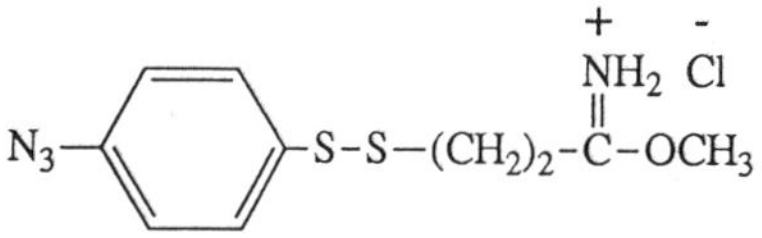

Ji, 1979
Vanin and Ji, 1981
Das et al., 1977

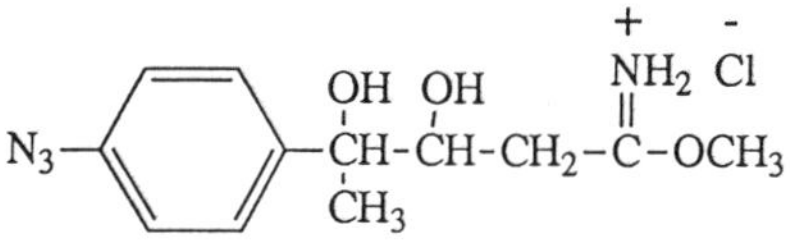

Rinke et al., 1980

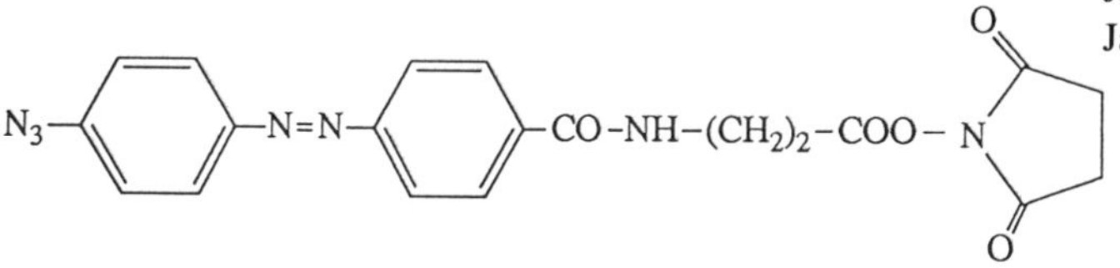

Jaffe et al., 1979
Jaffe et al., 1980

(continued on following page)

Disulfide Bridge Forming Reactions

Commercially available 3-(2-pyridyldithio)propionate (PDP) is the most commonly used group that is incorporated into peptide or protein ligands through its N-hydroxysuccinimide ester. The resulting highly stable affinity ligand undergoes very specific reaction at pH 6.0–6.5 with thiols to form a disulfide bond with the ligand. This is the method of choice for conjugating a ligand with another biomolecule or affinity matrix where reversibility is desired. The conjugated ligand can be easily cleaved off

Table 9   (continued)

Reagent	Reference

Thiol reactive groups:

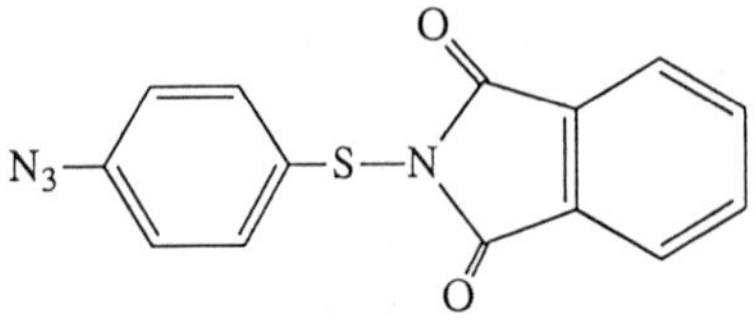

Vanin and Ji, 1981
Moreland et al.,
1982

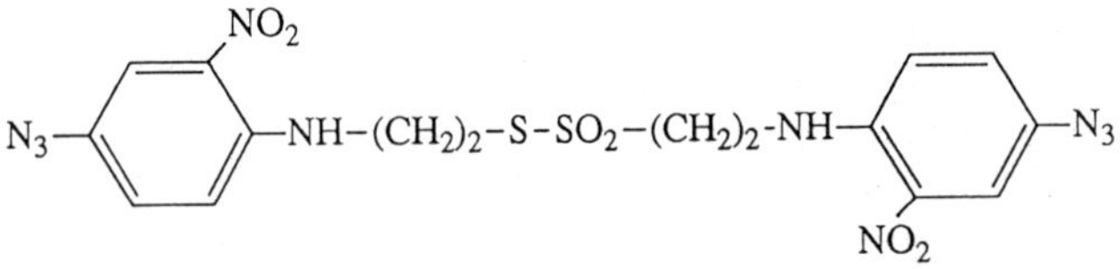

Huang and
Richards, 1977

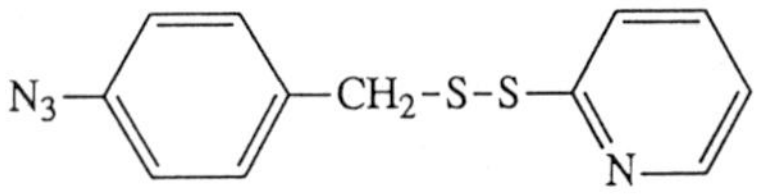

Henkin, 1977

by reducing the -S-S- bridge with a mild reducing agent, such as dithio-threitol (DTT).

Schwyzer and coworkers (Wunderlin et al., 1985a, 1985b) have established yet another facile technique for making a disulfide bridge between a ligand and another biomolecule. In this approach, as demonstrated by the conjugation of an α-melanotropin analogue with a biomolecule, a bromoalkyl group is first incorporated in the ligand by chemical methods of peptide synthesis. The treatment of the resulting stable derivative with sodium thiosulfate yields a "Bunte salt," which is treated in situ with a thiol-containing biomolecule, resulting in the establishment of the desired disulfide bridge.

Amine Reacting Derivatives

Carbonyl compounds, such as aldehydes and ketones, react with amino groups of lysine residues to form Schiff's bases that are reversible yet stable enough to allow further investigations into affinity labeled products. Imidate, isocyanate, or an isothiocyanate group attached to a ligand are the other most widely used modalities to target amino functionalities on the acceptor molecule. The synthesis of peptide ligands carrying these groups, however, is not straightforward. These groups, therefore, have found more use in bifunctional cross-linking reagents and are being widely utilized in such studies.

Photoaffinity Labels

Photoaffinity labeling of biological receptors for peptide and protein hormones and neurotransmitters is a most valuable means of receptor identification and isolation. In this approach, a chemically stable but photolabile group is conjugated to a potent hormonal ligand. After binding this ligand to its receptor site, the receptor-ligand complex is subjected to photolysis to generate highly reactive carbenes or nitrenes that may undergo hydrogen abstraction, insertion, or cycloaddition reactions with the adjacent chemical functionalities on the receptor molecule, thereby establishing a covalent bond between the receptor and the ligand (Bayley, 1983). Because of their high reactivity, the photoaffinity labels, in general, have proved to be of great advantage over the chemical affinity labels. They are stable in aqueous solutions, and, most importantly, the reactions exhibited upon photoactivation are not dependent on the presence of a nucleophilic center on the receptor for establishing chemical bonds. The use of a radiolabeled photoaffinity ligand in these studies helps to identify the covalently conjugated receptor-ligand complex.

In general, the use of five different groups of photolabels has been established in the literature. These groups are α-diazoketones, diazirines, α,β-unsaturated ketones, arylazides, and p-nitrophenyl. The singlet or the triplet state of the photogenerated species, as shown in Figure 9, tend to determine the type of reaction preferred in ligand-receptor cross-linking. Singlet states, in general, exhibit electrophilic reactions, such as insertions, additions, and so on. Triplet states exhibit radical reactions that lead to abstraction of a hydrogen atom (Turro, 1980). Nitrenes, due to longer half-lives than carbenes, tend to exist in low-energy triplet states that can produce insertion into C—H bonds (McRobbie et al., 1976). In electrophilic reactions, the singlet nitrenes prefer an O—H or N—H bond over a C—H bond and thereby are suggested to be less suitable than carbenes for labeling receptor areas rich in lipids or hydrophobic groups (Bayley and Knowles, 1978a, 1978b).

The α-diazoketones that give rise to α-keto carbenes were the first photoaffinity labels ever used. The active site of chymotrypsin was labeled by a derivative of this class of photolabels by Westheimer and coworkers (Singh, et al., 1962; Chowdhry and Westheimer, 1979). In the case of peptides, however, the use of this type of label has been rather limited. The main reason for it has been the inherent chemical reactivity

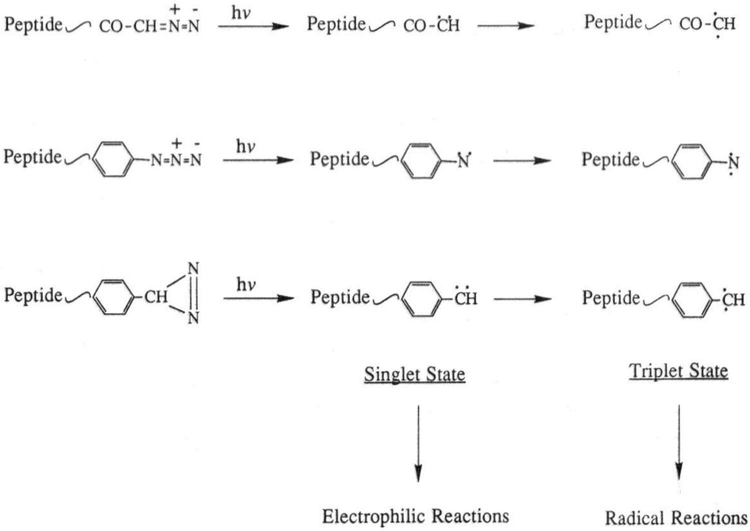

FIGURE 9   Alpha-diazocarbonyl, aryl azide, and aryl diazirine groups and their respective photolyzed reactive singlet and triplet states. Adapted from Eberle (1983).

of the α-diazo carbonyl group. The diazo group is a good leaving group in nucleophilic displacement reactions (for example, reactions with water, —OH, —SH, or —NH$_2$ functionalities). Further, the α-keto carbene generated upon photolysis of α-diazoketone is likely to undergo Wolff rearrangement reactions, thereby lowering the labeling efficiency of the photoaffinity labeling of the receptor. Replacement of $^{\alpha}$C—H with $^{\alpha}$C—COOEt, or $^{\alpha}$C—CF$_3$, or $^{\alpha}$C—SO$_2$C$_6$H$_4$CH$_3$ has been reported to increase the stability of the resulting diazocarbonyl compound as well as suppress the Wolff rearrangement (Vaughan and Westheimer, 1969; Chowdhry et al., 1976; Chowdhry and Westheimer, 1978). This substitution, however, also increases its lability in aqueous medium. Absorption maxima for α-diazoketones are 250 nm (high ε) and 350 nm (low ε). Both wavelengths can be used for photolysis.

Diazirines, in comparison to α-diazoketones, are chemically inert and generate the same type of reactive carbenes (Bayley and Knowles, 1978b). However, on photolysis, unsubstituted diazirines may generate linear diazo compounds as intermediates that may also exhibit undesired chemical reactivity, as described previously. Brunner et al. (1980) substituted a C^{α}—CF$_3$ for a C^{α}—H group to successfully overcome this problem. The resulting derivative can be easily photolyzed at 350 nm. The use of such a derivative has been shown by the success in labeling membrane components (Brunner and Richards, 1980). The main reason for its limited use, however, stems from the tedious synthesis of these derivatives. At the same time, it has also been argued that the higher reactivity of carbenes than nitrenes capable of facile C—H insertions may be a disadvantage in itself. This high reactivity may cause more predominant nonspecific insertion reactions into cell membrane lipids than the use of nitrene derivatives.

A few examples on the use of α,β-unsaturated ketones for the photo-affinity labeling of peptides are known. 4-Acetylbenzoyl pentagastrin was covalently coupled to albumin by irradiation at 320 nm (Galardy et al., 1974). In this case the photolysis produces a biradical triplet state that has selectivity in the abstraction of C—H hydrogens compared with O—H hydrogens (Turro, 1965). It has been predicted that this photo-label, due to its high lipophilicity, may considerably alter the binding and biological characteristics of short peptide hormones. That alteration is likely to make this type of photolabel unsuitable for peptide hormones.

Arylazides are the best-known labels used in combination with peptides, and they are stable to most of the acid and base procedures used in peptide synthesis. That makes them compatible with the synthesis of peptide derivatives. These groups, however, are not stable under reducing conditions, such as hydrogenation and exposure to thiols. Photolysis of arylazides yields highly reactive nitrenes. The absorption maxima for unsubstituted arylazides is around 260 nm. Introduction of an electron withdrawing group, for example, a nitro group, in the position *meta* to the azide group allows one to accomplish photolysis at higher wavelengths.

The *p*-nitrophenyl group has been used as a photolabel in the form of *p*-nitrophenylalanine derivatives incorporated into peptides. This technique has been successfully used for labeling the active site of α-chymo-trypsin (Escher and Schwyzer, 1974), receptor binding sites for angiotensin II (Escher and Guillemette, 1978), and bradykinin (Escher et al., 1981). However, it failed to label receptors for α-MSH, which could be labeled by using a *p*-azidophenyl group containing an MSH ligand (Eberle, 1984). The reason for this failure may lie in the mechanism of photolabeling by the nitrophenyl group. This photochemistry is poorly understood and is believed to proceed by means of a radical mechanism (Escher and Guillemette, 1978).

Incorporation of Photolabile Groups in the Peptide Hormones and Their Analogues

Much work on structure-activity relationships for peptide hormones strongly demonstrates that introduction of even a minor structural change in the peptide can affect its potency as well as its selectivity toward a particular receptor subtype. This factor requires careful consideration of the site where a potential group can be incorporated. While not com-

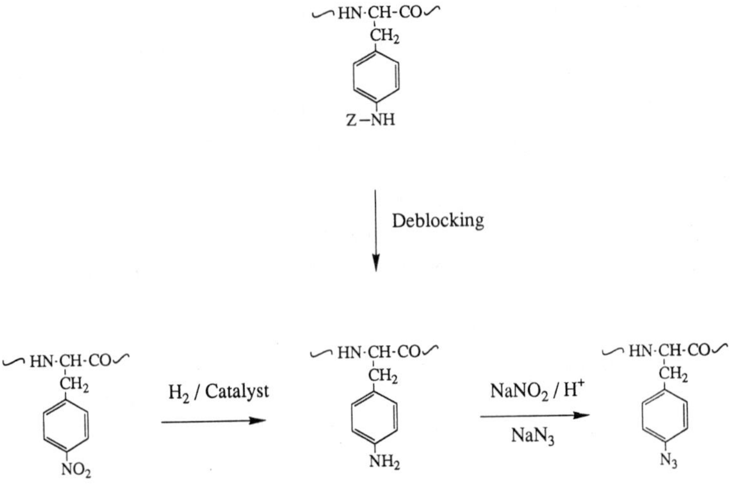

FIGURE 10   Alternate synthetic routes for *p*-azidophenylalanine (Pap) substituted peptides.

promising the potency and selectivity toward its biological receptors, the photoaffinity peptide analogues should also exhibit the ability to cross-link with the receptors. In general, the methods of incorporating photo-labile groups into peptides can be divided into two main categories: (a) de novo synthesis of a particular photoaffinity peptide analogue and (b) derivatization of peptides to photoaffinity peptides.

De Novo Synthesis of Photoaffinity Peptides

This approach is based on the introduction of *p*-nitrophenylalanine (Pnp) residue into the peptide during its de novo synthesis. Pnp can be substituted for an amino acid residue, preferably phenylalanine or tyrosine, that is not crucial for the potency and reactivity of a peptide. The finished peptide containing a Pnp usually acts as a photoaffinity ligand, as discussed earlier.

Alternatively, a Pnp residue, incorporated in a peptide, can be converted to a *p*-azidophenylalanine (Pap) residue which, as discussed earlier, is a more efficient photolabile group. This conversion can be achieved by hydrogenating the Pnp-containing peptide (Figure 10) to give the corresponding *p*-aminophenylalanine substituted peptide, which is then diazotized with sodium azide to form the Pap-substituted peptide analogue. As shown in Figure 10, a more direct method for obtaining *p*-amino phenylalanine substituted peptide derivative is by introducing a carbobenzyloxy (Z) protected p-NH_2-phenylalanine residue in the peptide during its synthesis. Photoreactive analogues containing Pap have been reported for angiotensin II (Escher and Guillemette, 1978), bradykinin (Escher et al., 1981), α-MSH (Eberle and Schwyzer, 1976; Eberle et al., 1981), a neurophysin analogue (Klausner et al., 1978), and substance-P (Escher et al., 1982).

Derivatization of Peptides to Yield Photoaffinity Peptides

A number of heterobifunctional reagents containing a photolabile p-azidophenyl group have been developed over last two decades that can be incorporated into finished peptides under rather mild conditions. The second chemically reactive species in these reagents is designed to react either with an amino or a carboxyl functionality that commonly occurs at the N or C terminus or on certain side-chain functionalities of some amino acids in the peptide. Some reagents that react with the side-chain functionalities of arginine, cysteine, tryptophan, and histidine residues also have been developed. Most of these reagents (Tables 7 and 8) have been used for constructing photoaffinity labels from peptide ligands, and they are commercially available form various sources.

Reagents for derivatizing amino groups, such as acid chlorides, acid bromides, or p-nitro phenyl esters of p-azido benzoic acid or p-azido phenylacetic acid, can lead to quantitative modifications. So do reagents such as fluoro-arylazides, p-azidophenylisothiocyanates, and imidates, which, in addition, allow the maintenance of the cationic characteristics of the modified amino group. Some of the reagents, such as those containing an N-hydroxysuccinimide ester, also offer the choice of varying the length of the spacer between the peptide and the photoreactive azidophenyl moiety. The carboxyl modifying reagents contain an amino group that can be coupled to the COOH group on the peptide by any of the well-established methods of peptide bond formation. In this case, however, any amino group that may be present in the peptide itself might have to be selectively protected prior to this reaction.

An arginine-specific reagent, p-azidophenyl glyoxal, reacts with the guanidino group at pH 7.0–7.5. The product formed, however, is unstable at alkaline pH, regenerating the original guanidino group. Cysteine-specific reagents that contain α-haloketones or maleimide groups, can react quantitatively with the thiol function. However, most cysteine residues are found in peptide hormones in the oxidized disulfide form. The tryptophan-modifying reagent 2-nitro-4-azido phenylsulfenyl chloride reacts at position 2 of the indole ring in tryptophan. This reagent is also capable of reacting with free cysteine (thiol) groups. The azo group in azido aryl azo derivatives reacts exclusively with the tyrosine or the histidine ring.

Three carbene-generating heterobifunctional reagents that modify an amino functionality also are available. These acid chloride or p-nitrophenyl ester type reagents are adequately stable for derivatization reactions.

A reversible or cleavable class of heterobifunctional reagents also has been developed. The use of these reagents (Table 9) offers the possibility of cleaving the peptide ligand from its receptor at will under mild conditions. The receptor thus separated from the ligand can be used in further studies, such as generation of receptor-specific antibodies, examination of structure, or cell-free reconstitution of the receptor. Normally, the cleavable heterobifunctional reagents contain, in addition to the usual two reactive groups, another group, such as a disulfide (—S—S—), glycol (—CHOH—CHOH—), or azo (—N=N—) group. These groups can be cleaved, respectively, by 10 to 100 mM 2-mercaptoethanol at pH 7–9 (mild reductive conditions), 15 mM sodium periodate for 4 to 10 hours (oxidative conditions), or sodium dithionite (Bayley, 1983).

RECEPTOR STUDIES—A BRIEF OVERVIEW OF STRUCTURE-ACTIVITY RELATIONSHIP

One of the most important applications of synthetic peptides is for the delineation of receptor types and subtypes and for providing analogues (radiolabeled and otherwise) that can be used for the examination of receptor systems in vivo. Clearly, receptor pharmacology is complicated by the fact that multiple types and subtypes of different receptors exist. Because there exist multiple receptor subtypes, one may assume that each subtype will have a particular functional role in vivo. Thus, the synthesis and design of peptides directed toward receptor subtype binding and kinetics as well as toward in vitro and in vivo pharmacology are a critical area of current synthetic peptide research.

One example of a receptor type determination that our group has been involved in is the δ-opioid receptor. When the concept of multiple opioid receptors began to surface in the mid-1970s (Hughes et al., 1975; Martin et al., 1976; Lord et al., 1976), the question arose concerning whether these individual subtypes might also be responsible for specific pharmacological and physiological responses. Thus, speculation began to arise concerning whether the therapeutic responses of the alkaloid opiates (for example, morphine) such as analgesia, vascular dilation, euphoria, and so on, as well as the adverse reactions, such as respiratory depression, decreased gut motility, dysphoria, and so on, were elicited by specific receptor types or subtypes.

Current thinking has established the existence for at least three types of opioid receptors; μ, κ, and δ, with some evidence for two additional receptor types referred to as σ and ε. In addition, there now is considerable evidence for subtypes of these receptors.

One method of peptide design that has been instrumental in defining these receptor types has been the technique of conformational constraint including cyclization (Hruby, 1982; Hruby et al., 1990a,b; for example, see Mosberg et al., 1983). The rationale for the use of conformational and topographical constraints has been to limit the number of conformations the peptide may assume. Decreasing the number of conformations will increase the likelihood that a peptide ligand will interact primarily at one receptor among several. This technique has been shown to be very fruitful in the discovery of the δ-selective peptide DPDPE (Mosberg et al., 1983), the μ-selective peptides CTOP (Pelton et al., 1985) and TCTAP (Kazmierski and Hruby, 1988; Kazmierski et al., 1989), and the κ-selective dynorphin analogues (Kawasaki et al., 1990).

Prior to describing the design and synthesis of receptor-selective peptides, however, a brief description of the bioassays to differentiate between the different receptor moieties is essential. As was indicated previously, Lord and coworkers (1976) demonstrated a difference in affinities of the alkaloid or morphine-like opiates versus the peptidic enkephalins. Utilizing tissues from the guinea pig ileum (GPI) and mouse vas deferens (MVD), Lord was able to show a preference of morphine-like opiates for μ receptors and a preference of enkephalins for δ receptors. This finding demonstrated that one could study ligands directed toward specific receptor types. It was later discovered that the MVD was approximately 80 percent δ receptors, the remainder being μ and κ, and the GPI was approximately 30 percent δ (Leslie et al., 1979). Even though the receptor populations are not homogeneous for δ (MVD) or μ (GPI), nevertheless, one can get a fairly accurate idea of *relative* ligand selectivity by calculating the GPI to MVD ratio of the corresponding IC_{50} values, where ratios < 1 exhibit μ selectivity and ratios > 1 exhibit δ selectivity. Of course, these ratios have to be based on absolute ligand standards, and today we use DPDPE (δ) and CTOP (μ).

In addition to these bioassays, binding assays generally utilize membranes from the brain, and ^3H or ^{125}I labeled highly receptor-selective ligands have been developed.

Development of δ-Selective Opioid Receptor Peptides

The prototypical peptide used as the original starting point for δ-selective peptide design begins with Leu[5]-enkephalin and Met[5]-enkephalin. As mentioned earlier, Lord and coworkers (1976) suggested a preference of both enkephalins for the δ receptor. Initial design principles that address either shortening or lengthening of the enkephalin pentapeptides met, in general, with little success regarding potency (versus Met[5] or Leu[5] enkephalin) or selectivity (Ling and Guillemin, 1976). The second design principle that was employed, the substitution of the corresponding L amino acid for a D amino acid, initially met with little success. Substitution of a D-Ala[2] for Gly[2] in Leu[5] and Met[5] enkephalin did not improve the binding character significantly, but it did result in an increase in

potency in both the MVD and GPI assay (Kosterlitz et al., 1980). Introduction of a D-Leu in position 5 (DADLE) likewise did not cause any
significant increase in affinity for the two binding sites; however, it increased the activity in the MVD and reduced it in the GPI such that there
was a fivefold increase in preference for the MVD over the GPI. Thus, it
became clear that substitution of D amino acids at positions 2 and 5 were
beneficial for δ-receptor selectivity.

A major breakthrough in the design of δ-selective peptides evolved
from the work of Gacel et al. (1980) and Fournie-Zaluskie et al. (1981).
Starting from structure-activity relationships in the enkephalin series, the
hexapeptide Tyr-D-Ser-Gly-Phe-Leu-Thr (DSLET) was prepared. Although no more potent than DADLE in the MVD (0.58 nM±0.20 versus
0.54 nM±0.09), DSLET was about sevenfold less potent in the GPI versus DADLE (DSLET IC$_{50}$ GPI/MVD = 620). Further modifications
based upon DSLET as the prototype eventually led to even more δ-selec-
tive peptides. Substitution of D-Thr^2 for D-Ser^2 increased the selectivity
for the δ receptor even more (Gacel et al., 1988). This was followed by
the development of the O-tert-butylated analogues [D-Ser^2(O-tert-butyl),
Leu^5, Thr^6]enkephalin (DSTBULET) and [D-Ser^2(O-tert-butyl), Leu^5,
Thr^6 (O-tert-butyl)]enkephalin (BUBU), which further improved upon
selectivity (Delay-Goyet et al., 1988; Gacel et al., 1988).

A design theme that proved to be extremely successful in our laboratory (Hruby, 1982; Hruby et al., 1990a) has been the use of conformational constraints to study receptor selectivity and, even more importantly, receptor-site topography. The concept takes advantage of the fact that
conformational constraints not only can help to generate receptor selectivity but also can aid in determining receptor topography, because now
the number of possible conformations is substantially reduced compared
to their unconstrained counterparts. Hence, a decreased number of degrees of conformational freedom gives one the added advantage of being
able to indirectly study the receptor topography by examining the conformational properties of the ligand in solution.

Work that led to the development of

[D-Pen^2, D-Pen^5]enkephalin (DPDPE)

(Mosberg et al., 1983) was based upon our knowledge of pseudo-
isosteric cyclization of side-chain groups in our previous studies with
melanotropins. In this case, Sawyer et al. (1982) showed that replacing
the Met^4 and Gly^{10} residues with Cys^{4,10} moieties, followed by subsequent oxidation, led to a compound that assumes a similar steric space
to its linear counterpart, but now it becomes constrained. This occurrence
led to a superpotent analogue. Similarly, it was realized that the Gly^2 and

Met[5] side chains of enkephalin would be pseudo-isosteric if cysteine substitutions were made at these positions. From the patent literature, Sarantakis had found that a

[D-C̄ys^2, L-C̄ys^5]enkephalin

was an effective analgesic in vivo, but no information was known about its selectivity until Schiller and coworkers resynthesized the compound and found it to be potent but not highly selective (Schiller et al., 1981). The suggestions from other work that D amino acids could be utilized in both the 2 and 5 positions encouraged Mosberg et al. to synthesize both

[D-P̄en^2, D-C̄ys^5]enkephalin (DPDCE)

and

[D-P̄en^2, L-C̄ys^5]enkephalin (DPLCE)

(Mosberg et al., 1983a). These were found to be the most selective δ ligands to date. A final permutation led to the synthesis of

[D-P̄en^2, D-P̄en^5]enkephalin (DPDPE),

which has served as the pharmacological standard for studying δ-opioid receptor properties.

The development of DPDPE by Mosberg and coworkers also led to extensive biophysical studies that helped give us a clue to the topography about the δ-opioid receptor. Hruby and coworkers (1988) conducted extensive 2D NMR studies in solution, including nuclear Overhauser enhanced NMR spectroscopy (NOESY) in 90% HOD and pH = 3.1 and 6.0. These studies revealed some rather interesting dipolar interactions between the aromatic moieties of Tyr[1] and Phe[4] and the β,β-dimethyl groups of D-Pen[2]. This direct evidence for transannular interactions led Hruby et al. (1988) to suggest that an amphiphilic conformation may be among the most stable in solution. Starting primarily with coupling con-

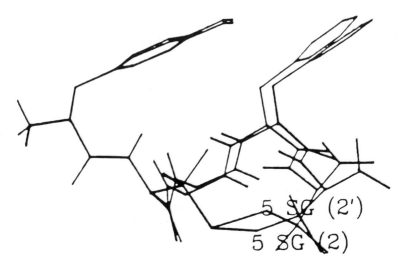

FIGURE 11   Superimposed structures of the two low-energy conformations found most consistent with ^{1}H NMR NOESY constraints and vicinal coupling constant data. A series of turn structures (type I, Iβ, II, IIβ, III, IIIβ, IV, IVβ, etc.) were used as starting points prior to energy minimization. Note the difference in helicity of the disulfide bridge.

stants derived by NMR determinations, combinations of all reasonable phi angles consistent with a 14-membered ring were calculated. These starting parameters were, in turn, energy-minimized and finally compared once again with the NMR data. The extensive energy minimization led to the finding of two pairs of low-energy conformations that differed primarily by the chirality of the disulfide (Figure 11). Of the two, one of the low-energy conformers satisfied all the NMR criteria. The conformation was distinguished by the existence of a type IV β-turn and the transannular interactions described previously. In addition, the molecule contained overall amphipathic topographical features. Thus, because of the constrained nature of this peptide ligand, one has a general idea of the δ-opioid receptor topography by virtue of the ligand that is specific for it. An additional approach, using quenched dynamics, was conducted by Pettitt and coworkers (Pettitt et al., 1991), which yielded a similar set of low-energy conformers that all exhibited the same amphiphilic character, which was evident in the earlier work. CPK models of the nine lowest-energy conformers can be seen in Figure 12.

In conclusion, using the δ-opioid receptor as an example, one can see how the development of δ-receptor selective analogues has developed from (1) lengthening and shortening the sequence, (2) systematic substitution of D amino acids for native L amino acids, (3) imposition of

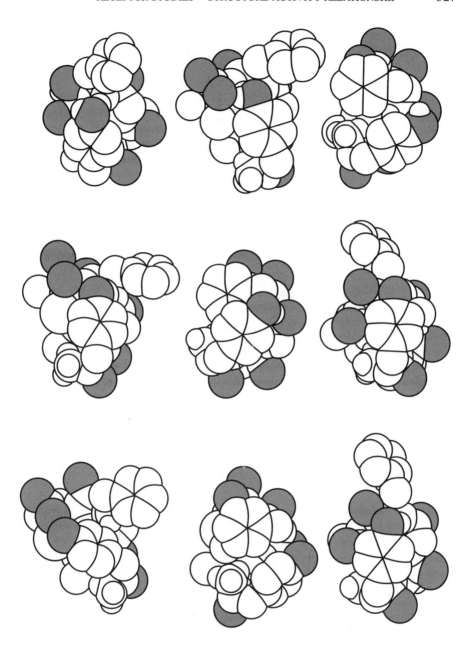

FIGURE 12   CPK models of the nine lowest energy conformers derived from quenched dynamics studies on [D-Pen^2, D-Pen^5]enkephalin (Pettitt, et al., 1991). Note that all nine conformers exhibit similar positioning of the aromatic residues above the plane at the backbone ring, while the carbonyl moieties (oxygens are shaded) are pointed toward the opposite side of the backbone ring, giving the molecules their amphiphilic character.

conformation constraints by pseudo-isosteric replacement, and, finally, (4) solution phase NMR and (5) subsequent molecular mechanics and molecular dynamics calculations in conjunction with molecular modeling of the appropriate conformationally constrained analogues. All have helped to develop a "picture" of the requirements for the δ-opioid receptor that can be tested. Clearly, more work needs to be done to refine the picture. However, future endeavors incorporating all the above design principles as well as molecular biology and molecular pharmacology techniques may one day soon allow us to isolate and to view the δ-opioid receptor itself.

SUMMARY AND FUTURE PERSPECTIVES

In this overview of possible applications of synthetic peptides to modern biology and medicine, we have tried to emphasize the need for both physical/chemical and biological considerations in peptide ligand design as well as the application of modern synthetic chemistry, including asymmetric syntheses and macrocyclic design and synthesis. Because macromolecular recognition and selectivity are essential features of most biological systems, it is critical that highly purified peptide ligands be utilized to examine these systems, whether the examination is in the form of a binding, in vitro, or in vivo bioassay, or whether it is in the form of a biophysical or biochemical measurement. It also is clear that peptides and their analogues, pseudopeptides, and peptidomimetics are likely to be the major drugs of the future. This finding raises many issues in need of further study. As already outlined herein, the discovery process for obtaining peptide ligands that are more or less selective for a particular receptor (or acceptor) type or subtype, a specific antibody or antigen, a specific enzyme, and so on, has been greatly accelerated in recent years by the tremendous advances in both synthetic peptide chemistry and the development of multiple, high-throughput binding and bioassay systems. The availability by synthesis of large (millions) peptide libraries that can be used to screen for binding and biologically relevant activities; the ability to design peptide derivatives and analogues with constrained conformations and topographies and specific two- and three-dimensional properties; the utilization of modern synthetic chemistry to prepare designed peptides, pseudo-peptides, and peptide mimetics; the ability to design peptide analogues that are stable to biodegradation and have good bioavailability; and the development of computational methods using modern fast computers to examine the conformations, topographies, and dynamics of peptides all provide powerful tools for future developments and will be critically important to future developments. Further work is needed in these areas as well as greatly accelerated efforts to improve methods of peptide delivery in biological systems, to find methods for passing peptides and peptide mimetics through various membrane barriers, to develop further rapid-throughput multiple assays systems, and to obtain more reliable methods for evaluating peptide conformations and dynamic properties. Most importantly, there is a need to find ways to facilitate the proper collaboration between chemists, physicists, biologists, and medical doctors that will be necessary to optimize progress in this area.

The prospects for important scientific advances and for their application to the treatment of disease are exceptionally good. The extent to which these prospects will be realized depends increasingly on a comprehensive approach to solving the problems. It is hoped that this chapter will help to facilitate this process.

LIST OF ABBREVIATIONS

α-MSH α-melanotropin, α-melanocyte stimulating hormone

ACTH   adrenocorticotropin hormone

CCK   cholecystokinin

Cp β-cyclopentamethylene-β-mercaptopropionic acid

CTOP   D-Phe-Cys-Tyr-D-Trp-Orn-Thr-Pen-Thr-NH$_2$

DADLE   [D-Ala^2,D-Leu^5]enkephalin

Dep β,β-diethyl-β-mercaptopropanoic acid

Dmp β,β-dimethyl-β-mercaptopropanoic acid

DPDPE   [D-Pen^2,D-Pen^5]enkephalin

DSLET   Tyr-D-Ser-Gly-Phe-Leu-Thr

DTT   dithiothreitol

EGF   epidermal growth factor

ELISA   enzyme-linked immunosorbent assay

FMDV   foot and mouth disease virus

GPI   guinea pig ileum

GnRH   gonadotropin-releasing hormone

HPAC   high-performance affinity chromatography

IL-2   interleukin-2

MAP   multiple antigenic peptide

MCH   melanocyte-concentrating hormone

Mpa β-mercaptoacetic acid

MVD   mouse vas deferens

NMR   nuclear magnetic resonance

nOe   nuclear Overhauser enhancement

NOESY   nuclear Overhauser enhancement spectroscopy

OXT   oxytocin

Pap   p-azidophenylalanine

Pen   penicillamine, β,β-dimethyl cysteine

Pip   L-Pipecolic acid

TASP   template assembled synthetic protein

TCTAP   D-Tic-Cys-Tyr-D-Trp-Arg-Thr-Pen-Thr-NH$_2$

TF   transferrin

Thr-ol   threonal, the reduced form of threonine, 2-amino-1,3-butanediol

Tic   1,2,3,4-tetrahydroisoquinoline carboxylate

ACKNOWLEDGMENTS

This work was supported by grants from the U. S. Public Health Service and the National Institute of Drug Abuse. We thank Cheryl McKinley for her work in typing and editing this manuscript.

REFERENCES

Abood, L.G., Michael, G.J., Xin, L., and Knigge, K.M. 1989. Interaction of putative vasopressin receptor proteins of rat brain and bovine pituitary gland with an antibody against a nanopeptide encoded by the reverse message of the complementary mRNA to vasopressin. J. Recept. Res. 9:19-25.

Al-Obeidi, F., Hadley, M.E., Pettitt, M.B., and Hruby, V.J. 1989a. Design of a new class of superpotent cyclic α-melanotropins based on quenched dynamic simulations. J. Am. Chem. Soc. 111:3413-3416.

Al-Obeidi, F., Castrucci, A.M., Hadley, M.E., and Hruby, V.J. 1989b. Potent and prolonged acting cyclic lactam analogues of α-melanotropin: design based on molecular dynamics. J. Med. Chem. 32:2555-2561.

Al-Obeidi, F., Hruby, V.J., Sharma, S.D., Hadley, M.E., and Castrucci, A.M. Antisense peptides of melanocyte stimulating hormone (MSH): Surprising results. In Peptides: Chemistry, Structure and Biology, J.E. Rivier and G.R. Marshall, eds. ESCOM, Leiden, The Netherlands, 1990, pp. 530-532.

Allen, J.S., Coligan, J.E., Barin, F., McLane, M.F., Sodraski, J.G., Rosen, C.A., and Haseltine, W.A. 1985. Major glycoprotein antigens that induce antibodies in AIDS patients are encoded by HTLV-III. Science 228:1091-9094.

Amit, A.G., Mariuzza, R.A., Phillips, S.E.V., and Paljak, R.J. 1986. Three-dimensional structure of an antigen-antibody complex at 2.8Å resolution. Science 233:747-753.

Anderson, K.P., Lucas, C., Hanson, C.V., Londe, H.F., Izu, A., Gregory, T., Amman, A., Berman, P.W., and Eichberg, J.W. 1989. Effect of dose and immunization schedule on immune response of baboons to recombinant glycoprotein 120 of HIV-1. J. Infect. Dis. 160:960-965.

Aota, S., Gojobori, T., Ishibashi, F., Marvyama, T., and Ikamvea, T. 1988. Codon usage tabulated from the gen bank genetic sequence data. Nucleic Acids Res. 16: Suppl., r315-391.

Audibert, F., Jolivet, M., Chedid, L., Alanf, J.E., Boquet, P., Ruaille, P., and Siffert, O. 1981. Active antitoxic immunization by a diphtheria toxin synthetic oligopeptide. Nature 289:593-599.

Bankowski, K., Manning, M., Halder, J., and Sawyer, W.H. 1978. Design of potent antagonist of the vasopressor response to arginine vasopressin. J. Med. Chem. 21:850-853.

Barin, F., McLane, M.F., Allan, J.S., and Lee, T.H. 1985. Virus Envelope protein of HTLV-III represents major target antigen for antibodies in AIDS patients. Science 228:1094-1095.

Barlow, D.J., Edwards, M.S., and Thornton, J.M. 1986. Continuous and discontinuous protein antigenic determinants. Nature 322:744-751.

Barret, A.D.T., and Gould, E.A. 1986. Antibody-mediated early death in vivo after infection with yellow fever virus. J. Gen. Virol. 67:2539-2549.

Bates, R.B., Gin, S.L., Hassen, H.A., Hruby, V.J., Janda, K.M., Kriek, G.R., Michaud, J.-P., and Vine, D.B. 1984. Synthesis of cyclo-N-methyl-L-Tyr-N-methyl-L-Tyr-D-Ala-L-Ala-O,N-dimethyl-L-Tyr-L-Ala, a cyclic hexapeptide related to the antitumor agent deoxybouvardin. Heterocycles 22:785-790.

Bauer, W., Boiner, U., Doepfner, W., Haller, R., Huguenin, R., Marbach, P., Petcher, T.J., and Pless, J. Structure-Activity Relationships of Highly Potent

Octapeptide analogues of Somatostatin. In Peptides 1982, K. Blaha and P. Malon, eds. de Gruyter, Berlin 1983.

Bayley, H., and Knowles, J.R. 1978a. Photogenerated reagents for membrane labeling. 1. Phenylnitrene formed within the lipid bilayer. Biochem. 17:2414-2419.

Bayley, H., and Knowles, J.R. 1978b. Photogenerated reagents for membrane labeling. 2. Phenylcarbene and adamentylidene formed within the lipid bilayer. Biochemistry 17:2420-2423.

Bayley, H. Photogenerated reagents in biochemistry and molecular biology. Elsevier, Amsterdam, 1983.

Birkbeck, T.H., and Penn, C.W. Antigenic Variation in Infectious Disease, Vol. 19. IRL Press, Oxford.

Biro, J. 1981a. Comparative analysis of specificity in protein-protein interactions. Part I: A theoretical and mathematical approach to specificity in protein-protein interactions. Med. Hypotheses 7:969-979.

Biro, J. 1981b. Comparative analysis of specificity in protein-protein interactions. Part II: The complementary coding of some proteins as the possible source of specificity in protein-protein interactions. Med. Hypotheses 7:981-993.

Biro, J. 1981c. Comparative analysis of specificity in protein-protein interactions. Part III: Models of the gene expression based on the sequential complementary coding of some pituitary proteins. Med. Hypotheses 7:995-1007.

Bittle, J.L., Houghton, R.A., Alexander, H., Shinnick, T.M., Sutcliffe, J.G., Lerner, R.A., Rowlands, D.J., and Brown, F. 1982. Protection against foot-and mouth disease by immunization with a chemically synthesized peptide predicted from the viral nucleotide sequence. Nature 298:30-34.

Bizzini, B., Blass J., Turpin, A., and Raymond, M. 1970. Chemical characterization of tetanus toxin and toxiod. Eur. J. Biochem. 17:100-109.

Blalock, J.E., and Bost K.L. 1988. Ligand receptor characteristics of peptides encoded by complementary nucleic acids: Implications for a molecular recognition code. Recent Prog. Horm. Res. 44:199-222.

Blalock, J.E., and Smith, E.M. 1984. Hydropathic anti-complementarity of amino acids based on the genetic code. Biochem. Biophys. Res. Commun. 121:203-207.

Blalock, J.E., and Bost, K.L. 1986. Binding of peptides that are specified by complementary RNAs. Biochem. J. 234:679-683.

Bodanszky, M., and du Vingeaud, V. 1959. Synthesis of a biologically active analog of oxytocin, with phenylalanine replacing tyrosine. J. Am. Chem. Soc. 81:6072-6075.

Bost, K.L., Smith, E.M., and Blalock, J.E. 1985. Regions of complementarity between the messenger RNAs for epidermal growth factor, transferrin, interleukin-2 and their respective receptors. Biochem. Biophys. Res. Commun. 128:1373-1380.

Bost, K.L., Smith, F.M., and Blalock, J.E. 1985. Similarity between the corticotropin (ACTH) receptor and a peptide encoded by an RNA that is complementary to ACTH mRNA. Proc. Natl. Acad. Sci. USA 82:1372-1375.

Bost, K.L., and Blalock, J.E. 1986. Molecular characterization of a corticotropin (ACTH) receptor. Mol. Cell. Endocrinol. 44:1-9.

Bost, K.L., and Blalock, J.E. 1989. Preparation and use of complementary peptides. Methods Enzymol. 168:16-28.

Brentani, R.R. 1988. Biological implications of complementary hydropathy of amino acids. J. Theor. Biol. 135:495-499.

Brentani, R.R., Ribeiro, S.F., Potocnjak, P., Pasqualini, R., Lopes, J.D., and Nakaie, C.R. 1988. Characterization of the cellular receptor for fibronectin through a hydropathic complementarity approach. Proc. Natl. Acad. Sci. USA 85:364-367.

Brown, F. The next generation of foot-and-mouth disease vaccines. In Synthetic Peptides: Approaches to Biological Problems, J.P Tam and E.T. Kaiser, eds. Alan R. Liss Inc., New York, 1989, pp. 127-142.

Brunner, J., Senn, H., and Richards, F.M. 1980. 3-Trifluoromethyl-3-phenyl diazirine. A new carbene containing group for photolabeling reagents. J. Biol. Chem. 255:3313-3318.

Brunner, J., and Richards, F.M. 1980. Analysis of membranes photolabeled with lipid analogues. Reaction of phospholipids containing a disulfide group and a nitrene or carbene precursor with lipids and with gramicidin A. J. Biol. Chem. 255:3319-3329.

Bryan, W.M., Callahan, J.F., Codd, E.E., Liemeux, C., Moore, M.L., Schiller, P.W., Walker, R.F., and Huffman, W.F. 1989. Cyclic enkephalin analogues containing α-amino-β-mercapto-β,β-cyclopentamethylenepropanoic acid at positions 2 and 5. J. Med. Chem. 32:302-304.

Budker, V.G., Knorre, D.G., Kravchenko, V.V., Lavrik, O.I., Nevinsky, G.A., and Teplova, M. 1974. Photoaffinity reagents for modifications of amino acyl-tRNA synthetases. FEBS Lett. 49:159-162.

Burgermeister, W., Nassal, M., Wieland, T., and Helmreich, E.J.M. 1983. A carbene-generating photoaffinity probe for beta-adrenergic receptors. Biochem. Biophys. Acta 729:219-228.

Carr, D.J.J., Bost, K.L., and Blalock, J.E. 1986. An antibody to a peptide specified by an RNA that is complementary to γ-endorphin mRNA recognizes an opiate receptor. J. Neuroimmunol. 12:329-337.

Chan, W.Y., and Kelly, N. 1969. A pharmacologic analysis on the significance of the chemical functional groups of oxytocin on its oxytocic activity and on the effect of magnesium on the in vitro and in vivo oxytocic activity of neurohypophyseal hormones. J. Pharmacol. Exp. Ther. 156:150-158.

Chanh, T.C., Dreesman, G.R., Konda, P., Lisette, G.P., Sparrow, J.T., Ho, D.D., and Kennedy, R.C. 1986. Induction of anti-HIV neutralizing antibodies by synthetic peptides. EMBO J. 5:3065-3072.

Charpentier, B., Durieux, C., Menant, E., and Roques, B.P. 1987. Investigation of peripheral cholecystokinin receptor homogeneity by cyclic and related linear analogues of CCK_{26-33}: Synthesis and biological properties. J. Med. Chem. 30:962-968.

Chowdhry, V., and Westheimer, F.H. 1979. Photoaffinity labeling of biological systems. Annu. Rev. Biochem. 48:293-325.

Chowdhry, V., and Westheimer, F.H. 1978. p-Toluenesulfonyl diazoacetates: Reagents for photoaffinity labeling. Bioorg. Chim. 7:189-205.

Chowdhry, V., Vaughan, R., and Westheimer, F.H. 1976. 2-Diazo-3,3,3, trifluoropropionyl chloride: Reagents for photoaffinity labeling. Proc. Natl. Acad. Sci. USA 73:1406-1408.

Coy, D.H., Horvath, A., Nekola, M.V., Coy, E.J., Erchegyi, J., and Schally, A.V. 1982. Peptide antagonists of LH-RH: Large increases in antiovulatory activities produced by basic D-amino acids in the six position. Endocrinology 110:1445-1447.

Cwirla, S.E., Peters, E.A., Barrett, R.W., and Dower, W.J. 1990. Peptides on phage: A vast library of peptides for identifying ligands. Proc. Natl. Acad. Sci. USA 87:6378-6382.

Das, M., Miyakawa, T., Fox, C.F., Pruss, R.M., Aharonov, A., and Herschman, H.R. 1977. Specific radiolabeling of a cell surface receptor for epidermal growth factor. Proc. Natl. Acad. Sci. USA 74:2790-2794.

Das, M., and Fox, C.F. 1978. Molecular mechanism of nitrogen action: Processing of receptor induced by epidermal growth factor. Proc. Natl. Acad. Sci. USA 75:2644-2648.

Dayhoff, M.O. Atlas of Protein Sequence and Structure, Vol. 5, Suppl. 2. NRB Foundation, Washington, D.C., 1976.

De Gasparo, M., Whitebread, S., Einsle, K., and Heussen, C. 1989. Are the antibodies to a peptide complementary to angiotensin II useful to isolate the angiotensin II receptor? Biochem. J. 261:310-311.

De Grado, W.F. 1988. Design of peptides and proteins. Adv. Protein Chem. 39:51-124.

De Lisi, C., and Berzofsky, J.A. 1986. T-cell antigenic sites tend to be amphipathic structures. Proc. Natl. Acad. Sci. USA 82:7048-7059.

Delay-Goyet, P., Sequin, C., Gacel, G., and Roques, B.P. 1988. [^{3}H][D-Ser^2 (O-tert-butyl), Leu^5]enkephalyl-Thr^6 and [D-Ser^2 (O-tert-butyl), Leu^5]enkephalyl-Thr^6 (O-tert-butyl). Two new enkephalin analogs with both a good selectivity and a high affinity toward δ-opioid binding sites. J. Biol. Chem. 263:4124-4130.

Demoliou, C.D., and Epand, R.M. 1980. Synthesis and characterisation of a heterobifunctional photoaffinity reagent for modification of tryptophan residues and its application to the preparation of a photoreactive glucagon derivative. Biochemistry 19:4539.

Dennis, S., Wallace, A., Hofsteenge, J., and Stone, S.R. 1990. Use of fragments of hirudin to investigate thrombin-hirudin interactions. Eur. J. Biochem. 188:61-66.

Devlin, J.J., Panganiban, L.C., and Devlin, P.E. Random peptide libraries: A source of specific protein binding molecules. Science 249:404-406.

Dyckes, D.F., Nestor, J.J., Jr., Ferger, M.F., and du Vigneaud, V. 1974. [1-β-Mercapto-β,β-diethylpropanoic acid]-8-lysine-vasopressin, A potent inhibitor of 8-lysine-vasopressin and oxytocin. J. Med. Chem. 17:250-252.

Dyson, H.J., Rance, M., Houghton, R.A., Lerner, R.A., and Wright, P.E. 1988. Folding of immunogenic peptide fragments of proteins in water solution. J. Mol. Biol. 201:161-201.

Eberle, A.N., and Schwyzer, R. 1976. Syntheses von [D-alanin^1, 4'-azido-3',5'-ditritio-L-phenylalanine^2, norvalin^4]-α-Melanotropins als <Photoaffinittsprobe> für Hormon-Rezeptor-Wechselwirkungen. Helv. Chim. Acta 59:2421-2431.

Eberle, A.N., Hübscher W., and Schwyzer, R. 1977. Synthesis von radioaktiv markierten Bromacetyl- und Diazoacetyl-α-Melanotropin-Derivaten zum Studium von kovalenten Hormone-Makromolekül-Komplexen. Helv. Chim. Acta 60:2895-2910.

Eberle, A.N., de Graan, P.N.E., and Hübscher, W. 1981. Synthesis and biological properties of p-azidophenylalanine[13]-α-melanotropin, a potent photoaffinity label for MSH receptors. Helv. Chim. Acta. 64:2645-2653.

Eberle, A.N. 1983. Photoaffinity labeling of peptide hormone receptors. J. Recept. Res. 3:313-316.

Eberle, A.N. 1984. Photoaffinity labeling of MSH receptors in Anolis melanophores: Irradiation techniques and MSH photolabels for irreversible stimulation. J. Receptor Res. 4:315-329.

Eberle, A.N., and de Graan, P.N.E. 1985. General principles for photoaffinity labeling of peptide hormone receptors. Methods Enzymol. 109:129-156.

Eberle, A.N., and Huber, M. 1991. Antisense peptides: Tools for receptor isolation? Lack of antisense MSH and ACTH to interact with their sense peptides and to induce receptor-specific antibodies. J. Recept. Res. 11:13-43.

Elton, T.S., Dion, L.D., Bost, K.L., Oparil, S., and Blalock, J.E. 1988a. Purification of an angiotensin II binding protein by using antibodies to a peptide encoded by angiotensin II complementary RNA. Proc. Natl. Acad. Sci. USA 85:2518-2522.

Elton, T.S., Oparil, S., and Blalock, J.E. 1988b. The use of complementary peptides in the purification of an angiotensin II binding proteins. J. Hypertens. 6:S404-S407.

Emmett, J., ed. Comprehensive Medicinal Chemistry, Vol. 3. Pergamon Press, Oxford, 1990.

Escher, E.H.F., Robert, H., and Guillemette, G. 1979. 4-Azidoaniline, a versatile protein and peptide modifying agent for photoaffinity labeling. Helv. Chim. Acta 62:1217-1222.

Escher, E., Couture, R., Champagne, G., Mizrahi, J., and Regoli, D. 1982. Synthesis and biological activities of photoaffinity labeling analogues of substance P. J. Med. Chem. 25:470-475.

Escher, E., and Schwyzer, R. 1974. p-Nitrophenylalanine, p-azidophenylalanine, m-azidophenylalanine, and o-nitro-p-azidophenylalanine as photoaffinity labels. FEBS Lett. 46:347-350.

Escher, E., and Guillemette, G. 1978. Photoaffinity labeling of the angiotensin II receptor. 3. Receptor inactivation with photolabile hormone analogue. J. Med. Chem. 22:1047-1050.

Escher, E., Laczko, E., Guillemette, G., and Regoli, D. 1981. Biological activities of photoaffinity labeling analogues of kinins and their irreversible effects on kinin receptors. J. Med. Chem. 24:1409-1413.

Fanning, D.W., Smith, J.A., and Rose, G.D. 1986. Molecular cartography of globular proteins with application to antigenic sites. Biopolymers 25:863-882.

Fassina, G., Roller, P.P., Olson, A.D., Thorgeirsson, S.S., and Omichinski, J.G. 1989a. Recognition properties of peptides hydropathically complementary to residues 356-375 of the c-raf protein. J. Biol. Chem. 264:11252-11257.

Fassina, G., Zamai, M., Brigham-Burke, M., and Chaiken, I.M. 1989b. Recognition properties of antisense peptides to Arg[8]-vasopressin/bovine neurophysin II biosynthetic precursor sequence. Biochemistry 28:8811-8818.

Ferrier, B.M., Jarvis, D., and du Vigneaud, V. 1965. Deamino-oxytocin. Its isolation by partition chromatography on sephadex and crystallization from water and its biological activities. J. Biol. Chem. 240:4264-4266.

Filatova, M.P., Kri, N.A., Komarova, O.M., Orfkchovich, V.N., Reiss, S. Liepinya, I.T., and Nikiforovitch, G.V. 1986. Synthesis and studies of conformationally restricted analogues of peptide inhibitors of the angiotensin-converting enzyme. Bioorg. Khim. 12:59-70.

Finnegan, A., Smith, M.A., Smith, J.A., Berzofsky, J., Sachs, D.H., and Hades, R.J. 1986. The T-cell repertoire for recognition of a phylogenetically distant protein antigen: Peptide specificity and MHC restriction of staphylococcal nuclease-specific T cell clones. J. Exp. Med. 164:897-903.

Fleet, G.W.J., Porter, R.R., and Knowles, J.R. 1969. Affinity labeling of antibodies with aryl nitrene as reactive group. Nature 224:511-512.

Fleet, G.W.J., Knowles, J.R., and Porter, R.R. 1972. The antibody binding site. Labeling of a specific antibody against the photoprecursor of an aryl nitrene. Biochem. J. 128:499-508.

Fodor, S.P., Read J.L., Pirrung, U.C., Stryer, L., Lu, A.T., and Salas, D. 1991. Light directed, spatially addressable parallel chemical synthesis. Science 251:767-771.

Fournie-Zaluski, M-C., Gacel, G.W., Maigret, B., Premilat, S., and Roques, B.P. 1981. Structural requirements for specific recognition of μ or δ opiate receptors. Mol. Pharmacol. 20:484-491.

Francis, M.J., Hastings, G.Z., Syred, A.D., McGinn, B., Brown, F., and Rowlands, D.J. 1987. Nonresponsiveness to a foot-mouth-disease virus peptide overcome by addition of foreign helper T-cell determinants. Nature 330:168-170.

Freidinger, R.M., and Veber, D.F. Design of novel cyclic hexapeptide somatostatin analogs from a model of the bioactive conformation. In Conformationally Directed Drug Design: Peptides and Nucleic Acids as Templates and Targets. J.A. Vida and M. Gordon, eds. ACS Symposium Series 251, Washington, D.C., 1984.

Freidinger, R.M., Colton, C.D., Randall, W.C., Pitzenberger, S.M., Veber, D.F., Saperstein, R., Brady, E.J., and Arison, B.H. A cyclic hexapeptide LH-RH antagonist. In Peptides: Structure and Function, C.M. Deber, V.J. Hruby, and K.D. Kopple, eds. Pierce Chemical Co., Rockford, Illinois, 1985.

Gacel, G.W., Fournie-Zaluski, M-C., and Roques, B.P. 1980. Tyr-D-Ser-Gly-Phe-Leu-Thr, a highly preferential ligand for δ opiate receptors. FEBS Lett. 18:245-247.

Gacel, G.W., Daugé, V., Breuze, P., Delay-Goyet, P., and Roques, B.P. 1988. Development of conformationally constrained linear peptides exhibiting a high affinity and pronounced selectivity for δ opioid receptors. J. Med. Chem. 31:1891-1897.

Gacel, G.W., Zajac, J.M., Delay-Goyet, P., Daugé, V., and Roques, B.P. 1988. Investigation of the structural parameters involved in the μ and δ opioid receptor discrimination of linear enkephalin-related peptides. J. Med. Chem. 31:374-383.

Galardy, R.E., Craig, L.C., Jamieson, J.D., and Printz, M.P. 1974. Photoaffinity labeling of peptide hormone binding sites. J. Biol. Chem. 249:3510-3518.

Geysen, H.M., Meloen, R.H., and Bartelling, S.J. 1984. Use of peptide synthesis to probe viral antigens for epitopes to a resolution of a single amino acid. Proc. Natl. Acad. Sci. USA 81:3998-4002.

Geysen, H.M., Rodda, S.J., and Mason, T.J. 1986. A priori delineation of a peptide which mimics a discontinuous antigenic determinant. Mol. Immunol. 23:709-715.

Ghiso, J., Saball, E., Leoni, J., Rostagano, A., and Frangione, B. 1990. Binding of cystatin C to C4: The importance of sense-antisense peptides in their interaction. Proc. Natl. Acad. Sci. USA 87:1288-1291.

Gierasch, L.M., Deber, C.M, Madison, V., Niu, C.H., and Blout, E.R. 1981. Conformations of (X-L-Pro-Y)$_2$ cyclic hexapeptides. Prefered β-turn conformers and implications for β-turns in proteins. Biochemistry 20:4730-4736.

Gilliand, D.G., and Callier, R.J. 1980. A model system involving anti-concanavalin A for antibody targeting of diphtheria toxin fragment A. Cancer Res. 40:3564-3571.

Goldstein, A., and Brutlag, D.L. 1989. Is there a relationship between DNA sequence encoding peptide ligands and their receptors? Proc. Natl. Acad. Sci. USA 86:42-45.

Goodfriend, T.L., Levine, L., and Fasman, G.D. 1964. Antibodies to bradykinin and angiotensin: Use of carbodiimides in immunology. Science 144:1344-1372.

Görcs, T.J., Gottschall, P.E., Coy, D.H., and Arimura, A. 1986. Possible recognition of the GnRH receptor by an antiserum against a peptide encoded by nucleotide sequence complementary to mRNA of a GnRH precursor peptide. Peptides 7:1137-1145.

Granthan, R., Gautier, C., and Gouy, M. 1980. Codon frequencies in 119 individual genes confirm consistent choices of degenerate bases according to genome type. Nucleic Acids Res. 8:1893-1912.

Guire, P., Fligen, D., and Hodgson, J. 1977. Photochemical coupling of enzymes to mammalian cells. Pharmacol. Res. Commun. 9:131-141.

Guillemette, G., Boulay, G., Gaynon, S., Bosse, R., and Escher, E. 1989. The peptide encoded by angiotensin II complementary RNA does not interfere with angiotensin II action. Biochem. J. 261:309.

Habeeb, A.F., and Hiramoto, R. 1968. Reaction of proteins with glutaraldehyde. Arch. Biochem. Biophys. 126:16-22.

Halstead, S.B. 1979. In vivo enhancement of dengue virus infection in Rhesus monkeys by passively transferred antibody. J. Infect. Dis. 140:527-536.

Hartfield, D., Mathew, C.R., and Rice, M. 1979. Aminoacyl-transfer RNA populations in mammalian cells: Chromatographic profiles and patterns of codon recognition. Biochim. Biophys. Acta. 564:414-423.

Hartfield, D., and Rice, M. 1986. Amino acyl-tRNA (anticodon): Codon adaptation in human and rabbit reticulocytes. Biochem. Int. 13:835-842.

Henkin, J. 1977. Photolabeling reagents for thiol enzymes. Studies on rabbit muscle creatine kinase. J. Biol. Chem. 252:4293-4297.

Hill, P.S., Smith, D.D., Slaninova, J., and Hruby, V.J. 1990. Bicyclization of a weak oxytocin agonist produces a highly potent oxytocin antagonist. J. Am. Chem. Soc. 112:3110-3113.

Hinds, M.G., Welsh, J.H., Biennoud, D.M., Fisher, J., Glennie, M.G., Richards, N.J.R., Turner, D.L., and Robinson, J.A. 1991. Synthesis, conformational properties and antibody recognition of peptides containing β-turn mimetics based on α-alkyl proline derivatives. J. Med. Chem. 34:1777-1789.

Hixson, S.H., and Hixson, S.S. 1975. p-Azidophenacyl bromide, a versatile photolabile bifunctional reagent. Reaction with glyceraldehyde-3-phosphate dehydrogenase. Biochemistry 14:4251-4254.

Ho, P.C., Mutch, D.A., Winkel, K.D., Saul, A.J., Jones, G.J., Daron, T.J., and Rzrpczyk, C.M. 1990. Identification of two promiscuous T cell epitopes from tetanus toxin. Eur. J. Immunol. 20:477-483.

Hopp, T.P. 1986. Protein surface analysis. Methods for identifying antigenic determinants and other interaction sites. J. Immunol. Methods 88:1-10.

Hopp, T.P., and Woods, K.R., 1981. Prediction of protein antigenic determinants from amino acid sequences. Proc. Natl Acad. Sci. USA 78:3824-3839.

Hruby, V.J. Structure and conformation related to the activity of peptide hormones. In Perspectives in Peptide Chemistry, A. Eberle, R. Geiger and T.Wieland, eds. S. Karger, Basel, Switzerland, 1981a.

Hruby, V.J. Relation of conformation to biological activity in oxytocin, vasopressin, and their analogues. In Topics in Molecular Pharmacology, Vol. 1, A.S.V. Burgen and G.C.K. Roberts, eds. Elsevier/North-Holland Biomedical Press, Amsterdam, 1981b.

Hruby, V.J. 1982. Conformational restrictions of biologically active peptides via amino acid side chain groups. Life Sci. 31:189-199.

Hruby, V.J., Sawyer, T.K., Yang, Y.C.S., Bregman, M.D., Hadley, M.E., and Heward C.B. 1980. Synthesis and structure-function studies of melanocyte stimulating hormone (MSH) analogues modified in the 2 and 4(7) position: Comparison of activities on frog skin melanophores and melanoma adenylate cyclase. J. Med. Chem. 23:1432-1437.

Hruby, V.J., and Smith, C.W. Structure-function studies of neurohypophyseal hormones. In The Peptides: Analysis, Synthesis, Biology, Vol. 8: Chemistry, Biology and Medicine of Neurohypophyseal Hormones and Their Analogs, C.W. Smith, ed. Academic Press, Inc., New York, 1987, pp. 77-207.

Hruby, V.J., Al-Obeidi, F., and Kazmierski, W. 1990a. Emerging approaches in the molecular design of receptor selective peptide ligands: Conformational, topographical and dynamic considerations. Biochem. J. 268:249-262.

Hruby, V.J., Kazmierski, W., Kawasaki, A.M., and Matsunaga, T. Synthetic peptide chemistry and the design of peptide based drugs. In Peptide Pharmaceuticals: Approaches to the Design of Novel Drugs, D. J. Ward, ed. Open University Press, London, 1990b, pp. 135-184.

Hruby, V.J., Fang, S., Knapp, R., Kazmierski, W.M., Lui, G.K., and Yamamura, H.I. 1990. Cholecystokinin analogues with high affinity and selectivity for brain membrane receptors. Int. J. Pept. Protein Res. 35:566-573.

Huang, C.K., and Richards, F.M. 1977. Reaction of a lipid soluble, unsymmetrical, cleavable cross-linking reagent with muscle aldolase and erythrocyte membrane protein. J. Biol. Chem. 252:5514-5521.

Hughes, J., Kosterlitz, H.W., and Leslie, F.M. 1975. Effect of morphine on adrenergic transmission in the mouse vas deferens. Assessment of agonist and antagonist potencies of narcotic analgesics. Br. J. Pharmacol. 53:371-381.

Jacobs, C.O., Sela, M., and Arnon, R. 1983. Antibodies against synthetic peptides of the B subunit of cholera toxin: Crossreaction and neutralization of the toxin. Proc. Natl Acad. Sci. USA 80:7611-7619.

Jaffe, C.L., Lis, H., and Sharon, N. 1979. Identification of peanut agglutinin receptors on human erythrocyte ghosts by affinity crosslinking using a cleavable heterobifunctional reagent. Biochem. Biophys. Res. Commun. 91:402-409.

Jaffe, C.L., Lis, H., and Sharon, N. 1980. New cleavable photoreactive heterobifunctional crosslinking reagent for studying membrane organization. Biochemistry 19:4423-4429.

Ji, T.H. 1977. A novel approach to the identification of surface receptors. The use of photosensitive hetero-bifunctional cross-linking reagent. J. Biol. Chem. 252:1566-1570.

Ji, T.H. 1979. The application of chemical crosslinking for studies on cell membranes and the identification of surface receptors. Biochem. Biophys. Acta 559:39-69.

Ji, T.H., and Ji, I. 1982. Macromolecular photoaffinity labeling with radioactive photoactivatable heterobifunctional reagents. Anal. Biochem. 121: 286-289.

Jones, D.S. 1972. Polypeptides. Part XIII. Peptides related to the C-terminal tertapeptide sequence of gastrins by complementary reading of the genetic message. J. Chem. Soc. Perkin I. 1407-1415.

Jost, K. Handbook of Neurohypophyseal Hormone Analogues, Vol. 1, Part 2, K. Jost, M. Lebl, and F. Brtnik, eds. CRC Press, Boca Raton, 1977.

Karplus, P.A., Schultz, G.E. 1985. Prediction of chain flexibility in proteins. Naturwissenschaften 72:212-222.

Kaurov, O.A., Martynov, V.F., Milhailov, Y.D., and Auna, Z.P. 1972. Synthesis of the new analogues of oxytocin, modified in position 2 (in Russian). Zh. Obshch. Khim. 42:1654.

Kayser, S. In Applied Therapeutics: The Clinical Use of Drugs, B.S. Katcher, L.Y. Young, and M.A. Koda-Kimble, eds. Applied Therapeutics, Inc., San Francisco, 1983, pp. 333-360.

Kazmierski, W.M., Wire, W.S., Lui, G.K., Knapp, R.J., Shook, J.E., Burks, T.F., Yamamura H.I., and Hruby, V.J. 1988. Design and synthesis of somatostatin analogues with topographical properties that lead to highly potent and specific μ opioid receptor antagonists with greatly reduced binding at somatostatin receptors. J. Med. Chem. 31:2170-2177.

Kazmierski, W.M., and Hruby, V.J. 1988. A new approach to receptor ligand design: Synthesis and conformation of a new class of potent and highly selective μ opioid antagonists utilizing tetrahydroisoquinoline carboxylic acid. Tetrahedron, 44:697-710.

Kennedy, R.C., Henkel, R.D., Pauletti, J.S., Allan, J.S, Lee, T.H., Essex, M., and Dreesman, G.R. 1986. Antiserum to synthetic peptide recognizes the HTLV-III envelope glycoprotein. Science 231:1556-1558.

Kitagawa, T., and Aikawa, T. 1976. Enzyme coupled immunoassay of insulin using a novel coupling reagent. J. Biochem. 79:233-236.

Klausner, Y.S., McCormick, W.M., and Chaiken, I.M. 1978. Design of a photoaffinity label for the hormone binding site of neurophysin. Int. J. Pept. Protein Res. 11:82-90.

Kobbs-Conrad, S., Gerdon, A., and Kaumaya, P.T.P. Multivalent B-and T-cell epitope vaccine design. In Proceedings of the Twelfth American Peptide Symposium. J. Smith and J. Rivier, eds. ESCOM, Leiden, 1991.

Kosterlitz, H.W., Lord, J.A.H, Paterson, S.J., and Waterfield, A.A. 1980. Effects of changes in the structure of enkephalins and of narcotic analgesic drugs on their interactions with μ- and δ-receptors. Br. J. Pharmacol. 68:333-342.

Krstenansky, J.L., and Mao, S.J.T. 1987. Antithrombin properties of C-terminus of hirudin using synthetic unsulfated N^α-acetyl-hirudin$_{45-65}$. FEBS Lett. 211:1 10-16.

Kruszynski, M., Lammek, B., Manning, M., Seto, J., Haldar, J., and Sawyer, W.H. 1980. [1-(β-Mercapto-β,β-cyclopentamethylenepropanoic acid), 2(O-methyl)tyrosine]arginine-vasopressin and [1-(β-mercapto-β,β-cyclopentamethylenepropanoic acid)]arginine-vasopressin, two highly potent antagonists of the vasopressor response to arginine-vasopressin. J. Med. Chem. 23:364-368.

Kyte, J., and Doolittle, R.F. 1982. A simple method for displaying hydropathic character of a protein. J. Mol. Biol. 157:105-132.

Lam, K.S., Salmon, S.E., Hersh, E.M., Hruby, V.J., Al-Obeidi, F., Kazmierski, W.M., and Knapp, R.J. The selectide process: Rapid generation of large synthetic peptide libraries linked to identification and structure determination of acceptor binding ligands. In Peptides: Chemistry and Biology, J. A. Smith and J. E. Rivier, eds., ESCOM, Leiden, 1992, pp. 492-495.

Lam, K.S., Salmon, S.E., Hersh, E.M. Hruby, V.J., Kazmierski, W.M., and Knapp, R.J. 1991. A new type of synthetic peptide library for identifying ligand binding activity. Nature, 354:82-84.

Lebl, M., Barth, T., Servitova, L., Slaninova, J., and Jost, K. 1985. Oxytocin analogues with inhibitory properties, containing in position 2 a hydrophobic amino acid of D-configuration. Collec. Czech. Chem. Commun. 50:132-145.

Lerner, R.A., 1982. Tapping the immunological repertoire to produce antibodies of predetermined specificity. Nature 299:592-595.

Lerner, R.A. 1984. Antibodies of predetermined specificity in biology and medicine. Adv. Immunol. 36:1-46.

Leslie, F.M., Chavkin, C., and Cox, B.M. 1980. Opioid binding properties of brain and peripheral tissues: Evidence for heterogeneity in opioid ligand binding sites. J. Pharmacol. Exp. Ther. 214:395-402.

Levy, D. 1973. Preparation of photo-affinity probes for the insulin receptor site in adipose and liver cell membranes. Biochim. Biophys. Acta 322:329-336.

Lin, Y., Trivedi, D., Siegel, M., and Hruby, V.J. Conformationally constrained glucagon analogues: new evidence for the conformational features important to glucagon-receptor interactions. In Peptides: Chemistry and Biology, J.A. Smith and J.E. Rivier, eds., ESCOM, Leiden, 1992, pp. 439-440.

Lord, I.A.H., Waterfield, A.A., Hughes, J., and Kosterlitz, H.W. 1977. Endogenous opioid peptides: Multiple agonists and receptors. Nature 267:495-499.

Lomant, A.J., and Fairbanks, G. 1976. Chemical probes of extended biological structures: Synthesis and properties of the cleavable protein cross-linking reagent (^{35}S)dithiobis-(succinimidyl propionate). J. Mol. Biol. 104:243-261.

Lowbridge, J., Manning, M., Seto, J., Haldar, J., and Sawyer, W.H. 1979. Synthetic antagonists of in vivo responses by the rat uterus to oxytocin. J. Med. Chem. 22:565-569.

Lu, F.X., Aiyar, N., and Chaiken I. 1991. Affinity capture of [Arg^8]vasopressin-receptor complex using immobilized antisense peptide. Proc. Natl. Acad. Sci. USA 88:3642-3646.

Malley, A., Saka, A., and Halliday, W.J. 1965. Immunochemical studies of hemocyanin from the gaint keyhole limpet and horseshoe crab. J. Immunol. 95:141-149.

Mann, K.G. 1987. The assembly of blood clotting complexes on membranes, Trends Biochem. Sci. 12:229-233.

Mao, S.J.T., Yates, M.T., Owen, T.J., and Krstenansky, J.L. 1988. Interaction of hirudin with thrombin: Identification of a minimal binding domain of hirudin that inhibits clotting activity. Biochemistry 27:8170-8173.

Martin, F.J., and Papahadjopoulos, D. 1982. Irreversible coupling of immunoglobulin fragments to preformed vesicles. An improved method for liposome targeting. J. Biol. Chem. 257:286-288.

Martin, W.R., Eades, C.G., Thompson, J.A., Huppler, R.E., and Gilbert, P.E. 1976. The effects of morphine and nalorphine-like drugs in the nondependent and morphine-dependent chronic spinal dog. J. Pharmacol. Exp. Ther. 197:517-532.

Matsunaga, T.O., de Lauro Castrucci, A.M., Hadley, M.E., and Hruby, V.J. 1989. Melanin concentrating hormone (MCH): Synthesis and bioactivity studies of MCH fragment analogues. Peptides 10:349-354.

McRobbie, M., Meth-Cohn, O., and Suschitzky, H. 1976a. Competitive cyclizations of singlet and triplet nitrenes. Part I. Cyclization of 1-(2-nitrophenyl)pyrazoles. Tetrahedron Lett. 925-928.

McRobbie, M., Meth-Cohn, O., and Suschitzky, H. 1976b. Competitive cyclizations of singlet and triplet nitrenes. Part II. Cyclization of 2-nitrophenyl-, thiophens-, benzo thiazoles-, and benzimidazoles. Tetrahedron Lett. 929-932.

Means, G.E., and Feeney, R.E. Chemical Modification of Proteins. Holden-Day, San Francisco, 1971, pp. 254-281.

Mekler, L.B. 1969. On specific selective interactions between amino acid residues of poly peptide chain. Biophys. USSR (Engl. Trans.) 14:613-617.

Melin, P., Vilhardt, H., and Akerlund, M. Uteronic oxytocin and vasopressin antagonists with minimal structure modifications. In Peptides: Structure and Function, V.J. Hruby, and D.H. Rich, eds. Pierce Chemical Co., Rockford, Illinois, 1983.

Meraldi, J.P., Yamamoto, D., Hruby, V.J., and Brewster, A.I.R. Conformational studies of neurohypophyseal hormones and inhibitors using high resolution NMR spectroscopy. In Peptides: Chemistry, Structure and Biology, R. Walter and J. Meinhofer, eds., Ann Arbor Sci. Publ., Ann Arbor, 1975, pp. 803-814.

Meraldi, J.P., Hruby, V.J., and Brewster, A.I.R. 1977. Relative conformational rigidity in oxytocin and [1-penacillamine]-oxytocin: A proposal for the relationship of conformational flexibility to peptide hormone agonism and antagonism. Proc. Natl. Acad. Sci. USA 74:1373-1377.

Merrifield, R.B. 1963. Solid phase peptide synthesis. I. The synthesis of a tetrapeptide J. Am. Chem. Soc. 85:2149-2152.

Mitchell, A.R., Kent, S.B.H., Engelhard, M., and Merrifield, R.B. 1978. A new synthetic route to tert-butyoxycarbonylaminacyl-4-(oxymethyl) phenylacetamidomethyl-resin, an improved support for solid-phase peptide synthesis. J. Org. Chem. 43:2845-2852.

Mizuno, T., Chou, M.-Y., and Inouye, M. 1984. A unique mechanism regulating gene expression: Translational inhibition by a complementary RNA transcript (micRNA). Proc. Natl. Acad. Sci. USA 81:1966-1970.

Moreland, R.B., Smith, P.K., Fujimoto, E.K., and Dockter, M.E. 1982. Synthesis and characterization of N-[4-azido phenylthio] phaliimidate: A cleavable photoactivable crosslinking reagent that reacts with sulfhydryl groups. Anal. Biochem. 121:321-326.

Morgan, B.A., Bowers, J.D., Guest, K.P., Handa, B.K., Metcalf, G., and Smith, C.F.C. Structure-activity relationships of enkephalin analogues. In Peptides, M. Goodman and J. Meienhofer, eds. John Wiley & Sons, New York, 1977, and references therein.

Mosberg, H.I., Hurst, R., Hruby, V.JU., Galligan, J.J., Burks, T.F., Gee, K., and Yamamura, H.I. 1983. Conformationally constrained cyclic enkephalin analogs with pronounced delta opioid receptor agonist selectivity. Life Sci. 32:2565-2569.

Mosberg, H.I., Hurst, R., Hruby, V.J., Gee, K., Yamamura, H.I., Galligan, J.J., and Burks, T.F., 1983b. Bis-penicillamine enkephalins possess highly improved specificity toward d opioid receptors. Proc. Natl. Acad. Sci. USA 80:5871-5874.

Mosberg, H.I., Hurst, R., Hruby, V.J., Gee, K., Yamamura, H.I., Galligan, J.J., and Burks, T.F. 1983. Bis-penicillamine enkephalins possess highly improved specificity toward delta opioid receptors. Proc. Natl. Acad. Sci. USA 80:5871-5874.

Mulchahey, J.J., Neill, J.D., Dion, L.D., Bost, K.L., and Blalock, J.E. 1986. Antibodies to the binding site of the receptor for luteinizing hormone-releasing hormone (LHRH): Generation with a synthetic decapeptide encoded by an RNA complementary to LHRH mRNA. Proc. Natl. Acad. Sci. USA 83:9714-9718.

Muramoto, K., and Ramachandran, J. 1980. Photoreactive derivative of corticotropin 2. Preparation and characterisation of 2-nitro-4(5)-azido phenylsulfenyl derivatives of corticotropin. Biochemistry 19:3280-3286.

Mutter, M. 1988. Synthetic proteins with a new three-dimensional architecture. In Proceedings of the Tenth American Peptide Symposium. G.R. Marshall, ed. ESCOM, Leiden, pp. 349-353.

Nassal, M. 1983. 4(1-Azi-2,2,2,trifluoroethyl)benzoic acid, a highly photolabile carbene generating label readily fixable to biochemical agents. Liebigs Ann. Chem. 1510-1523.

Nizbet, A.D., Saundry, R.H., Mari, A.J., Fothergil L.A., and Fothergill, J.E. 1981. The complete amino acid sequence of hen ovalbumin. Eur. J. Biochem. 155:335-339.

Ngo, T.T., Yam, C.F., Lenhoff, H.M., and Ivy, I.I. 1981. p-Azidophenylglyoxal. A heterobifunctional photoactivatable cross-linking reagent selective for arginyl residues. J. Biol. Chem. 256:11313-11318.

Omichinski, J.G., Olson, A.D., Thorgeirsson, S.S., and Fassina, G. Computer assisted design of recognition peptides. In Techniques in Protein Chemistry, T.E. Hugli, ed. Academic, San Diego, 1989, pp. 430-438.

Pelton, J.T., Gulya, K., Hruby, V.J., Duckles, S.P., and Yamamura, H.I. 1985. Conformationally restricted analogues of somatostatin with high μ-opiate receptor specificity. Proc. Natl. Acad. Sci. USA 82:236-239.

Pfaff, E., Mussgay, M., Bohm, H.O., Schulz, G.E., and Scaller, H. 1982. Antibodies against a preselected peptide recognize and neutralize foot and mouth disease virus. EMBO J. 1:869-874.

Piercy, M.F., Schroeder, A., Einspahr, F.J., Folkers, K., Xu, J.-C.,and Horig, J. Behavioral evidence that substance P may be a spinal cord nociceptor neurotransmitter. In Peptides: Synthesis-Structure-Function, D.H. Rich and E. Gross, eds. Pierce Chemical Co., Rockford, Illinois, 1981.

Putney, S.D., Lynn, D., Javaherion, K., Fadey, J., and Mueller, W.T. 1989. Recombinant proteins of human T-cell lymphotropic virus III (HTLV-III) and uses thereof. S. African ZA 8803,261. 25 Jan 1989. US Appl. 107,231. 09 Oct 1987; 94 pp.

Rasmussen, U.B., and Hesch, R.-D. 1987. On antisense peptides: The parathyroid hormone as an experimental example and a critical theoretical view. Biochem. Biophys. Res. Commun. 149:930-938.

Richman, S.J., Thomas, V., Sharma, P., Flint, J., Ardeshir, F., Grass, M., Silverman, C., and Reese, R.T. Construction of carrier-free synthetic peptide antigens capable of stimulating biostable antibody responses to a 75kDa malarial parasite protein. In Synthetic Peptides: Approaches to Biological Problems, J.P. Tam and E.T. Kaiser, eds. Alan R. Liss, Inc., New York, 1989, pp. 181-196.

Rinke, J., Meinke, M., Brimacombe, R., Fink, G., Rommel, W., and Fasold, H. 1980. The use of azidoarylimidoesters in RNA-protein cross-linking studies with *Escherichia coli* ribosomes. J. Mol. Biol. 137:301-3143.

Robinson, W.E., Montefiori, D.C., and Mitchell, W.M. 1988. Antibody-dependent enhancement of HIV infection. Lancet I:1285-1286.

Robinson, W.E., Kauamura, T., Gamry, M.K., Lake, D., Xu, J.-X., Matsumato, Y., Sugano, T., Markho, Y., Mitchell, W.M., Herzh, E., and Zoller-Pazner, S. 1990. Human monoclonal antibodies to the human immunodeficiency virus type I (HIV-I) transmembrane glycoprotein gp41 enhance HIV-I infection in vitro. Proc. Natl. Acad. Sci. USA 87:3185.

Rodda, S.J., Geysen, H.M., Manson, I.J, and Schaafs, P.G. 1986. The antibody response to myoglobin-I. Systematic synthesis of myoglobin peptides reveals location and substructure of species dependent continuous antigenic determinants. Mol. Immunol. 23:603-610.

Rodriguez, M., Lignon, M.-F., Galas, M.-C., and Martinez, J. Highly selective cyclic cholecystokinin analogs for central receptors. In Peptides: Chemistry, Structure, and Biology, J.E. Rivier and G.R. Marshall, eds. ESCOM, Leiden, 1990.

Rothbard, J.B., and Taylor, W.R. 1988. A sequence pattern common to T-cell epitopes. EMBO J. 7:93-101.

Rudinger, J., Pliska, V., and Krejci, I. 1972. Oxytocin analogues in the analysis of some phases of hormone action. Recent Prog. Hormone Res. 28:131-172.

Sawyer, T.K., Sanfilippo, P.J., Hruby, V.J., Engel, M.H., Heward, C.B., Burnett, C.B., and Hadley, M.E. 1980. [Nle^4, D-Phe^7]-α-Melanocyte stimulating hormone: A highly potent α-melanotropin with ultralong biological activity. Proc. Natl. Acad. Sci. USA 77:5754-5758.

Sawyer, T.K., Hruby, V.J., Darman, P.S., and Hadley, M.E. 1982. [half-Cys^4, half-Cys^{10}]-α-melanocyte stimulating hormone: A cyclic α-melanotropin exhibiting superagonist biological activity. Proc. Natl. Acad. Sci. USA 79:1751-1755.

Schiller, P.W., Eggimann, B., DiMaio, J., Lemieux, C., and Nguyen, T.M.-D. 1981. Cyclic enkephalin analogs containing a cystine bridge. Biochem. Biophys. Res. Commun. 101:337-343.

Schiller, P.W., Nguyen, T.M.-D., Maziak, L.A., and Lemieux, C. 1985. Novel cyclic opioid peptide analog showing high preference for μ-receptors. Biochem. Biophys. Res. Commun. 127:558-564.

Schulz, H., and du Vigneaud, V. 1966. Synthesis of 1-L-penacillamine-oxytocin, 1-D-penacillamine-oxytocin, and 1-deamino-penacillamine-oxytocin, potent inhibitors of oxytocic response to oxytocin. J. Med. Chem. 9:647-650.

Schwartz, I., and Ofengand, J. 1974. Photoaffinity labeling of tRNA binding site in macromolecules. I. Linking of the phenacyl-p-azide of 4-thiouridine in (*Escherichia coli*) valyl-tRNA to 16S RNA at the ribosomal P site. Proc. Natl. Acad. Sci. USA 71:3951-3955.

Schwyzer, R. 1977. ATCH: A short introductory review. Ann. N.Y. Acad. Sci. 297:3-26.

Scott, J.K., and Smith, G.P. 1990. Searching for peptide ligands with an epitope library. Science 249:386-390.

Seela, F., and Rosemeyer, H. 1977. 5-Azido-ω-bromo-2-nitroacetophenone. A cross-linking reagent with groups of selective reactivity. Hoppe Seyler Z. Physiol. Chem. 358:129-131.

Segerstéen, U., Nordgren, H., and Biro, J.-C. 1986. Frequent occurence of short complementary sequences in nucleic acids. Biochem. Biophys. Res. Commun. 139:94-101.

Sette, A., Doria, G., and Adauin, L. 1986. A microcomputer program for hydrophilicity and amphipathicity analysis of protein antigens. Mol. Immunol. 23:807-815.

Shai, Y., Brunck, T. K., and Chaiken, I.M. 1989. Antisense peptide recognition of sense peptides: Sequence simplification and evaluation of forces underlying the interaction. Biochemistry 28:8804-8811.

Shai, Y., Flashner, M., and Chaiken, I.M. 1987. Anti-sense peptide recognition of sense peptide: Direct quantitative characterization with the ribonuclease S-peptide system using analytical high-performance affinity chromatography. Biochemistry 26:669-675.

Siemion, I.Z., Szewezuk, Z., Herman, Z.S., and Stachura, Z. 1980. To the problem of the biologically active conformation of enkephalin. Mol. Cell. Biochem. 34:23-29.

Sigrist, H., and Zahler, P. 1980. Heterobifunctional crosslinking of bacteriorhodopsin by azidophenylisothiocyanate. FEBS Lett. 113:307-311.

Singh, A., Thornton, E. R., and Westheimer, F.H. 1962. The photolysis of diazoacetyl chymotrypsin. J. Biol. Chem. 237:PC3006-PC3008.

Smith, J.A., Hunnell J.G.R., and Leach, S.J. 1977. A novel method for delineating antigenic determinants: Peptide synthesis and radioimmunoassay using the same solid support. Immunochemistry 14:565-572.

Smith, J.A. Synthetic peptides: Tools for elucidating mechanism of protein antigen processing and presentation. In Synthetic Peptides: Approaches to Biological Problems, J.P. Tom and E.T. Kaiser, eds. Alan R. Liss, Inc., New York, 1989, pp. 31-42.

Smith, L.R., Bost, K.L., and Blalock, J.E. 1987. Generation of idiotypic and anti-idiotypic antibodies by immunization with peptides encoded by complementary RNA: A possible molecular basis for the network theory. J. Immunol. 138:7-9.

Smith, R.A.G., and Knowles, J.R. 1974. The utility of photoaffinity labels as mapping reagents. Biochem. J. 141:51-56.

Sonder, S.A., and Fenton, J.W. 1984. Proflavin binding within the fibrinopeptide groove adjacent to the catalytic site of human α-thrombin. Biochemistry 23:1818.

Stone, S.R., and Hofsteenge, J. 1986. Kinetics of the inhibition of thrombin by hirudin. Biochemistry 25:4622.

Strohmaier, K., Franze, R., and Adam, K.H. 1982. Localization and characterization of the antigenic portion of the foot-and-mouth disease virus protein. J. Gen. Virol. 59:295-305.

Struthers, R.S., Rivier, J., and Hagler, A. 1984. Molecular dynamics and minimum energy of GnRH and analogs: A methodology for computer-aided drug design. Ann. N.Y. Acad. Sci. 439:81-96.

Stryer, L. Biochemistry, W.H. Freeman & Co., San Francisco, 1975.

Sutcliffe, J.G., Shinnick, T.M., Green, N., Liu, F.-T., Niman, H.L., and Lerner, R.A. 1980. Chemical synthesis of a polypeptide predicted from nucleotide sequence allows detection of a new retroviral gene product. Nature 287:801-805.

Sutcliffe, J.G., Shinnick, T.M., Green, N., Lerner, R.A. 1983. Antibodies that react with predetermined sites on proteins. Science 219:660-663.

Tam, J.P. Multiple antigenic peptide system: A novel design for synthetic peptide vaccines and immunoassay. In Synthetic Peptides: Approaches to Biological Problems. J.P. Tam and E.T. Kaiser, eds. Alan R. Liss, Inc., New York, 1989, pp. 3-18.

Thornton, J.M., Edward, M.S., Taylor, W.R., and Barlow, D.J. 1986. Location of 'continuous' antigenic determinants in the protruding regions of proteins. EMBO J. 5:409-415.

Toniolo, C. 1990. Conformationally restricted peptides through short-range cyclizations. Int. J. Pept. Protein Res. 35:287-300.

Torchiana, M.L., Cook, P.G., Weise, S.R., Saperstein, R., and Veber, D.F. 1978. Subcutaneous administration of somatostatin analogs as a major factor in the enhancement of drug duration of action. Arch. Int. Pharmacol. Ther. 235:170-176.

Trommer, W.E., and Hendrick, M. 1973. The formation of maleimides by a new mild cyclization process. Synthesis 484-485.

Turro, N.J. In Molecular Photochemistry, Benjamin, New York, 1965.

Turro, N.J. 1980. Structure and dynamics of important reactive intermediates involved in photobiological systems. Ann. N. Y. Acad. Sci. 346:1-17.

Unanue, E.R., and Allen, P.M. 1987. The basis for the immunoregulatory role of macrophages and other accesory cells. Nature 236:551-557.

van Regenmortel, M.H.V., Briand, J.P., Muller, S., and Plaue, S. Synthetic polypeptides as antigens. In Laboratory Techniques in Biochemistry and Molecular Biology, R.H. Burdon and P.H. von Knippenberg, eds. Elsevier, Amsterdam, 1988.

Vanin, E.F., Burkhard, S.J., and Kaiser, I.I. 1981. p-Azidophenylglyoxal: A heterobifunctional photosensitive reagent. FEBS Lett. 124:89-92.

Vanin, E.F., and Ji, T.H. 1981. Synthesis and applications of cleavable photoactivable heterobifunctional reagents. Biochemistry 20:6754-6760.

Vaughan, R.J., and Weotheimer, F.H. 1969. A method for marking the hydrophobic binding sites of enzymes. An insertion into the methyl group of an alanine residue of trypsin. J. Am. Chem. Soc. 91:217-218.

Vavrek, R.J., Ferger, M.F., Allen, G.A., Rich, D.H., Blomquist, A.T., and du Vigneaud, V. 1972. Synthesis of three oxytocin analogs related to [1-deaminopenacillamine]oxytocin possessing antioxytocic activity. J. Med. Chem. 15:123-126.

Veber, D.F., Freidinger, R.M., Prelow, D.S., Poleveda, W.J., Holly, F.W., Strachan, R.G., Nutt, R.F., Arison, B.H., Homnick, C., Randall, W.C., Glitzer, M.S., Saperstein, R., and Hirchmann, R. 1981. A potent cyclic hexapeptide of somatostatin. Nature (London) 292:55-57.

Veronese, F.D., De Vico, A.L., Copeland, T.D., Orozzcan, S., Gallo, R.C., and Sangadhazan, M.G. 1985. Characterization of gp41 as the HTLV-III/LAV envelope gene. Science 229:1402-1404.

Walker, B., Wikstrom, P., and Shaw, E. 1985. Evaluation of inhibitor constants and alkylation rates for a series of thrombin affinity labels. Biochem J. 230:245.

Walter, G. 1986. Production and use of antibodies against synthetic peptides. J. Immunol. Methods 88:149-157.

Walter, R. 1977. Identification of sites in oxytocin involved in uterine receptor recognition and activation. Fed Proc. 36:1872-1878.

Ward, D.J., ed. Peptide Pharmaceuticals: Approaches to the Design of Novel Drugs. Open University Press, London, 1990.

Weigent, D.A., Hoeprich, P.D., Bost, K.L., Brunck, T.K., Reiher III, W.E., and Blalock, J.E. 1986. The HTLV-III envelope protein contains a hexapeptide homologous to a region of interleukin-2 that binds to the interleukin-2 receptor. Biochem. Biophys. Res. Commun. 139:367-374.

Wunderlin, R., Minakakis, P., Tun-Kyi, A., Sharma, S.D., and Schwyzer, R. 1985a. Melanotropin receptors. I. Synthesis and biological activity of N^{α}-(5-bromovaleryl)-N^{α}-desacetyl-α-melanotropin. Helv. Chim. Acta 68:1-11.

Wunderlin, R., Sharma, S.D., Minakakis, P., and Schwyzer, R. 1985b. Melanotropin receptors. II. Synthesis and biological activity of α-melanotropin/tobacco mosaic virus disulfide conjugates. Helv. Chim. Acta 68:12-22.

Yip, C.C., Yeung, C.W.T., and Moule, M.L. 1980. Photoaffinity labeling of insulin receptor proteins of liver plasma membrane preparations. Biochemistry 19:70-76.

Yoshitake, S., Yamada Y., Ishikawa, E., and Masseyeff, R. 1979. Conjugation of glucose oxidase from *Aspergillus niger* and rabbit antibodies using N-hydroxysuccinimide ester of N-(4-carboxy cyclohexylmethyl)-maleimide. Eur. J. Biochem. 101:395-399.

6

Setup of a Core Facility

Ruth Hogue Angeletti

Setting up a core peptide synthesis laboratory is usually a singular opportunity for an institution. A set of decisions needs to be made, from the early stages of setting objectives, to implementing these goals, to ensuring in the long term that the technology continues to serve the changing scientific environment. In this chapter, the problem is approached by posing specific questions that must be answered throughout the process of planning a core laboratory (Table 1). Most of these questions will apply both to setting up large and small full-service core facilities and to creating a small, shared peptide synthesis laboratory.

Table 1   Questions to Ask Before Setting Up a Core Peptide Synthesis Laboratory

How do we get started?

Who will be responsible for the laboratory?

Who will oversee the operations?

What are the expenses involved in operating a core peptide synthesis laboratory?

How will we pay for day-to-day operations?

Can our institution subsidize the costs?

Should there be user fees?

What is the ideal balance of institutional support and user fees?

What will happen when there is a budget shortfall?

How do we avoid one?

How can we economize?

How do we plan for future needs?

What services do we want?

How rapidly must these services be provided?

What is the best synthesizer?

How much quality control is actually needed?

What is the best equipment for analysis of the synthetic peptides?

What equipment is needed for peptide purification and other post-synthesis procedures?

How do we take care of our expensive equipment?

How many people are needed to provide the desired services?

What are their qualifications?

How do we find and select these people?

How can they be trained?

What kind of laboratory is needed for this type of work?

What safety concerns must be met?

What is the best way to interact with investigators?

How can we help them understand their data and improve the the effectiveness of their use of the laboratory?

How do we deal with complaints and problems?

AN INSTITUTIONAL CORE LABORATORY

Setting Goals, Taking Responsibility

How Do We Get Started?

The person or committee charged with setting up the core facility must first decide what services are needed and then designate a scientist to be responsible for organizing and operating the laboratory. They must secure funding for renovating the laboratory and purchasing the equipment as well as obtain commitments for supporting the day-to-day operations. In early planning stages, it is likely that a few mistakes will be made. Problems with the physical design of the laboratory can be corrected, perhaps at some cost. However, the most critical decisions to be made, and the most difficult problems to remedy, are often those that deal with day-to-day operations or long-term scientific goals. For the final setup to be successful, the scientific goals must meet the expectations of the user group. Thus, there must be a willingness to change it, if needed. There must also be an awareness that a peptide synthesis laboratory is a scientific facility, not just a technical operation. Finally, a good peptide synthesis laboratory is expensive. If plans are not made to ensure the financial stability of operation, meeting concerns about costs of the chemicals, trained personnel, and maintenance, then the scientific goals will founder.

Who Will Be Responsible for the Laboratory?
Who Will Oversee the Operations?

There are two common ways of managing a core facility. The first and most common solution is to hire a scientist who not only will manage the technical operations of the facility but also will be accountable for all aspects of scientific and financial planning of the laboratory. The other alternative is to find a scientist who will manage the day-to-day operations of the facility and who will participate in hands-on technical work. In this case, a senior scientist or faculty member is then designated to be responsible for the upper-level management and for interactions with the institution's administration. Although the paired situation can work very well, it is imperative that a good working relationship develop between the two scientists. Therefore, it is usually desirable for the senior faculty member, or scientist, to be involved in hiring the peptide chemist.

It is helpful for the search committee to draft a job description for potential candidates. This action is important not only to make certain that the right candidate is chosen for local needs but to ensure that there is a mutual understanding about what is expected and what is necessary for the core laboratory to be successful. The director will be responsible for overall peptide synthesis, analysis, quality control, and purification;

hiring and training technical personnel; purchase and maintenance of equipment; interacting with users to help them plan their synthetic work; preparing educational materials and technical sessions to aid in this effort; keeping logs of syntheses, customers and their affiliations, and expenditures; establishing a mechanism for the preparation and mailing of invoices and collecting payments, if fees are to be assessed; preparing detailed budgets; providing materials for grant applications to promote investigators' research efforts and to support the core facility's operations; writing grant applications for new equipment; and planning future directions. If the director of the core laboratory will report to a scientific director, then these responsibilities are divided among the two scientists. The search committee will need to assess the depth of the scientific expertise of the candidate and judge the experience and ability of the candidate to handle the complexity of the scientific problems that will arise in the institution. The candidates will themselves be judging whether the scientific environment will be interesting and challenging and if the recruiting institution has a real scientific and financial commitment to its core facilities. The institution will be most attractive to a peptide chemist if there are other local laboratories interested in protein structure and function. However, a group that has a rich variety of interesting problems that might require synthetic peptides will also offer an attractive environment for a peptide chemist.

What Are the Expenses Involved in Operating a Core Peptide Synthesis Laboratory?

No matter how expensive the setup of a core laboratory is, the routine, recurring operational expenses are usually more difficult to cover. Although equipment for a full-service synthesis laboratory will cost about $300,000 to $500,000 in 1992, it will be a one-time expense. For a laboratory with a single peptide synthesizer, an HPLC for purification, and necessary small equipment, the setup cost can reach $125,000 to $150,000. The yearly costs could be as much as one-half to three-quarters of the original cost of the equipment. Routine expenses for personnel, chemicals, equipment maintenance, and small-equipment purchases must be considered in annual budgets. Certain expenses, such as small amounts of funds for personnel to attend technical meetings or workshops, may seem frivolous, but they help to keep a laboratory successful by providing the employees the opportunity to learn and to exchange ideas with their peers.

A facility that synthesizes and purifies about 75 peptides per year, without performing the necessary analytical chemistry, will spend $25,000 to $40,000 per year on supplies. Service contracts usually cost about 10 percent of the initial cost of the instrument per year. If the instruments are new and employees are experienced, service contracts need not be taken. However, if the instruments are older, or if employees are not very experienced, service contracts are an economy.

Personnel costs will vary from one region to another. If the laboratory director also performs hands-on service work, then a laboratory with one synthesizer and resultant purification needs could manage with a single technician, although two would be more efficient. A director who does not perform actual laboratory work will require two technicians.

How Will We Pay for the Day-to-Day Operations?

The issues involved in setting up and financing a core peptide synthesis facility are both practical and philosophical. The basic purpose of the laboratory is to make peptides for local scientists at a higher quality, lower cost, and shorter turnaround time than those that are available from outside laboratories. Beyond this fundamental need, however, is the potential to enhance the creativity of the scientific environment. If the core laboratory can provide peptides at a lower cost, then the local investigators can design experiments incorporating synthetic peptides with less concern about minimizing their use and cost. The availability of low-cost peptides can be among those elements that foster the experimental risk-taking that is necessary for scientific progress. Each institution will have to decide the mixture of user fees and subsidy that will achieve their goals.

Can Our Institution Subsidize the Costs?

A financial commitment from the institution will lower the cost of peptides to investigators and give financial stability to the core laboratory. It is particularly important that funding be provided during the start-up and training years, when costs cannot be recovered by reasonable fees. The benefit to the institution will be the higher productivity of its scientists, which will ultimately be reflected in higher levels of research project funding. Furthermore, financial commitment of an institution to its core facilities helps these laboratories to obtain outside funding for new equipment.

Should There Be User Fees? What Is the Ideal Balance of Institutional Support and User Fees?

If user fees are to be charged, quick calculations will show that they will need to be quite high to cover full operational costs. High fees would negate many of the beneficial effects of having an institutional core laboratory. Yet a completely subsidized laboratory that makes peptides available at no cost is not necessarily a good solution. Free work is not always valued. Peptides requested in a mood of excitement may sit unused. However, research budgets are tight and are likely to remain so. It is shortsighted to expect that researchers will be able to support a state-of-the-art core laboratory entirely from user fees. In seeking a balance between cost effectiveness, budget consciousness, and scientific productivity, a good solution is for the institution or center grants to provide support for salary and benefits of the core laboratory personnel. The user fees could then be calculated simply to reimburse for the costs of consumable supplies, equipment maintenance, the cost of resynthesizing intractable peptides, and other daily expenditures. In an environment where most of the salaries are subsidized, user fees may range from $20 to $50 (in 1992) per amino acid residue, depending upon the type of synthesis and quality control procedures performed.

What Will Happen When There Is a Budget Shortfall? How Do We Avoid One?

Just as in a small business, a core facility must have financial support during its start-up period, when instruments are being installed, technicians trained, and the user interactions developed. This period usually lasts about two years, ending when everyone is more experienced and the director, the technical staff, and the scientists have developed effective interactions. However, the use of the core laboratory may still vary with time, depending upon experimental exigencies. The lower the level of institutional commitment, the greater the financial problems will be. Terminating a skilled staff member because of a temporary financial shortfall would be an unfortunate loss. Even with strong institutional support, there must be a mechanism for absorbing occasional losses. However, when there are continued financial problems, the organization and operating conditions of the laboratory will need to be examined by a group of concerned users. This scrutiny will undoubtedly yield some good ideas for cost cutting, some of which are detailed in the next section. It is also likely that this group will become aware that they have not been sufficiently supporting the laboratory.

How Can We Economize?

Some economies can keep the annual costs of operation down. Salary and benefits for the trained technical personnel should not be too much lower than industrial scales. However, academic institutions can compensate for lower salaries by offering educational benefits, longer vacations, and a campus environment. Organic solvents, reagents, and chemical intermediates can be bought in large quantities, often with special purchase agreements. This is particularly true for synthesizers that use prepackaged amino acid intermediates, which are provided at great cost by the manufacturer. These containers can be filled with bulk intermediates at dramatic savings. Maintenance costs can be kept at a minimum by properly caring for the instruments. One of the principal unforeseen expenses is the need for peptide resynthesis. The sophistication and apparent ease of automated peptide synthesis can create the illusion that peptide synthesis is a rote process. As well documented in other chapters in this book, each synthesis is unique, and many represent true challenges. For short antigenic peptides, the cost of one resynthesis effort may have to be absorbed by the core facility. Before beginning synthesis of longer peptides, it is important that investigators understand that these syntheses are experiments and that they will need to pay for the cost of resynthesis, if necessary. For shorter peptides, the resynthesis efforts should be only a small percentage of the laboratory's total syntheses. If the percentage of failed syntheses rises too high, then the director will need to make a serious examination of equipment operation and maintenance, purity of intermediates and solvents, efficacy of canned synthetic programs, and cleavage procedures. It will also be helpful to keep records of difficult couplings, so that they can be recognized in future peptides and special synthetic procedures can be developed. A final way to economize is for the core facility to eliminate time-consuming ancillary services that are easily performed in investigators' own laboratories. The service laboratory can then focus on the synthesis and quality control efforts that it is uniquely qualified to perform.

How Do We Plan for Future Needs?

Even in a well-planned core facility, there will be needs for capital equipment purchases that were not originally anticipated. The technology for automated synthesis and quality control procedures is continuing to evolve and must be incorporated into the core facility's operation. Older equipment will eventually need replacement. A facility that was originally planned as a small laboratory may become heavily used and require expansion. There are many ways in which funds can be assembled for equipment purchases. Some amount of savings for future equipment purchases can be built into the user fees. A similar mechanism is to arrange lease-purchase agreements for equipment, with payback funds built into the facility fee structure. Institutional funds, donations, program projects, or core grants can provide all, or a portion, of the money needed. A large group of investigators, each of whom contributes a modest amount, will be able to fund even an expensive piece of equipment. Finally, an academic institution can apply to a government agency for equipment funds. In order to qualify for this type of funding, a core facility director must prepare a carefully organized grant application, justifying the need for the equipment and detailing the expertise, effective organization, and functioning of the current operation. All of this effort will be relatively straightforward if the core facility has kept good records of its past activities. A final critical component of requesting government funds for equipment purchase is the level of institutional commitment, which is demonstrated by the level of financial support of the core facility. Final purchase of the desired equipment may require a combination of funding sources and the cooperation of a number of groups. This type of effort is one in which both the commitment of the institution to the core facility and the ability of the director to engage a cooperative effort will be tested.

Services/Equipment

What Services Do We Want?

For the purposes of our example, the equipment needed for establishing a medium to large-sized full-service peptide synthesis laboratory will be described first. This laboratory will be able to make a large number of peptide antigens, biologically active peptides, and longer peptides for structural model studies. There will be occasional needs for peptides with unusual amino acid derivatives or with post-translational type modifications. Structure-function analyses or epitope definition studies will require the synthesis of peptides with multiple substitutions at each site. Peptides may be purified and, perhaps, peptide antigens may be coupled to carrier proteins.

The array of services determines both the selection of instruments and the number of instruments required (Table 2). The user group may also prefer to perform its own post-cleavage sample workup. In these cases, a smaller investment in equipment will be necessary.

How Rapidly Must These Services Be Provided?

There is a physical limit to how rapidly a peptide can be planned, synthesized, cleaved, analyzed, and purified. Beyond this limit, however, the turnaround time depends primarily on the backlog of synthesis requests and the number of people working in the laboratory. If economy of time is desired, then more funds will have to be spent, primarily on technical personnel. If operating funds are limiting and investigators can be somewhat patient, then fewer technicians can be hired to carry out the functions of the core facility (see the following discussion). When a high level of synthesis is required and a rapid turnaround time is also desirable, instruments and technicians can be added. Alternatively, the core laboratory can limit its work to synthesis and quality assurance. Investigators can clean up their own peptides, prepare their own antigens, and perform other ancillary technical procedures if the quality of the synthetic peptides is guaranteed. This is the most economical way to increase turnaround times, because it will free the core facility to perform the primary synthetic work, which the users are unable to do.

Table 2   Equipment for a Large Peptide Core Laboratory

Synthesis	
Routine antigens	Automated synthesizer with canned chemistry and some programming capability
Experimental peptides	Automated synthesizer with flexibility in reaction vessel size and programming
	Manual shaker
Multiple replacement	RAMPS™ apparatus or robotic instrument
	Manual pin or paper synthesis
Separations	HPLC with interchangeable heads for both medium- and large-scale purifications or two HPLCs (one analytical and one preparative), HPLC columns
	Fraction collector
	Columns for reverse-phase chromatography and gel filtration
Quality control	Amino acid analyzer
	Automated sequencer availability
	Single quadrupole mass spectrometer
	Analytical HPLC and/or capillary electrophoresis
Small equipment	HF cleavage apparatus
	Tabletop centrifuge
	Rotary evaporator with vacuum pump
	Peptide hydrolysis setup
	Fraction collectors

What Is the Best Synthesizer?

The best automated synthesizer may not be the same for different purposes. The demand for antigens, active peptides, structural motifs, and multiple replacement synthesis calls for a flexibility in synthetic approaches. One way of dealing with this diversity is to have one automated synthesizer for production of peptide antigens, another automated instrument for more complex syntheses, and some type of manual synthesizer for special needs. The ability to choose either tBoc- or Fmoc-based syntheses, depending on peptide design, will also be helpful in covering most of these needs.

When there is a need for large number of peptide antigens 10 to 20 residues in length, it is helpful to have one workhorse instrument that can be optimized to produce the best possible immunogens, with very few repeat syntheses and with the fastest turnaround times. This need can be met by a number of instruments, but there are compromises that will have to be made. To minimize problems with antibody production, it is critical to avoid the synthesis of deletion peptides, which may be difficult to separate from the correct sequence and can potentially elicit antibodies against incorrect structures. For this reason, acetylation (or "capping") after each coupling is helpful—a procedure that is easier to perform with some instruments than with others. Some difficult sequences can be made more straightforward by the investigator by altering the design of the sequence to be made. Taking note of reports in the literature, manufacturer's advice, and one's own laboratory records, it is possible to circumvent many problems by recoupling at sites predicted to be difficult. Most synthesizers have one or more "canned" programs that have been optimized for the instrument. Some of these programs can provide very rapid coupling times, permitting a larger number of peptides to be synthesized. However, it has been reported that some of the abbreviated procedures produce a smaller percentage of high-purity peptides (Stevens et al., 1991). As long as careful attention is paid to quality control, as discussed in Chapter 4, most of the instruments can be used to advantage.

The choice of a second instrument can be made on the basis of instrument flexibility and defined needs. There are several synthesizers that have different-sized reaction vessels, providing peptides from micro-scale to even kilogram quantities. Although a number of machines permit either tBoc or Fmoc synthesis to be performed, a few offer the possibility of selecting different derivatives within an individual synthesis. This feature provides interesting chemical opportunities. A manual shaker for specialized reactions would complement the repertoire of automated synthetic equipment.

Multiple replacement peptide synthesis, used for studies of structure-activity relationships or for defining antigenic epitopes in protein sequences, can be accomplished in several ways. The RAMPS^{TM} apparatus is a simple, inexpensive system for manual synthesis of up to 25 peptides at a time. Alternatively, robotic automated synthesizers can perform the same tasks on a faster time scale. For either approach, careful record keeping is critical. If there are no financial restrictions, a robotic instrument could be chosen for the routine antigen synthesis. With this type of synthesis, careful quality control of the peptides will be essential. Furthermore, it is a daunting prospect to cleave, analyze, and purify dozens of peptides every few days. The synthesis of small amounts of multiple peptides on polyethylene pins, as devised by Geysen and co-workers (1984), offers an experimental technique that is not replaced by either of the two methods discussed. Large stretches of sequence can be synthesized in overlapping fragments, furnishing sufficient material for biological testing. A similar procedure, which utililizes cellulose paper as the solid support for synthesis and can produce somewhat larger amounts of peptides, has recently been developed (Frank and Doring, 1988; Frank et al., 1991). Although these are relatively simple manual procedures, they are expensive, and careful attention to detail is required. However, such methods can yield biological data that are not possible to obtain in any other way (Van Bleek and Nathenson, 1990).

Facilities for both hydrogen fluoride cleavage of tBoc-based syntheses and trifluoracetic acid cleavage of Fmoc-based syntheses need to be provided, depending on which method of synthesis is being used. The latter is technically very simple, requiring special glassware and a table-top centrifuge. HF cleavage is more complex and potentially more lethal. The standard apparatus provides sealed reaction chambers that need to be cleaned and o-rings that need to be changed and otherwise maintained. A vacuum pump is more efficient than a water pump for removing HF. A calcium oxide trap protects the pump and the environment and can prevent small accidents. The entire cleavage apparatus, including the vacuum pump, should be contained within a hood with a high-flow capacity. For a smaller full-service laboratory, it may be more convenient to use Fmoc-based syntheses, which require less equipment for the cleavage steps. Alternatively, there are commercial laboratories that will perform HF cleavage for a reasonable fee (see Appendix).

How Much Quality Control Is Actually Needed? What Is the Best Equipment for Analysis of the Synthetic Peptides?

It is tempting to bypass or minimize the analytical stages of peptide synthesis, placing faith in the quality of the synthetic equipment. This happens most often when there are restrictions on funds and personnel. Several anecdotes may be persuasive evidence of the need for quality control when important biological and structural hypotheses are being tested. In one case, a group was studying the biological activities of a CD4-related peptide that blocked HIV-induced cell fusion (Liefson et al., 1988). They found that the purified peptide lacked the activity found in the crude peptide mixture. Bioassay and mass spectral analysis of fractions from a reverse-phase separation of the starting material showed that the biological activity resided in a peptide that still contained the cysteine blocking group. In another example, an elegant study of the three-dimensional structure of a synthetic peptide model, as determined by X-ray diffraction, was presented (Hill et al., 1990). The data refinement revealed that the peptide was only 12 residues long, not the 16-residue structure that had been planned. A simple amino acid analysis prior to the lengthy X-ray analysis could have prevented this error.

The technology for assessing the success of peptide synthesis now includes a broad repertoire of approaches: amino acid analysis, HPLC, mass spectrometry, Edman degradation, and capillary electrophoresis. As evaluated in detail in Chapter 4, each of these methods has specific advantages. It is critical to be able to evaluate and distinguish problems derived from synthetic procedures or from cleavage of peptides from resins. A balanced analysis of all available data will help to assess the quality of each synthesis.

In the past, an amino acid analyzer was the one analytical instrument to which a peptide core laboratory needed access on a daily basis. Although some now doubt the efficacy of amino acid analysis for evaluating synthesis, this method can most easily reveal problems when capping is used to terminate incomplete couplings. In these cases, truncation peptides will be readily detected. Many of the HPLC-based systems will work well for amino acid analysis, although the instrument that provides the best separations of natural and modified amino acids, plus reliable quantitation, is still a system based on ion-exchange separations of free amino acids with post-column ninhydrin detection.

Mass spectrometry is the most valuable new tool for determining the presence of the desired peptide product. Mass spectrometry is not quantitative, and it does not necessarily give a representation of the proportions of products in a mixture. However, it is the best way to quickly analyze what went wrong with a bad synthesis, revealing problems with deprotection of amino acid side chains, peptides with deleted or oxidized amino acids, truncated peptides, or unusual peptide adducts (Smith et al., 1992; Kumagaye et al., 1991). Although expensive, a single quadrupole mass spectrometer, preferably with an electrospray gun, is a valuable investment for the core peptide synthesis laboratory. The instrument could be shared with a core protein sequencing laboratory. In this case, a triple quadrupole mass spectrometer, which will permit peptide sequencing, is the preferred instrument.

When mass spectrometry or amino acid analysis indicates that there is a problem with the synthesis, analytical HPLC can help in evaluating the proportions of the correct and incorrect peptide products. One must cautiously interpret the chromatograms, for a single peak may be a pure incorrect product. Capillary electrophoresis also has the potential for rapidly analyzing the complexity of the product mixture, but it has not yet been sufficiently used for this purpose (Crimmins, 1992). For either type of analysis, most of the commercially available analytical level instruments are suitable.

Analysis of synthetic peptide on an automated protein sequencer can be a useful supplement to these approaches. When a peptide is very long, a good signal may not be obtained in the mass spectrometer, and the amino acid analysis of a mixture of correct and incorrect sequences may not differ significantly from the analysis of the correct sequence alone. A careful quantitation of the preview in each cycle of Edman degradation will reveal which coupling failed, aiding in resynthesis efforts. However, if capping protocols have been used to block uncoupled amino termini, the sequence obtained will be only the correct one, with very little preview. A protein sequencer is not necessarily a good investment for a peptide synthesis laboratory alone. An agreement with a protein sequencing facility can provide this valuable method of analysis. Some other instruments, such as the amino acid analyzer, may also be shared with a protein sequencing laboratory.

What Equipment Is Needed for Peptide Purification and Other Postsynthesis Procedures?

The type of purification that is required for synthetic peptides depends on the final use of the peptide and level of purity that is indicated by the quality control. For peptide antigens that have been synthesized using capping protocols, and for which no significant truncation peptides are indicated, a simple desalting column will suffice for purification. Peptides needed for tissue culture or structure-function studies, or which have interfering truncation or undeprotected peptides, should be purified by HPLC. Almost any modular HPLC will suffice. However, if both small and large amounts of peptides may need to be purified, it is useful to select an instrument with interchangeable pump heads. If there is another HPLC in the laboratory for analytical separations (see previous discussion), then it is more convenient to purchase a second HPLC with high flow rate pump heads for large-scale purifications. The ability to detect at least two wavelengths, ca. 220 nm and 280 nm, will be useful in detecting peptides without and with aromatic residues, respectively. A selection of medium- and high-capacity columns as well as a fraction collector should be purchased. Using the methods outlined in Chapters 3 and 5, cross-linking of peptide antigens to carrier proteins can be performed with routine laboratory equipment.

How Do We Take Care of Our Expensive Equipment?

Equipment maintenance is one of the most important jobs in the core laboratory. An instrument that is kept clean and has routine maintenance performed by laboratory personnel will have far fewer operational problems. In addition, a synthesizer that is operated constantly, with only short lapse times between syntheses, will need less repair. Even major repairs can be taken care of in-house. However, if either people or experience is limiting, a service contract will be an economy. Whichever choice is made, detailed maintenance logs kept for each instrument will aid in both routine maintenance and troubleshooting.

Personnel and Training

How Many People Are Needed to Provide the Desired Services? What Are Their Qualifications? How Do We Find and Select These People? How Can They Be Trained?

All of these questions are interrelated. It is important to remember that, even in such a technology-oriented field, people are the most important component in the successful operation of a core facility.

In order to carry out the full array of services described previously, at least three trained people will be needed. As described in the section on economizing, fewer people will suffice if longer turnaround times can be tolerated. Finding technicians trained in peptide chemistry is usually not possible. It is often a good strategy to hire people with the best possible science education and to teach them peptide chemistry and separations. A peptide synthesis laboratory cannot function well with personnel who simply learn to operate a machine according to a manufacturer's manual or training course. It is in everyone's best interest for the technical staff to become as knowledgeable as possible. They need to be able to learn to make some judgments about synthesis and separations as well as to implement new techniques when requested. This type of training is a commitment, and usually it has to be done by the laboratory's director. Weekly group meetings can be used to review recent work as well as to discuss recent papers and new techniques and to bring everyone up to the same level on established techniques. Funds for the technicians to attend relevant scientific meetings, such as the American Peptide Symposium, the Protein Society Symposium, or specific technical meetings should be included in the annual budget if at all possible. These gatherings provide scientific perspective for the peptide synthesis work as well as technical presentations, and they afford an opportunity for staff to interact productively with colleagues in the same field. An additional benefit is increased interest and motivation.

Laboratory Design and Safety

What Kind of Laboratory Is Needed for This Type of Work?

In order to house one synthesizer, HPLCs for peptide purification, small equipment, hoods, adequate bench space, and personnel, a laboratory of 400 to 500 square feet, plus office space, will be adequate. Although it is technically possible to design a laboratory in less space, efficiency and safety may be sacrificed. If the peptide synthesis laboratory is not affiliated with an analytical chemistry laboratory, another 250 square feet of space will be needed for these functions.

What Safety Concerns Must Be Met?

In the design of a peptide synthesis laboratory, safety features have to be considered in addition to the routine concerns of bench space, equipment placement, and power lines. Most of the reagents and solvents used in peptide synthesis should be carefully stored, and they must be used in a hood. The automated synthesizers must be vented to hood lines, and manual synthesis should be performed in a fume hood. It is very useful to have one hood for manual synthesis and other manipulations and another for resin cleavage operations. The hoods must be maintained in optimal conditions of flow, particularly when HF cleavage is performed.

No cleavage reactions should be undertaken without first testing the hood's function. HF is lethal if breathed and can cause serious burns, so all possible safety precautions should be taken. HF cleavage should not be attempted without special training by personnel experienced in this technique. It is wise to have someone else present nearby to offer assistance in case of an HF accident. If the organic chemistry background of the laboratory personnel is not strong, then it is best to choose Fmoc synthesis as the routine approach, using trifluoracetic acid for cleavage. Trifluoracetic acid and trifluoromethanesulfonic acid are not as deadly as HF, but they are very hygroscopic, rapidly causing severe burns where they touch the skin.

Other chemicals used in peptide synthesis are also potentially dangerous. These chemicals can include the peptides themselves. Peptides are often lyophilized to fluffy powders after cleavage from the resin, exposing personnel to potential hazards from inhalation and skin contact. Therefore, it is useful to know whether a requested peptide is known to be biologically active or toxic, so that appropriate biohazard safety precautions can be taken.

Organic solvent use in a peptide synthesis laboratory is very high. Care must be taken in terms of both use and disposal; these chemicals should be used in a fume hood. Gloves, eye protection, laboratory coats, and other protective gear should be worn as recommended. Laboratory workers should be familiar with safety concerns, described in material data safety sheets, as well as organic and biological toxicity. It is helpful to review and to plan appropriate measures to take if a spill or other exposure should occur. Among the more commonly used solvents, dichloromethane is notable in that it is not flammable. However, it can be toxic, and it is a suspected carcinogen. Both N-methylpyrrolidone and N,N-dimethylformamide are combustible as well as toxic. Other acids, organic bases, and acetic anhydride are used in much smaller quantities, but they must still be handled carefully. Acetonitrile, employed in many

HPLC applications, is flammable, but it is normally used in aqueous solution. The organic, often sulfur-containing scavengers that are used to protect the peptide during cleavage also require cautious utilization. They must often be stored in sealed dessicators, even when in a fume hood, in order to avoid the release of noxious fumes into the laboratory environment. Finally, such compounds as dicyclohexylcarbodiimide can elicit an allergic response, which is usually noted by intense itching and red spots on the skin. There is also a potential danger that these agents can cross-link small amounts of the body's natural proteins, which might then stimulate an autoimmune response.

In addition to precautions in use, care must be taken in disposal. Disposal problems and expense are exacerbated by the large volumes of organic chemicals used. There are federal, state, and local ordinances that regulate organic waste disposal. Most institutions have a safety office that will take care of this problem. If the safety officers are consulted during laboratory planning and are also aware of the nature of peptide synthesis work, they will usually be very helpful and prompt in waste removal. They will also be prepared to help if a laboratory accident should occur.

Education and Training

What Is the Best Way to Interact with Investigators? How Can We Help Them Understand Their Data and Improve the Effectiveness of Their Use of the Laboratory?

The goal of a core peptide synthesis laboratory is to make peptides that will promote the success of basic biomedical research. If one stops at simply making high-quality peptides, the core facility will not reach its maximum effectiveness. Interactions with users can be fostered through direct discussions, detailed technical handouts, and technical workshops or seminars.

A first approach is to discuss both the peptide sequence and its planned use with the investigator. At this point, the core director can suggest changes in the peptide sequence that might make it more soluble or make the synthesis itself more successful. These discussions will improve the chances that the peptides synthesized will have the correct structure and will not require costly resyntheses. The final product will often be more amenable to the investigator's experimental conditions as well. Scientists who are not peptide chemists are usually not aware of the complexities of peptide chemistry. If they are guaranteed that the finished peptide will be correct and that it will have an assured level of purity, users will be cooperative in altering their peptide design. If they cannot change a difficult peptide sequence, they will, at least, be aware of potential problems. It will be helpful for the director to suggest ways of dealing with the properties of challenging peptides. When discussions with the investigators are accompanied by written handouts describing strategies in peptide synthesis, it will be easier to plan a useful peptide, with maximum likelihood for a successful synthesis and peptide solubility. It is also a common problem that the people doing the actual laboratory bench work with peptides do not have experience with quantitating peptides, cross-linking them to carrier proteins, or even packing or using a simple desalting column. A set of written protocols or other handouts with references will help to alleviate this problem and greatly reduce potential complaints. Finally, it may also help to hold occasional seminars or workshops on peptide synthesis and applications and the interpretation of quality control data.

How Do We Deal with Complaints and Problems?

When preparing peptides for use by other scientists who are knowledgeable only about the final experimental applications, complaints and problems frequently arise. An investment in the education of users will help the core laboratory as well as aid the investigators in successful applications. A common misconception about peptide synthesis is that it is as simple and straightforward as oligonucleotide synthesis. Although an occasional peptide may have a turnaround time of less than a week, including synthesis, cleavage, and quality control, this is a rare event. An average turnaround time may be 3 to 6 weeks, depending on the workload of the laboratory. Furthermore, intractable peptides, or those that have problems requiring resynthesis, can lead to dissension. Further misunderstanding can develop when the final peptides are difficult to work with under the desired experimental conditions. When someone complains that the peptides are "no good," he or she has rarely based that judgment on the quality control data; rather, he or she usually means that the peptides are insoluble or that there are other technical problems in using the peptides in experiments. A more fundamental issue arises when the peptide simply does not function in an experiment in the way the investigator predicted. One of the most frustrating problems with running a core facility is that the users do not really examine the analytical data given to them until there are such experimental difficulties. At these times, the director will need to carefully and calmly explain the quality control data obtained for an individual peptide and show similar data for other peptides produced by the facility. It may also help to prepare a handout describing how to interpret the data returned to the users with the peptide. After these discussions, the investigator can then reevaluate the experimental hypotheses, confident that the synthetic peptide used had the desired structure. Finally, the facility director will, at times, have to acknowledge an error and will need to compensate the investigator by performing additional synthetic work.

A SMALL, SHARED PEPTIDE SYNTHESIS LABORATORY

There can be several reasons why setting up a large core peptide synthesis facility is either not suitable or not desirable for a particular institution. It may be that in a small school or research institution, extensive needs for peptide synthesis are not supported by a large enough scientific base. Alternatively, the financial structure that could maintain a larger, shared facility may not be in place. Finally, the research community, although excellent, may not provide an attractive environment for a peptide chemist. There may be no colleagues in related fields with whom the potential facility director can interact; or the number of projects that require synthetic peptides may be small and, as such, may not be as challenging to a qualified peptide chemist. However, a small group of scientists with no experience in peptide chemistry may still have a need for synthetic peptides whose cost would be prohibitive if purchased from a commercial laboratory. These investigators must answer the same questions and face the same problems that a larger institution must deal with. Their unique question becomes, How do we set up a productive shared synthesis laboratory, in the face of our ignorance of peptide synthesis, and make reliable peptides of high quality, which will aid, not hinder, our research efforts?

For a small group of investigators, the most crucial common need is for high-quality peptide synthesis. Ancillary services need not be performed by the core laboratory but, instead, can be the responsibility of each individual laboratory. The investigators involved should decide from the beginning how to share the costs of equipment and operations. This type of arrangement demands trust from the participating group of scientists, because effective operation requires that one person be responsible for daily management and interactions with a technician. That person must be willing to make the effort to learn some peptide chemistry and to understand the basic analytical techniques.

The finances of a small, shared laboratory are relatively simple. It is unlikely that there will be institutional subsidy or fees. Instead, individual laboratories will contribute to the technicians' salaries and to the supplies budget. This is where dissension can easily arise. Careful logs of the synthetic work performed are the best basis for deciding the percentage of support that should be derived from each laboratory. Thought should be given to the problems that can develop if one of the support laboratories should temporarily lose its research support.

Separate laboratory space is not always necessary if a hood and sufficient bench space are already available. The type of equipment can be selected based on financial resources. For a small laboratory, Fmoc synthesis is easier to perform, and the cleavage procedures are somewhat less dangerous. However, HF cleavage services can be obtained from several commercial laboratories (see Appendix). If funds are not limited, one of the less expensive automated instruments can be purchased. With severely restricted funds, the RAMPS peptide apparatus is an effective system. However, the investigators and the technician will need to learn more about peptide chemistry if this more interactive type of small equipment is used.

All of the safety precautions mentioned in previous sections and in books on peptide chemistry should be carefully followed. These concerns are particularly important in a laboratory that is not used to organic chemical experiments. Although all steps of peptide synthesis require safe laboratory practice, the cleavage step is the most dangerous. Inexperienced personnel should not attempt to perform HF cleavage without training. Fmoc synthesis strategies instead of Boc strategies, which will require HF cleavage, should be strongly considered for small, inexperienced laboratories.

Quality control in a small laboratory is often neglected, or the success of the synthesis is taken in good faith. The anecdotes recounted in the previous section clearly demonstrate the risks. An amino acid analyzer is the minimum requirement for even a small laboratory. If such an instrument cannot be purchased, then it is imperative to contract out for the service. As discussed in a previous section, the single most useful instrument for evaluating the results of synthetic work is a mass spectrometer. This type of instrument is clearly beyond the reach of a small laboratory. However, there are mass spectrometry facilities that are willing to perform these analyses at reasonable fees (Appendix). The value of the fee will be more than compensated by the long-term quality of the research performed using well-defined peptides.

SUMMARY

1. Use the questions presented in Table 1 as a guide for setup and operation.

2. Seek support from your institution, pointing out the benefits to the institution as a whole. Seek a solution that will place value on your services and products but will not overburden the individual investigator's budgets.

3. Select a competent director, preferably with experience, but certainly with an interest and commitment to the success of the laboratory.

4. Do not neglect good quality control for convenience or speed. A poor product or incorrect peptide will not be a benefit to anyone and could set back a research project.

5. Familiarity of the users with the requirements and potential problems of peptide synthesis will promote smoother and more successful interactions. Devote some effort to instruction and consultation.

6. Safety is important. Peptide synthesis presents many potential hazards. Train your personnel well, and monitor conditions routinely.

REFERENCES

Crimmins, D.L. Comparative analysis of synthetic peptides by free-solution capillary electrophoresis (FSCE) and strong-cation exchange (SCX) HPLC. In Techniques in Protein Chemistry III, R. Hogue Angeletti, ed., Academic Press, Orlando, 1992, pp. 171-182.

Frank, R., and Doring, R. 1988. Simultaneous multiple peptide synthesis under continuous flow conditions on cellulose paper discs as segmental solid supports. Tetrahedron 44:6031-6040.

Frank, R., Guler, S., Krause, S., and Lindenmaier, W. In Peptides, E. Giralt and D. Andreu, eds., ESCOM, Leiden, the Netherlands, 1991.

Geysen, H.M., Meleon, R.H., and Barteling, S.J. 1984. Use of peptide synthesis to probe viral antigens for epitopes to a resolution of a single amino acid. Proc. Natl. Acad. Sci. USA 81:3998-4002.

Hill, C.P., Anderson, D.H., Wesson, L., DeGrado, W.F., and Eisenberg, D. 1990. Crystal structure of alpha-1: Implications for protein design. Science 249:543-546.

Kumagaye, K.Y., Inui, T., Nakajimi, K., Kimura, T., and Sakakibara, S. 1991. Suppression of a side reaction associated with N^{im}-benzyloxymethyl group during synthesis of peptides containing a cysteinyl residue at the N-terminus. Peptide Research 4:84-86.

Liefson, J.D., Hwang, K.M., Nara, P.L., Fraser, B., Padgett, M., Dunlog, N.M., and Eiden, L.E. 1988. Synthetic CD4 peptide derivatives that inhibit HIV infection and cytopathicity. Science 241:712-716.

Smith, R.L., Young, J.D., Carr, S.A., Marshak, D.R., Williams, L.C., and Williams, K.R. State-of-the-art peptide synthesis: Comparative characterization of a 16-mer synthesized in 31 different laboratories. In Techniques on Protein Chemistry III, R. Hogue Angeletti, ed., Academic Press, Orlando, 1992, pp. 219-232.

Stevens, R.L., Stevens, K.L., and Young, J.D. Comparison of several methods for synthesis of a difficult test peptide by BOC chemistry. In Techniques in Protein Chemistry II, J.J. Villafranca, ed., Academic Press, Orlando, 1991.

Van Bleek, G.M., and Nathenson, S.G. 1990. Isolation of an endogenously processed immunodominant viral peptide from the class I H-2k^{b} molecule. Nature 348:213-216.

Appendix

Services

The followling is a partial list of sources offering custom peptide synthesis or other analytical services on a contract or fee basis. The commericial sources were compiled largely from information gathered at the Twelfth American Peptide Symposium (Cambridge, Mass., 1991). The other sources, largely from academic laboratories, are compiled mainly from the authors' experience. These lists are not intended to be complete, and undoubtedly are not, but they are offered for the potential benefit of our readers. No claim regarding quality of service or products is implied.

PEPTIDE SYNTHESIS

Commercial

United States and Canada

Advanced Chemtech
5609 Fern Valley Road
Louisville, Kentucky 40228
(502) 969-0000
(800) 456-1403
FAX (502) 968-1000

Amino Tech
35 Antares Dr., Unit 2
Nepean, Ontario, Canada K2E 8B1
(613) 723-5843
(800) 567-8098
FAX (613) 723-5846

367

Applied Biosystems, Inc.
850 Lincoln Centre Dr.
Foster City, California 94404
(415) 570-6667
(800) 874-9868
FAX (415) 572-2743

Bachem California
3132 Kashiwa St.
Torrence, California 90505
(213) 539-4171
FAX (213) 530-1571

Bio-Synthesis, Inc.
P.O. Box 2067
Denton, Texas 76202
(214) 420-8505
FAX (214) 420-0442

Cambridge Research Biochemicals
Wilmington, Delaware 19897
(800) 327-0125
FAX (302) 886-2370

Clontech Laboratories, Inc.
4030 Fabian Way
Palo Alto, California 94303
(415) 424-8222
(800) 662-2566
FAX (415) 424-1352

Coast Scientific
P.O. Box 1142
LaJolla, California 92038-1142
(619) 459-8581
FAX (610) 459-3687

Immuno-Dynamics
P.O. Box 766
LaJolla, California 92038
(619) 452-1270
FAX (619) 452-3515

Multiple Peptide Systems
3550 General Atomics Court
San Diego, California 92121
(619) 455-3710
(800) 338-4965
FAX (619) 455-3713

Nova Biochem
10933 N. Torrey Pines Rd.
LaJolla, California 92037
(619) 450-5695
(800) 228-9622
FAX (800) 432-9622

Nuros Corporation
2020 Lundy Ave.
San Jose, California 95131
(408) 433-4850
FAX (408) 433-4850

Peptides International
P.O. Box 24658
Louisville, Kentucky 40224
(502) 425-8765
(800) 777-4779
FAX (502) 425-8771

Peptide Laboratory, Inc.
P.O. Box 8300
Berkeley, California 94707
(510) 262-0800
FAX (415) 262-9127

Synthecell
9620 Medical Center Dr.
Rockville, Maryland 20850
(301) 309-1910
(800) 336-7455
FAX (301) 309-1916

Synthetic Peptides, Inc.
355 Medical Sciences Building
University of Alberta
Edmonton, Alberta, Canada
T6G 2H7
(403) 492-3155
FAX (403) 492-7168

Europe

Bachem (UK), Ltd.
69 High St., Saffron Walden
Essex CB10 1AA, England
(0799) 26465
FAX (0799) 26351

Bissendorf Biochemicals GmbH
Feodor-Lynen-Strasse 14
3000 Hannover 61 Germany
(0511) 530080
FAX (0511) 5300888

Carlbiotech, Ltd.
Tagensvej 16 DK-2200
Copenhagen, Denmark
+45 31 37 21 00
FAX +45 31 37 27 94

Noncommercial

Beckman Center, B062
Stanford University Medical Center
Stanford, California 94305
(415) 723-1907
FAX (415) 725-0766

Biomolecular Resource Center
University of California-San Francisco
1235 Health Sciences East
San Francisco, California 94143
(415) 476-1839

Biopolymer Analysis Laboratory
Department of Biochemistry and
    Molecular Biology
Texas College of Osteopathic Medicine
3500 Camp Bowie Blvd.
Fort Worth, Texas 76107
(817) 735-2138
FAX (817) 735-2133

Biotechnology Laboratory
University of Oregon
Institute of Molecular Biology
Eugene, Oregon 97403
(503) 346-5151
FAX (503) 346-5891

541 Norris Cancer Center
University of Southern California
1441 Eastlake Ave.
Los Angeles, California 90033
(213) 224-6506
FAX (213) 224-6417

Harvard Microchemistry Facility
Harvard University
16 Divinity Ave.
Cambridge, Massachusetts 02138
(617) 495-4043
FAX (617) 495-1374

Laboratory for Macromolecular Analysis
Albert Einstein College of Medicine
1300 Morris Park Ave.
Bronx, New York 10461
(212) 430-3475
FAX (212) 823-5877

Macromolecular Sequence/Synthesis Core
Georgetown University Medical Center
Vincent T. Lombardi Cancer Research Center
3800 Reservoir Rd., N.W.
Washington, D. C. 20007
(202) 687-3570
(202) 687-7776

Peptide Synthesis Laboratory
Department of Biological Chemistry
UCLA School of Medicine
    Center for Health Sciences
10833 Le Conte Ave.
Los Angeles, California 90024
(213) 206-6866
FAX (213) 206-5272

Peptide Synthesis Laboratory
Department of Biochemistry
University of Kentucky
Lexington, Kentucky 40536
(606) 257-4932
FAX (606) 258-1037

Protein and Nucleic Acid Research Facility
Box 441, School of Medicine
University of Virginia
Charlottesville, Virginia 22908
(804) 924-2356

Protein Chemistry Facility
W. Alton Jones Cell Science Center
Lake Placid, New York 12946
(518) 523-1281
FAX (518) 523-1849

Protein Chemistry Laboratory
Washington University School of Medicine
660 S. Euclid St.
Campus Box 8231
St. Louis, Missouri 63110
(314) 362-3366
FAX (314) 362-4698

Protein Core Facility
Department of Biochemistry and
    Molecular Biophysics
Virginia Commonwealth University
Box 614
Richmond, Virginia 23298
(804) 225-3509
FAX (804) 786-1473

Protein/DNA Sequence/Synthesis Facility
University of Wisconsin
    Biotechnology Center
1710 University Avenue
Madison, Wisconsin 53705
(608) 262-1985
FAX (608) 262-6748

The Rockefeller University
Protein Sequencing/Biopolymer Facility
1230 York Ave.
P.O. Box 105
New York, New York 10021
(212) 570-8446
FAX (212) 570-7974

Yale University Peptide Synthesis Facility
Department of Molecular Biophysics
    and Biochemistry
Yale University School of Medicine
333 Cedar St., P.O. Box 3333
New Haven, Connecticut 06510
(203) 785-6718

CUSTOM HF CLEAVAGE

Bio-Synthesis, Inc.
P.O. Box 2067
Denton, Texas 76202
(214) 420-8505
FAX (214) 420-0442

Coast Scientific
P.O. Box 1142
La Jolla, California 92038-1142
(619) 459-8581
FAX (619) 459-3687

Immuno-Dynamics
P.O. Box 766
La Jolla, California 92038
(619) 452-1270
FAX (619) 452-3515

Multiple Peptide Systems
3550 General Atomics Court
San Diego, California 92121
(619) 455-3710
(800) 338-4965
FAX (619) 455-3713

Nuros Corporation
2020 Lundy Ave.
San Jose, California 95131
(408) 433-4850
FAX (408) 433-4850

Synthecell
9620 Medical Center Dr.
Rockville, Maryland 20850
(301) 309-1910
(800) 336-7455
FAX (301) 309-1916

Synthetic Peptides, Inc.
355 Medical Sciences Building
University of Alberta
Edmonton, Alberta, Canada T6G 2H7
(403) 492-3155
FAX (403) 492-7168

MASS SPECTROMETRY

Cambridge Research Biochemicals
Wilmington, Delaware 19897
(800) 327-0125
FAX (302) 886-2370

Multiple Peptide Systems
3550 General Atomics Court
San Diego, California 92121

(619) 455-3710
(800) 338-4965
FAX (619) 455-3713

M-Scan, Inc.
137 Brandywine Parkway
West Chester, Pennsylvania 19380
(215) 696-8210
(215) 696-8240
FAX (215) 696-8370

PE Sciex Instruments
55 Glen Cameron Rd.
Thornhill, Ontario L3T 1P2 Canada
(416) 881-9832

Synthetic Peptides, Inc.
355 Medical Sciences Building
University of Alberta
Edmonton, Alberta Canada T6G 2H7
(403) 492-3155
FAX (403) 492-7168

Synthecell
9620 Medical Center Dr.
Rockville, Maryland 20850
(301) 309-1910
(800) 336-7455
FAX (301) 309-1916

Index